Developments in Technologies for Human-Centric Mobile Computing and Applications

Joanna Lumsden
Aston University, UK

Managing Director:	Lindsay Johnston
Book Production Manager:	Jennifer Romanchak
Publishing Systems Analyst:	Adrienne Freeland
Managing Editor:	Joel Gamon
Development Editor:	Heather Probst
Assistant Acquisitions Editor:	Kayla Wolfe
Typesetter:	Travis Gundrum
Cover Design:	Nick Newcomer

Published in the United States of America by
Information Science Reference (an imprint of IGI Global)
701 E. Chocolate Avenue
Hershey PA 17033
Tel: 717-533-8845
Fax: 717-533-8661
E-mail: cust@igi-global.com
Web site: http://www.igi-global.com

Library of Congress Cataloging-in-Publication Data

Developments in technologies for human-centric mobile computing and applications / Joanna Lumsden, editor.
   p. cm.
 Includes bibliographical references and index.
 Summary: "This book is a comprehensive collection of knowledge and practice in the development of technologies in human -centric mobile technology, focusing on the developmental aspects of mobile technology; bringing together research-ers, educators and practitioners to encourage readers to think outside of the box"-- Provided by publisher.
 ISBN 978-1-4666-2068-1 (hardcover) -- ISBN 978-1-4666-2069-8 (ebook) -- ISBN 978-1-4666-2070-4 (print & perpetual access) 1. Mobile computing. I. Lumsden, Joanna.
 QA76.59.D44 2013
 004--dc23
                        2012013063

British Cataloguing in Publication Data
A Cataloguing in Publication record for this book is available from the British Library.

The views expressed in this book are those of the authors, but not necessarily of the publisher.

Table of Contents

Preface xv

Chapter 1
Evaluating the Visual Demand of In-Vehicle Information Systems:
The Development of a New Method 1
    Ainojie Alexander Irune, University of Nottingham, UK

Chapter 2
Classic and Alternative Mobile Search: A Review and Agenda 22
    Matt Jones, Swansea University, UK

Chapter 3
How Do People Use Their Mobile Phones? A Field Study of Small Device Users 38
    Tianyi Chen, University of Manchester, UK
    Simon Harper, University of Manchester, UK
    Yeliz Yesilada, Middle East Technical University, Northern Cyprus Campus, Turkey

Chapter 4
Evaluating the Readability of Privacy Policies in Mobile Environments 56
    R. I. Singh, University of Alberta, Canada
    M. Sumeeth, University of Alberta, Canada
    J. Miller, University of Alberta, Canada

Chapter 5
Remote and Autonomous Studies of Mobile and Ubiquitous Applications in Real Contexts 79
    Kasper Løvborg Jensen, Aalborg University, Denmark

Chapter 6
Nudging the Trolley in the Supermarket: How to Deliver the Right Information to Shoppers 99
    Peter M. Todd, Indiana University, Bloomington, USA
    Yvonne Rogers, The Open University, UK
    Stephen J. Payne, University of Bath, UK

Chapter 7
Speech for Content Creation 114
Joseph Polifroni, Nokia Research Center, USA
Imre Kiss, Nokia Research Center, USA
Stephanie Seneff, MIT CSAIL, USA

Chapter 8
3D Talking-Head Interface to Voice-Interactive Services on Mobile Phones 130
Jiri Danihelka, Czech Technical University in Prague, Czech Republic
Roman Hak, Czech Technical University in Prague, Czech Republic
Lukas Kencl, Czech Technical University in Prague, Czech Republic
Jiri Zara, Czech Technical University in Prague, Czech Republic

Chapter 9
feelabuzz: Direct Tactile Communication with Mobile Phones 145
Christian Leichsenring, Bielefeld University, Germany
René Tünnermann, Bielefeld University, Germany
Thomas Hermann, Bielefeld University, Germany

Chapter 10
Human-Centered Design for Development 155
Hendrik Knoche, EPFL IC LDM, Switzerland
PR Sheshagiri Rao, CK Trust, India
Jeffrey Huang, EPFL IC LDM, Switzerland

Chapter 11
A Festival-Wide Social Network Using 2D Barcodes, Mobile Phones and
Situated Displays 168
Jakob Eg Larsen, Technical University of Denmark, Denmark
Arkadiusz Stopczynski, Technical University of Denmark, Denmark

Chapter 12
Wearable Tactile Display of Landmarks and Direction for Pedestrian Navigation:
A User Survey and Evaluation 186
Mayuree Srikulwong, University of Bath, UK
Eamonn O'Neill, University of Bath, UK

Chapter 13
Good Times?! 3 Problems and Design Considerations for Playful HCI 204
Abdallah El Ali, University of Amsterdam, The Netherlands
Frank Nack, University of Amsterdam, The Netherlands
Lynda Hardman, Centrum voor Wiskunde en Informatica (CWI), The Netherlands

Chapter 14
A Comparison of Distribution Channels for Large-Scale Deployments of iOS Applications............222
    Donald McMillan, University of Glasgow, UK
    Alistair Morrison, University of Glasgow, UK
    Matthew Chalmers, University of Glasgow, UK

Chapter 15
WorldCupinion: Experiences with an Android App for Real-Time Opinion
Sharing During Soccer World Cup Games 240
    Robert Schleicher, Technical University of Berlin, Germany
    Alireza Sahami Shirazi, University of Duisburg-Essen and
        University of Stuttgart, Germany
    Michael Rohs, Technical University of Berlin and
        Ludwig-Maximilians-Universität München, Germany
    Sven Kratz, Technical University of Berlin, Germany
    Albrecht Schmidt, University of Duisburg-Essen and
        University of Stuttgart, Germany

Chapter 16
SGVis: Analysis of Data from Mass Participation Ubicomp Trials....................................258
    Alistair Morrison, University of Glasgow, UK
    Matthew Chalmers, University of Glasgow, UK

Chapter 17
Experimenting Through Mobile 'Apps' and 'App Stores' 277
    Paul Coulton, Lancaster University, UK
    Will Bamford, Lancaster University, UK

Chapter 18
My App is an Experiment: Experience from User Studies in Mobile App Stores............................294
    Niels Henze, University of Oldenburg, Germany
    Martin Pielot, OFFIS - Institute for Information Technology, Germany
    Benjamin Poppinga, OFFIS - Institute for Information Technology, Germany
    Torben Schinke, Worldiety GbR, Germany
    Susanne Boll, University of Oldenburg, Germany

Compilation of References 316

About the Contributors 347

Index...358

Detailed Table of Contents

Preface.. xv

Chapter 1

Evaluating the Visual Demand of In-Vehicle Information Systems:
The Development of a New Method.. 1
    Ainojie Alexander Irune, University of Nottingham, UK

In-vehicle information systems (IVIS) provide a variety of driver support and infotainment functionality; however, there is a growing concern that the resulting engagement with IVIS could present significant sources of distraction to drivers. This paper summarises the PhD thesis of Dr Ainojie Alexander Irune, which was awarded at the University of Nottingham in December 2009. The primary aims of the research were to develop a framework to aid the selection of an appropriate HF/HCI method, for use at particular stages in the design process, and to develop a novel method (with a focus on glance duration) for assessing the visual demand afforded by IVIS. Five empirical studies are reported in the thesis. In the first study, interviews were conducted with subject experts and the results were combined with the literature to provide guidance regarding the appropriate use of human factors methods. The remaining four studies present an iterative development of a novel method capable of predicting the visual demand imposed by an IVIS.

Chapter 2

Classic and Alternative Mobile Search: A Review and Agenda 22
    Matt Jones, Swansea University, UK

As mobile search turns into a mainstream activity, the author reflects on research that provides insights into the impact of current interfaces and pointers to yet unmet needs. Classic text dominated interface and interaction techniques are reviewed, showing how they can enhance the user experience. While today's interfaces emphasise direct, query-result approaches, serving up discrete chunks of content, the author suggests an alternative set of features for future mobile search. With reference to example systems, the paper argues for indirect, continuous and multimodal approaches. Further, while almost all mobile search research has focused on the 'developed' world, the paper outlines challenges and impact of work targeted at 'developing' world contexts.

Chapter 3

How Do People Use Their Mobile Phones? A Field Study of Small Device Users 38

Tianyi Chen, University of Manchester, UK
Simon Harper, University of Manchester, UK
Yeliz Yesilada, Middle East Technical University, Northern Cyprus Campus, Turkey

The usability evaluation of small devices (i.e., mobile phones and PDAs) is an emerging area of research. Compared with desktop computers, designing a usability evaluation for small devices is more challenging. Context of use, such as environmental disturbance and a user's physical activities affect the evaluation results. However, these parameters are usually ignored or excluded from simple and unnatural evaluation settings; therefore generating unrealistic results. This paper presents a field study that investigates the behaviour of small device users in naturalistic settings. The study consists of a series of unobtrusive remote observations and interviews. Results show that small device users normally use the device with just one hand, press the keys with thumb and make phone calls and send text messages while walking. They normally correct typing errors and use abbreviations. On average, small device users switch their attention between the device screen and the surrounding environment 3 times every 20 seconds, and this increases when they are walking.

Chapter 4

Evaluating the Readability of Privacy Policies in Mobile Environments 56

R. I. Singh, University of Alberta, Canada
M. Sumeeth, University of Alberta, Canada
J. Miller, University of Alberta, Canada

Recent work has suggested that the current "breed" of privacy policy represents a significant challenge in terms of comprehension to the average Internet-user. Due to display limitations, it is easy to represent the conjecture that this comprehension level should drop when these policies are moved into a mobile environment. This paper explores the question of how much does comprehension decrease when privacy policies are viewed on mobile versus desktop environments and does this decrease make them useless in their current format? It reports on a formal subject-based experiment, which seeks to evaluate how readable are privacy policy statements found on the Internet but presented in mobile environments. This experiment uses fifty participants and privacy policies collected from ten of the most popular web sites on the Internet. It evaluates, using a Cloze test, the subject's ability to comprehend the content of these privacy policies.

Chapter 5

Remote and Autonomous Studies of Mobile and Ubiquitous Applications in Real Contexts 79

Kasper Løvborg Jensen, Aalborg University, Denmark

As mobile and ubiquitous applications become increasingly complex and tightly interwoven into the fabric of everyday life it becomes more important to study them in real contexts. This paper presents a conceptual framework for remote and autonomous studies in the field and two practical tools to facilitate such studies. RECON is a remote controlled data capture tool that runs autonomously on personal mobile devices. It utilizes the sensing and processing power of the devices to capture contextual information together with general usage and application specific interaction data. GREATDANE is a tool for exploration and automated analysis of such rich datasets. The presented approach addresses some key issues of existing methods for studying applications in situ, namely cost, scalability and obtrusiveness to the user experience. Examples and experiences are given from remote and autonomous studies of two mobile and ubiquitous applications where the method and tools have been used.

Chapter 6
Nudging the Trolley in the Supermarket: How to Deliver the Right Information to Shoppers............ 99
    Peter M. Todd, Indiana University, Bloomington, USA
    Yvonne Rogers, The Open University, UK
    Stephen J. Payne, University of Bath, UK

The amount of information available to help decide what foods to buy and eat is increasing rapidly with the advent of concerns about, and data on, health impacts, environmental effects, and economic consequences. This glut of information can be overwhelming when presented within the context of a high time-pressure, low involvement activity such as supermarket shopping. How can we nudge people's food shopping behavior in desired directions through targeted delivery of appropriate information? This paper investigates whether augmented reality can deliver relevant 'instant information' that can be interpreted and acted upon in situ, enabling people to make informed choices. The challenge is to balance the need to simplify and streamline the information presented with the need to provide enough information that shoppers can adjust their behavior toward meeting their goals. This paper discusses some of the challenges involved in designing such information displays and indicate some possible ways to meet those challenges.

Chapter 7
Speech for Content Creation.. 114
    Joseph Polifroni, Nokia Research Center, USA
    Imre Kiss, Nokia Research Center, USA
    Stephanie Seneff, MIT CSAIL, USA

This paper proposes a paradigm for using speech to interact with computers, one that complements and extends traditional spoken dialogue systems: speech for content creation. The literature in automatic speech recognition (ASR), natural language processing (NLP), sentiment detection, and opinion mining is surveyed to argue that the time has come to use mobile devices to create content on-the-fly. Recent work in user modelling and recommender systems is examined to support the claim that using speech in this way can result in a useful interface to uniquely personalizable data. A data collection effort recently undertaken to help build a prototype system for spoken restaurant reviews is discussed. This vision critically depends on mobile technology, for enabling the creation of the content and for providing ancillary data to make its processing more relevant to individual users. This type of system can be of use where only limited speech processing is possible.

Chapter 8
3D Talking-Head Interface to Voice-Interactive Services on Mobile Phones 130
    Jiri Danihelka, Czech Technical University in Prague, Czech Republic
    Roman Hak, Czech Technical University in Prague, Czech Republic
    Lukas Kencl, Czech Technical University in Prague, Czech Republic
    Jiri Zara, Czech Technical University in Prague, Czech Republic

This paper presents a novel framework for easy creation of interactive, platform-independent voice-services with an animated 3D talking-head interface, on mobile phones. The Framework supports automated multi-modal interaction using speech and 3D graphics. The difficulty of synchronizing the audio stream to the animation is examined and alternatives for distributed network control of the animation and application logic is discussed. The ability of modern mobile devices to handle such applications is documented and it is shown that the power consumption trade-off of rendering on the mobile phone versus streaming from the server favors the phone. The presented tools will empower developers and researchers in future research and usability studies in the area of mobile talking-head applications (Figure 1). These may be used for example in entertainment, commerce, health care or education.

Chapter 9

feelabuzz: Direct Tactile Communication with Mobile Phones 145
Christian Leichsenring, Bielefeld University, Germany
René Tünnermann, Bielefeld University, Germany
Thomas Hermann, Bielefeld University, Germany

Touch can create a feeling of intimacy and connectedness. This work proposes feelabuzz, a system to transmit movements of one mobile phone to the vibration actuator of another one. This is done in a direct, non-abstract way, without the use of pattern recognition techniques in order not to destroy the feel for the other. The tactile channel enables direct communication, i. e. what another person explicitly signals, as well as implicit context communication, the complex movements any activity consists of or even those that are produced by the environment. This paper explores the potential of this approach, presents the mapping use and discusses further possible development beyond the existing prototype to enable a large-scale user study.

Chapter 10

Human-Centered Design for Development 155
Hendrik Knoche, EPFL IC LDM, Switzerland
PR Sheshagiri Rao, CK Trust, India
Jeffrey Huang, EPFL IC LDM, Switzerland

This paper describes the challenges faced in ICTD by reviewing the lessons learned from a project geared at improving the livelihood of marginal farmers in India through wireless sensor networks. Insufficient user participation, lack of attention to user needs, and a primary focus on technology in the design process led to unconvinced target users who were not interested in the new technology. The authors discuss benefits that ICTD can reap from incorporating human-centered design (HCD) principles such as holistic user involvement and prototypes to get buy-in from target users and foster support from other stakeholders and NGOs. The study's findings suggest that HCD artifacts can act as boundary objects for the different internal and external actors in development projects.

Chapter 11

A Festival-Wide Social Network Using 2D Barcodes, Mobile Phones and
Situated Displays 168
Jakob Eg Larsen, Technical University of Denmark, Denmark
Arkadiusz Stopczynski, Technical University of Denmark, Denmark

This paper reports on the authors' experiences with an exploratory prototype festival-wide social network. Unique 2D barcodes were applied to wristbands and mobile phones to uniquely identify the festival participants at the CO2PENHAGEN music festival in Denmark. The authors describe experiences from initial use of a set of social network applications involving participant profiles, a microblog and images shared on situated displays, and competitions created for the festival. The pilot study included 73 participants, each creating a unique profile. The novel approach had potential to enable anyone at the festival to participate in the festival-wide social network, as participants did not need any special hardware or mobile client application to be involved. The 2D barcodes was found to be a feasible low-cost approach for unique participant identification and social network interaction. Implications for the design of future systems of this nature are discussed.

Chapter 12
Wearable Tactile Display of Landmarks and Direction for Pedestrian Navigation:
A User Survey and Evaluation 186
Mayuree Srikulwong, University of Bath, UK
Eamonn O'Neill, University of Bath, UK

This research investigates representation techniques for spatial and related information in the design of tactile displays for pedestrian navigation systems. The paper reports on a user survey that identified and categorized landmarks used in pedestrian navigation in the urban context. The results show commonalities of landmark use in urban spaces worldwide. The survey results were then used in an experimental study that compared two tactile techniques for landmark representation using one or two actuators. Techniques were compared on 4 measures: distinguishability, learnability, memorability, and user preferences. Results from the lab-based evaluation showed that users performed equally well using either technique to represent just landmarks alone. However, when landmark representations were presented together with directional signals, performance with the one-actuator technique was significantly reduced while performance with the two-actuator approach remained unchanged. The results of this ongoing research programme can be used to help guide design for presenting key landmark information on wearable tactile displays.

Chapter 13
Good Times?! 3 Problems and Design Considerations for Playful HCI 204
Abdallah El Ali, University of Amsterdam, The Netherlands
Frank Nack, University of Amsterdam, The Netherlands
Lynda Hardman, Centrum voor Wiskunde en Informatica (CWI), The Netherlands

Using Location-aware Multimedia Messaging (LMM) systems as a research testbed, this paper presents an analysis of how 'fun or playfulness' can be studied and designed for under mobile and ubiquitous environments. These LMM systems allow users to leave geo-tagged multimedia messages behind at any location. Drawing on previous efforts with LMM systems and an envisioned scenario illustrating how LMM can be used, the authors discuss what playful experiences are and three problems that arise in realizing the scenario: how playful experiences can be inferred (the inference problem), how the experience of capture can be motivated and maintained (the experience-capture maintenance problem), and how playful experiences can be measured (the measurement problem). In response to each of the problems, three design considerations are drawn for playful Human-Computer Interaction: 1) experiences can be approached as information-rich representations or as arising from human-system interaction 2) incentive mechanisms can be mediators of fun and engagement, and 3) measuring experiences requires a balance in testing methodology choice.

Chapter 14
A Comparison of Distribution Channels for Large-Scale Deployments of iOS Applications 222
Donald McMillan, University of Glasgow, UK
Alistair Morrison, University of Glasgow, UK
Matthew Chalmers, University of Glasgow, UK

When conducting mass participation trials on Apple iOS devices researchers are forced to make a choice between using the Apple App Store or third party software repositories. In order to inform this choice, this paper describes a sample application that was released via both methods along with comparison of user demographics and engagement. The contents of these repositories are examined and compared, and statistics are presented highlighting the number of times the application was downloaded and the user retention experienced with each. The results are presented and the relative merits of each distribution method discussed to allow researchers to make a more informed choice. Results include that the application distributed via third party repository received ten times more downloads than the App Store application and that users recruited via the repository consistently used the application more.

Chapter 15

WorldCupinion: Experiences with an Android App for Real-Time Opinion
Sharing During Soccer World Cup Games 240

Robert Schleicher, Technical University of Berlin, Germany

Alireza Sahami Shirazi, University of Duisburg-Essen and
  University of Stuttgart, Germany

Michael Rohs, Technical University of Berlin and
  Ludwig-Maximilians-Universität München, Germany

Sven Kratz, Technical University of Berlin, Germany

Albrecht Schmidt, University of Duisburg-Essen and
  University of Stuttgart, Germany

Mobile devices are increasingly used in social networking applications and research. So far, there is little work on real-time emotion or opinion sharing in large loosely coupled user communities. One potential area of application is the assessment of widely broadcasted television (TV) shows. The idea of connecting non-collocated TV viewers via telecommunication technologies is referred to as Social TV. Such systems typically include set-top boxes for supporting the collaboration. In this work the authors investigated whether mobile phones can be used as an additional channel for sharing opinions, emotional responses, and TV-related experiences in real-time. To gain insight into this area, an Android app was developed for giving real-time feedback during soccer games and to create ad hoc fan groups. This paper presents results on rating activity during games and discusses experiences with deploying this app over four weeks during soccer World Cup. In doing so, challenges and opportunities faced are highlighted and an outlook on future work in this area is given.

Chapter 16

SGVis: Analysis of Data from Mass Participation Ubicomp Trials 258

Alistair Morrison, University of Glasgow, UK

Matthew Chalmers, University of Glasgow, UK

The recent rise in popularity of 'app store' markets on a number of different mobile platforms has provided a means for researchers to run worldwide trials of ubiquitous computing (ubicomp) applications with very large numbers of users. This opportunity raises challenges, however, as more traditional methods of running trials and gathering data for analysis might be infeasible or fail to scale up to a large, globally-spread user base. SGVis is a data analysis tool designed to aid ubicomp researchers in conducting trials in this manner. This paper discusses the difficulties involved in running large scale trials, explaining how these led to recommendations on what data researchers should log, and to design choices made in SGVis. The authors outline several methods of use and why they help with challenges raised by large scale research. A means of categorising users is also described that could aid in data analysis and management of a trial with very large numbers of participants. SGVis has been used in evaluating several mass-participation trials, involving tens of thousands of users, and several use cases are described that demonstrate its utility.

Chapter 17

Experimenting Through Mobile 'Apps' and 'App Stores' 277

Paul Coulton, Lancaster University, UK

Will Bamford, Lancaster University, UK

Utilizing App Stores as part of an 'in-the-large' methodology requires researchers to have a good understanding of the effects the platform has in the overall experimental process if they are to utilize it effectively. This paper presents an empirical study of effects of the operation an App Store has on an App lifecycle through the design, implementation and distribution of three games on the WidSets platform which arguably pioneered many of the features now seen as conventional for an App Store. Although these games achieved in excess of 1.5 million users it was evident through their App lifecycle that very large numbers of downloads are required to attract even a small number of active users and suggests such Apps need to be developed using more commercial practices than would be necessary for traditional lab testing. Further, the evidence shows that 'value added' features such as chat increase not only the popularity of an App but also increase the likelihood of continued use and provide a means of direct interaction with users.

Chapter 18

My App is an Experiment: Experience from User Studies in Mobile App Stores............................ 294

Niels Henze, University of Oldenburg, Germany

Martin Pielot, OFFIS - Institute for Information Technology, Germany

Benjamin Poppinga, OFFIS - Institute for Information Technology, Germany

Torben Schinke, Worldiety GbR, Germany

Susanne Boll, University of Oldenburg, Germany

Experiments are a cornerstone of HCI research. Mobile distribution channels such as Apple's App Store and Google's Android Market have created the opportunity to bring experiments to the end user. Hardly any experience exists on how to conduct such experiments successfully. This article reports on five experiments that were conducted by publishing Apps in the Android Market. The Apps are freely available and have been installed more than 30,000 times. The outcomes of the experiments range from failure to valuable insights. Based on these outcomes, the authors identified factors that account for the success of experiments using mobile application stores. When generalizing findings it must be considered that smartphone users are a non-representative sample of the world's population. Most participants can be obtained by informing users about the study when the App had been started for the first time. Because Apps are often used for a short time only, data should be collected as early as possible. To collect valuable qualitative feedback other channels than user comments and email have to be used. Finally, the interpretation of collected data has to consider unpredicted usage patterns to provide valid conclusions.

Compilation of References 316

About the Contributors 347

Index 358

Preface

Celebrating its third year in print in 2011, the *International Journal of Mobile Human Computer Interaction* (IJMHCI) continued to see the coming together of innovation, high quality research, and thought provoking and challenging articles to create exciting issues of the journal.

This Advances book is a compendium of articles from the third volume of the *International Journal of Mobile Human Computer Interaction (IJMHCI)*. The mission of the IJMHCI is to provide an international forum for researchers, educators, and practitioners to advance knowledge and practice in all facets of design and evaluation of human interaction with mobile technologies; to encourage readers to think out of the box to ensure that novel, effective user interface design and evaluation strategies continue to emerge and, in turn, the true potential of mobile technology is realized whilst being sensitive to the societal impact such technologies may have. The IJMHCI brings together a comprehensive collection of research articles from international experts on the design, evaluation, and use of innovative handheld, mobile, and wearable technologies; it also considers issues associated with the social and/or organizational impacts of such technologies. Emerging theories, methods, and interaction designs are included and are complemented with case studies which demonstrate the practical application of these new ideas. The aim of the journal is to increase exposure to, and heighten awareness of, the complexity of current and future issues concerning mobile human-computer interaction. In its infant years, and over the life of the journal, articles have presented (and will undoubtedly and indeed hopefully continue to present) alternative points of view for some of the field's hotly debated topics. Such variance is not only stimulating but also essential in terms of encouraging readers to think to the future and embrace the challenge of new paradigms both for interaction design and evaluation.

HIGHLIGHTING DIVERSITY IN MOBILE HCI

The third volume of the IJMHCI comprised 4 issues which, collectively, continue to highlight the diversity encompassed by the field of Mobile HCI. The first four articles included in this publication appeared in issue 3(1) of the IJMHCI. Spanning research on In-Vehicle Information Systems, discussion on mobile search, and studies to highlight how people actually use mobile phones as well as to investigate the comprehensibility of privacy policies on mobile devices, these comprehensive yet diverse articles exemplify the broad spectrum of research encompassed by the field of Mobile HCI and thereby captured within the journal.

The first article launched Volume 3 by showcasing another of the latest talents to join our field. A lot of PhD work is published in a somewhat disjointed fashion as a series of papers, each covering distinct sub-components of the bigger PhD research picture. The result is that the collective whole of a PhD is often un- or under-appreciated. The IJMHCI invites researchers who have recently graduated with a PhD in a mobile HCI-related topic to submit a comprehensive overview article of their PhD research. Such articles provide recent PhD graduates with a unique opportunity to showcase to a broad reader base, in a comprehensive and centralized fashion, the "big picture" of their research. In turn, it allows the mobile HCI research community to become aware of, to appreciate, and to take pride in the achievements of the newest members of our research family. This is certainly possible in terms of the first article of Volume 3 and, thereby, this compendium. In his article entitled "Evaluating the Visual Demand of In-Vehicle Information Systems: The Development of a New Method," Ainojie Alexander Irune provides a comprehensive overview of his PhD research (supervised by Dr. Gary Burnett, Irune's PhD was awarded at the University of Nottingham in December 2009). In this article, Irune notes the paradox between the variety of driver support and infotainment functions provided by in-vehicle information systems (IVIS) and the increasing concern about the distraction potential presented by such systems. He highlights the need to be able to assess and predict drivers' visual attention to ensure the safe operation of IVIS. In response, Irune presents a novel method, focusing on glance duration, for predicting the visual demand imposed by an IVIS, and discusses the series of studies that led to its development.

As acknowledged before, the IJMHCI would not be possible without the invaluable input of its advisory board members. As the first in a series of invited articles designed to both recognize the contribution of the journal's advisors and to intellectually challenge IJMHCI readers, I am honored to present an invited paper by Dr. Matt Jones, a member of the International Advisory Board of the IJMHCI. In "Classic and Alternative Mobile Search: A Review and Agenda," Matt reflects on *research that provides insights into the impact of current interfaces and pointers to yet unmet needs* (this volume). Recognizing the role of classic text-dominated interfaces and interaction mechanisms, Matt suggests an alternative set of search features to meet the needs of future mobile search activities. Arguing for indirect, continuous, and multimodal approaches to mobile search, Matt highlights the challenges associated with work directed specifically at "developing" world contexts.

In the third article, entitled "How Do People Use Their Mobile Phones? A Field Study of Small Device Users," Tianyi Chen, Simon Harper, and Yeliz Yesilada reflect on the challenges we face when designing usability evaluations for small devices. They note that contextual factors – including users' physical activities – are often omitted from evaluation settings which, as a result, return results that lack meaning. Contributing to the ongoing debate about how best to approach the evaluation of mobile technologies, and providing empirical behavioral information which we can all draw on when designing such devices, Chen *et al.* discuss a field study which they conducted to investigate the behavior of users of small devices in naturalistic settings.

In the fourth article – *"Evaluating the Readability of Privacy Policies in Mobile Environments"* – Ravi Singh, Manasa Sumeeth, and James Miller discuss the issue of presentation of privacy policies on mobile devices to support maximum comprehension across average internet users. It is sensible to assume that the limited screen resources of most mobile technologies will lead to reduced comprehension when policies drafted for review on desktop-based systems are moved, without substantial adaptation, to a mobile environment. In their article, Singh *et al.* empirically illustrate the extent to which this assumption holds true and explore the question of whether privacy policies are useless in their current format on mobile devices.

CELEBRATING THE BEST

Issues 3(2) and 3(3) saw the launch of a new initiative for the IJMHCI. Collectively, these issues comprised a two part themed issue designed to showcase the best papers from each of the workshops run during the 12th International Conference on Human-Computer Interaction with Mobile Devices and Services (MobileHCI'2010) in Lisbon, Portugal. MobileHCI'2010 saw a record number of workshop proposals, from which seven full day and two half day workshops were selected. These workshops covered topics ranging from the use of audio and haptics for spatial information delivery through the social mobile web to playful experiences in Mobile HCI! Across the two parts of this themed issue, the IJMHCI highlighted the best papers from eight of the nine[1] exciting and varied workshops.

- **Workshop on Tool-support for Mobile and Pervasive Application Development:**
  - **Organizers:** Ilhan Aslan (University of Salzburg, Austria), Paul Holleis (DOCOMO Communications Laboratories Europe, Germany), Karin Leichtenstern (Augsburg University, Germany), Christoph Stahl (Saarland University, Germany), Rainer Wasinger (University of Sydney, Australia).
  - **Best Paper:** *Remote and Autonomous Studies of Mobile and Ubiquitous Applications in Real Contexts* by Kasper Løvborg Jensen (Aalborg University, Denmark).

Few of today's available mobile and pervasive applications consider users' preferences, situational context, and other devices in their environment. Developing such mobile applications requires a thorough understanding of the complex interplay of users, personal devices, and their environment. The workshop on Tool-Support for Mobile and Pervasive Application Development (TSMPAD) focuses on software tools which support the designer of a mobile or pervasive application in the development process – e.g., task and requirements analysis, conceptual design, prototyping, and evaluation. In order to effectively provide support for the development of a mobile or pervasive application it is important for a tool to: (i) provide representations and concepts to describe the users and their needs in mobile settings; (ii) realize the design as a mobile application; and (iii) aid the designer in evaluating the developed applications with target users. The overall goal of the workshop was to reflect the state of the art of available tools, discuss current and emerging issues (e.g., tool-support for cross-platform development) and identify future problems for researchers and designers of mobile and pervasive applications. [Overview by organizer Ilhan Aslan, University of Salzburg, Austria].

In his paper, Jensen presents a conceptual framework for remote and autonomous studies in the field which he then applies via the delivery of two practical tools (RECON and GREATDANE) to facilitate such studies. Jensen's approach addresses key issues such as cost, scalability, and obtrusiveness to the user experience associated with existing methods for studying applications in situ, and he reflects on his experience of using these methods and tools for the remote and autonomous study of two mobile and ubiquitous environments.

- **NIMD'10: First International Workshop on Nudge & Influence Through Mobile Devices:**
  - **Organizers:** Parisa Eslambolchilar (Swansea University, UK), Max L. Wilson, (Swansea University, UK), and Andreas Komninos (Glasgow Caledonian University, UK).

- ○ **Best Paper:** *Nudging the trolley in the supermarket: How to deliver the right information to shoppers* by Peter M. Todd (Indiana University, USA), Yvonne Rogers (The Open University, UK) and Stephen J. Payne (University of Bath, UK).

The aim of Nudge and Influence through Mobile Devices (NIMD) workshop was to provide a focal point for research and technology dedicated to persuasion and influence on mobile platforms. We were inspired to establish a scientific network and community dedicated to emerging technologies for persuasion using mobile devices. This workshop was an excellent opportunity for interaction designers and researchers in this area to share their latest research and technologies on 'nudge' methods with the Mobile HCI community. Patterns of consumption such as drinking and smoking are shaped by the taken-for-granted practices of everyday life. However, these practices are not fixed and are 'immensely malleable'. Consequently, it is important to understand how the habits of everyday life change and evolve. Our decisions are inevitably influenced by how the choices are presented. Therefore, it is legitimate to deliberately 'nudge' people's behavior in order to improve their lives. Mobile devices can play a significant role in shaping normal practices in three distinct ways: (1) they facilitate the capture of information at the right time and place; (2) they provide non-invasive and cost effective methods for communicating personalized data that compare individual performance with relevant social group performance; and (3) social network sites running on the device facilitate communication of personalized data that relate to the participant's self-defined community. Among the issues addressed by the workshop were: What opportunities do mobile interventions provide?; Is persuasion ethical?; and How can we extend the scale of intervention in a society using mobile devices? [Overview by organizer Parisa Eslambolchilar, Swansea University, UK].

In response to the plethora of information now available that is intended to help consumers decide what food to buy and eat, Todd *et al.* explore the potential to nudge people's food shopping behaviour during a visit to the supermarket such that they are better able to make informed purchasing decisions despite the time pressures inherent in visits to the supermarket. Exploring the potential for augmented reality to deliver 'instant information' that can assist consumers in making informed choices in situ, Todd *et al.* discuss some of the challenges involved in designing such information displays and indicate some possible ways to meet those challenges.

- **SiMPE: 5th Workshop on Speech in Mobile and Pervasive Environments:**
  - ○ **Organizers:** Amit A. Nanavati (IBM India Research Laboratory, India), Nitendra Rajput (IBM India Research Laboratory, India), Alexander I. Rudnicky (Carnegie Mellon University, USA), Markku Turunen (University of Tampere, Finland), Andrew Kun (University of New Hampshire, USA), Tim Paek (Microsoft Research, USA), and Ivan Tashev (Microsoft Research, USA).
  - ○ Best Paper #1: *Speech for Content Creation* by Joseph Polifroni (Nokia Research Center, USA) Imre Kiss (Nokia Research Center, USA) and Stephanie Seneff (MIT CSAIL, USA).
  - ○ Best Paper #2: *3D Talking-Head Interface to Voice-Interactive Services on Mobile Phones* by Jiri Danihelka (Czech Technical University in Prague, Czech Republic), Roman Hak (Czech Technical University in Prague, Czech Republic), Lukas Kencl (Czech Technical University in Prague, Czech Republic), and Jiri Zara (Czech Technical University in Prague, Czech Republic).

With the proliferation of pervasive devices and the increase in their processing capabilities, client-side speech processing has been emerging as a viable alternative. The SiMPE workshop series (http://research.ihost.com/SiMPE/) started in 2006 with the goal of enabling speech processing on mobile and embedded devices to meet the challenges of pervasive environments (such as noise) and leveraging the context they offer (such as location). SiMPE 2010, the 5th in the series, continues to explore issues, possibilities, and approaches for enabling speech processing as well as convenient and effective speech and multimodal user interfaces.

The workshop started with a keynote speech by Albrecht Schmidt on "Trends and Challenges in Mobile Interaction." He opined that classical computing limitations (such as memory and processing) will not play a significant role (they haven't in the past) and that speech can be used for implicit interactions. The rest of the workshop was divided into 4 sessions: position papers, early results, demos and full papers. The sessions were really interactive and the participants were enthusiastically engaged during the entire workshop. In the last hour, there was a round table discussion on modes and topics for collaboration. The workshop summary is available on the SiMPE wiki (http://simpe.wikispaces.com/SiMPE+2010) [Overview by organizer Amit A Nanavati, IBM India Research Laboratory, India].

In the first of two best papers awarded by this workshop, Polifroni *et al.* propose the paradigm of speech for content creation. They argue that it is time to use mobile devices to create content on-the-fly and propose the use of speech to generate uniquely personalisable data in mobile contexts. In their paper, Polifroni *et al.* discuss a prototype system they have developed to enable the speech-based creation of restaurant review content.

In the second of the workshop's best papers, Danihelka *et al.* present a novel framework – which supports multi-modal interaction using speech and 3D graphics – for easy creation of interactive, platform-independent voice-services incorporating an animated 3D talkinghead interface, for mobile phones. They suggest that their tools have the capacity to empower developers and researchers who wish to use mobile talkinghead applications for mobile entertainment, m-commerce, mobile healthcare delivery, and/or mobile education.

- **International Workshop on Mobile Social Signal Processing:**
  - **Organizers:** Alessandro Vinciarelli (University of Glasgow, UK/Idiap Research Institute, Switzerland), Rod Murray-Smith (University of Glasgow, UK), and Herve' Bourlard (Idiap Research Institute, Switzerland/EPFL, France).
  - **Best Paper:** *feelabuzz – Direct Tactile Communication with Mobile Phones* by Christian Leichsenring (Bielefeld University, Germany), René Tünnermann (Bielefeld University, Germany) and Thomas Hermann (Bielefeld University, Germany).

Conversation is the primordial site of social interaction, and mobile phones, allowing one to talk with virtually anybody at virtually any moment, have, not surprisingly, pervaded our everyday life more quickly and deeply than any previous technology.

However, while becoming a preeminent form of social interaction, mobile phone conversations have been the subject of limited investigation from both a psychological and technological points of view.

The reason is not only that the diffusion of mobile phones is a relatively recent phenomenon, but also that phone conversations have traditionally been considered nothing more than particular cases of face-to-face conversations, characterized by speech being the only information at disposition, in contrast with actual face-to-face conversations where humans are known to exchange not only words, but also a wide spectrum of nonverbal behavioral cues accounting for social, affective, and relational phenomena.

This leaves open a major gap in the moment where two important phenomena take place in the scientific and technological landscape. The first is that nowadays, standard mobile phones contain a large number of sensors (e.g., GPS, accelerometers, magnetometers, capacitive touch, and, in the near future, pressure sensing). Also the increasing processing power and the potential to use server-side processing allows the use of algorithms previously considered only possible on powerful PCs, capturing, with unprecedented depth and precision, context and behavior of their users (e.g., position, movement, hand grip behavior, proximity to social network members, gait type, auditory context). This behavior can also potentially be compared with large numbers of other users, to categories the style of interaction. The second is that automatic analysis, synthesis, and understanding of verbal and nonverbal communication, typically captured with multiple sensors, is one of the hottest topics in the computing community. This applies in particular to Social Signal Processing (SSP), the new, emerging domain aimed at bringing social intelligence in machines.

The goal of the International Workshop on Mobile Social Signal Processing was to bridge the gap mentioned above by gathering, for the first time, researchers active in the communities that have dealt, so far separately, with the two phenomena described earlier. The participants included representatives from both industry (Nokia, Sony-Ericsson, and Google) and academia. Presentations and discussions focused on the potential that Social Signal Processing can have for the improvement of mobile phones, possibly leading to new applications, and on the value of cellular phones as a tool for studying and understanding social interactions. The outcomes of the workshop will be published in a volume of the Springer LNCS Series [Overview by organizer Alessandro Vinciarelli, University of Glasgow, UK].

Recognizing the potential for touch to elicit feelings of connectedness, Leichsenring *et al.* present feelabuzz, a system to transmit movements of one mobile phone to the vibration actuator of another one. Having used a direct, non-abstract approach (without the use of pattern recognition techniques) in order not to destroy the feeling at the receiving end, Leichsenring *et al.* explore the potential of their approach, present the mapping they used and discuss further possible development beyond the existing prototype to enable a large-scale user study.

- **Mobile HCI and Technical ICTD: A Methodological Perspective:**
  - **Organizers:** Jörg Dörflinger (SAP Research, Germany), Tom Gross (Bauhaus-University Weimar, Germany), Gary Marsden (University of Cape Town, South Africa), Matt Jones (Swansea University, UK), and Mark Dunlop (University of Strathclyde, UK).
  - **Best Paper:** *Human-Centered Design for Development* by Hendrik Knoche (EPFL IC LDM, Switzerland), PR Sheshagiri Rao (CK Trust, India) and Jeffrey Huang (EPFL IC LDM, Switzerland).

Mobile HCI has a great set of methods supporting effective research in all phases of mobile research. Methodologies for data collection, theory building, testing of hypothesis, and framework creation are represented with survey research, applied research, basic research, and normative writings. Methodologies are available for laboratory experiments supporting controlled experiments and theory testing in artificial environments. Generation, testing, and description of hypothesis and theory as well as studying of current practices and evaluation of new practices in natural settings are supported by methodologies like case studies, field studies, and action research. Due to the huge opportunities of mobile computing in developing countries Technical Information and Communication Technologies for Development (ICTD) could benefit from this complete set of Mobile HCI methodologies supporting all phases of a mobile research lifecycle.

Moreover, not only could technical ICTD benefit from Mobile HCI but so too could Mobile HCI benefit from technical ICTD. A review of Mobile HCI publications[2] revealed that there is very little research done using field methods or studies (action research, case studies, field studies) in natural settings. Most Mobile HCI studies focused on the technical aspects of prototyping and performing evaluations, if at all, in laboratory settings only. This is where Mobile HCI could learn from technical ICTD because field studies in natural settings and real world evaluations using action research and participatory design form the foundation of good technical ICTD research. The utilization of the rarely used Mobile HCI research methods in technical ICTD could foster a valuable cross fertilization.

The goal of this workshop (http://www.uctictd2010.org/) was to elaborate on the application of Mobile HCI methods for technical ICTD. Researchers from the ICTD research field presented their work and experiences with Mobile HCI research methods. Presentations covered topics including: experiences and lessons learned in technical ICTD research; success stories and failures of technical ICTD research; utilization and combination of Mobile HCI research methods in technical ICTD research; and utilization of action research, user centered design and participatory design in technical ICTD research.

In discussions during the workshop it became clear that there is a benefit in using Mobile HCI research methods in ICTD research, specifically technical ICTD. As an initial workshop result, however, the participating researchers revealed some points of adaptation of Mobile HCI research methods necessary for their efficient utilization in the ICTD context – e.g., cultural adaptation, ethics, end user focus, trust building, end user incentive models, focus on real business case, user centered research approach, and capacity building, to name only a few. The workshop participants will continue the discussion and further refine the necessary adaptations of Mobile HCI research methods to support technical ICTD research [Overview by organizer Jörg Dörflinger, SAP Research, Germany].

In reviewing the lessons learned from a project geared at improving the livelihood of marginal farmers in India through wireless sensor networks, Knoche *et al.* highlight the challenges faced by ICTD. They argue that a lack of focus on users and principal attention to technology in the design process resulted in lack of interest and acceptance of their technology by their target population. Knoche *et al.* discuss and argue for the benefits to ICTD of embracing human-centred design principles.

- **Social Mobile Web (SMW'10):**
  - ○ **Organizers:** Karen Church (Telefonica Research, Spain), Josep M. Pujol, (Telefonica Research, Spain), Barry Smyth (University College Dublin, Ireland), Noshir Contractor (Northwestern University, USA).
  - ○ **Best Paper:** *A Festival-Wide Social Network using 2D Barcodes, Mobile Phones, and Situated Displays* by Jakob Eg Larsen (Technical University of Denmark, Denmark) and Arkadiusz Stopczynski (Technical University of Denmark, Denmark).

The mobile space is evolving at an astonishing rate. At present there are close to 5 billion mobile subscribers worldwide and with continued advances in devices and services the mobile web looks set to inspire a new age of anytime, anywhere information access. The nature of the information being accessed is also shifting from traditional content towards social content. Online social networking sites such as Facebook, Twitter, and Foursquare continue to experience huge increases in usage, with more and more users seeking novel ways of interacting with their friends and family, e.g. sharing pictures, status updates, and providing recommendations.

The combination of the social and the mobile spaces are going to define the future of information consumption and communication. This edition of the Social Mobile Web workshop explored this novel and soon to be prevalent space. The contributions we received addressed different aspects of this challenging domain. For example, enabling novel social interactions via cell-phones, leveraging context-awareness in social settings beyond location, social-based recommendations while on the move and challenges in evaluating social mobile services in the wild [Overview by organizers Karen Church and Josep M. Pujol, Telefonica Research, Spain].

In their paper, Larsen and Stopczynski report on their experiences of applying unique 2D barcodes on wristbands and mobile phones to uniquely identify festival participants in order to develop an exploratory festival-wide social network prototype. They describe a set of social network applications and reflect on their experiences from initial use of these applications. They discuss how their novel approach enabled mass participation in the festival-wide social network due to the lack of any requirement for participants to have any special hardware or mobile client applications. On the basis that they found the 2D barcodes represented a feasible low-cost approach for unique participant identification and social network interaction, Larsen and Stopczynski reflect on the implications for the design of future systems of this nature.

- **Using Audio and Haptics for Delivering Spatial Information via Mobile Devices:**
  - ○ **Organizers:** Margarita Anastassova (CEA, LIST, France), Charlotte Magnusson (Lund University, Sweden), Martin Pielot (OFFIS Institute for Information Technology, Germany), Gary Randall (BMT Group Ltd., UK), and Ginger B. Claassen (C-Lab Siemens, Germany).
  - ○ **Best Paper:** *Wearable Tactile Display of Landmarks and Direction for Pedestrian Navigation: A User Survey and Evaluation* by Mayuree Srikulwong (University of Bath, UK) and Eamonn O'Neill (University of Bath, UK).

Orientation and navigation are very important skills for getting along in daily life. The acquisition and use of these skills is based on the processing of visual, auditory and sensorimotor/kinesthetic information, denoting the relations between objects, places, and people. With the recent availability of global positioning method, of comprehensive GIS systems, of powerful mobile computers and of advanced interaction techniques, multisensory spatial information could now be presented in a personalized, context-aware and intuitive manner. However, it is still not completely clear how to design, and how and when to present multisensory (audio, visual, haptic) spatial information on mobile devices. This question was the main focus of the workshop we organized at MobileHCI 2010.

Fourteen papers and demonstrations were accepted and presented at the workshop. All of them were highly relevant to the workshop topics. The following questions were discussed:

The workshop was very successful. The participants had very different backgrounds (e.g., computer science, engineering, design, psychology, marketing). They were coming both from academia and industry. This diversity in our educational and institutional backgrounds resulted in very rich, interesting and active discussions and demonstrations.

The workshop was organized in the framework of the EU Haptimap project. We are grateful to the European Commission which co-funds it (FP7-ICT-224675) [Overview by organizer Margarita Anastassova, CEA, LIST, France].

Srikulwong and O'Neill report on their investigation of representation techniques for spatial and related information in the design of tactile displays for pedestrian navigation systems. Srikulwong and O'Neill compared – on the basis of distinguishability, learnability, memorability and user preferences – two tactile techniques for landmark representation using either one or two actuators. They suggest their results – which suggest that when landmark and directional information is presented simultaneously, their two actuator approach was better at enabling users to maintain levels of performance – may be able to help guide the design of presentation of key landmark information on wearable tactile displays.

- **Please Enjoy!? Workshop on Playful Experiences in Mobile HCI:**
  - ○ **Organizers:** Ylva Fernaeus (SICS, Sweden), Henriette Cramer (SICS, Sweden), Hannu Korhonen (Nokia Research, Finland), and Joseph 'Jofish' Kaye (Nokia Research, USA).
  - ○ **Best Paper:** *Good Times?! 3 Problems and Design Considerations for Playful HCI* by Abdallah El Ali (University of Amsterdam, The Netherlands), Frank Nack (University of Amsterdam, The Netherlands) and Lynda Hardman (CWI, The Netherlands).

Designing for play and playful experiences has become a central theme in how people use and value interaction with mobile devices, and is also a focus in several research projects. This workshop aimed to gather experiences and explore different approaches and challenges to this particular sub-field of mobile HCI.

We believe that mobile devices are especially interesting to study in terms of playful experiences. First, mobile devices include a range of interesting sensors and media facilities (GPS, Sensors, Camera) which potentially enable extended possibilities for playful interaction. The widespread use of mobile devices also means that they are used in many different contexts, including social settings where playful activities emerge easily. With respect to this, mobile devices are used as mediators in social interaction both remotely and locally. Moreover, people carry mobile devices with them, making for very personal, smooth, and habitual practices, integrating some play in daily routines, in transitional 'non-places', and while waiting. A simplistic example is how people no longer have to 'stop' what they are doing and go away to participate in, for example, an online social network; instead such activities may run in parallel and on top of other activities. These are all aspects that have been extensively studied and addressed in, for example, ubiquitous gaming.

Researchers, designers and developers with interest in this theme were welcomed to participate in a full day activity of presentations and discussions. Nine position papers were presented in the morning session and thereafter focus was on discussing examples of own experiences playfulness with mobile technology. An overarching theme was discussions around what kinds of experiences can be considered playful, especially as there are many uncertainties as to how playful experiences can be addressed in design as well as in research [Overview by organizer Ylva Fernaeus, SICS, Sweden].

In their paper, El Ali *et al.* analyze how "fun or playfulness" can be studied and designed for in mobile and ubiquitous environments. They discuss the notion of a 'playful experience' and highlight three problems that arise in realizing the scenario: the inference problem, the experience-capture maintenance problem, and the measurement problem. In response, they suggest that experiences can be approached as information-rich representations or as arising from human-system interaction, incentive mechanisms can be mediators of fun and engagement, and measuring experiences requires a balance in testing methodology choice.

RESEARCH IN THE LARGE

Directing our attention to the opportunities and associated challenges of utilizing mobile app stores and markets as research vehicles, issue 3(4) of the IJMHCI was a special issue on *Research in the Large* guest-edited by Henriette Cramer, Mattias Rost, and Frank Bentley. As the guest editors noted, although mobile app stores and markets provide researchers with a huge opportunity to gather research data from the public at large, evaluation and research methods need to be adapted to this new context. Given a lack of information about successful strategies and ways to overcome the methodological challenges inherent to large-scale deployment for research purposes, the guest editors organized a workshop dedicated to this topic, from which the special issue of the IJMHCI arose. Interested readers should refer to IJMHCI Vol 3., Issue 4 to read the guest editors' preface which provides an overview of strategies and opportunities in 'research in the large', as well as providing an introduction to the challenges with associated research ethics and validity. In collating this special issue of the IJMHCI, the guest editors hoped "*that these articles will help to form a beginning of a community around large-scale deployments and that the lessons from these authors can be taken forward to help improve the quality of future work in this area*" and that "*as more research is completed in this domain, best practices and updated methods will emerge that can more reliably lead to valid and repeatable results.*"

The articles included in this special issue ascribe to three main themes – distribution, data analysis, and validity – with some articles spanning the themes. In the first article, entitled "A Comparison of Distribution Channels for Large-Scale Deployments of iOS Applications," Donald McMillan, Alistair Morrison, and Matthew Chalmers provide an analysis of two ways to distribute an iOS application: through the official Apple App Store or through third-party repositories. This analysis highlights decision criteria that research teams should analyze if deciding to deploy a new application for the iOS platform.

In an article entitled "WorldCupinion: Experiences with an Android App for Real-Time Opinion Sharing During Soccer World Cup Games," Robert Schleicher, Alireza Sahami Shirazi, Michael Rohs, Sven Kratz, and Albrecht Schmidt present their experiences with releasing a research application on the Android Market. They highlight important issues that researchers must take into account when conducting research in the large, in addition to giving accounts to what they learned about their application, WorldCupinion.

This time focusing on data-analysis, Alistair Morrison and Matthew Chalmers describe (in "SGVis: Analysis of Data from Mass Participation Ubicomp Trials") a tool to visualize real-time results from a large-scale deployment as well as a way to see patterns of use across individual users. This system shows great promise in identifying patterns and combining more qualitative methods with the quantitative data that is being collected.

The final two articles focus on validity of results. In "Experimenting through Mobile Apps and App Stores," Paul Coulton and Will Bamford give accounts of a longitudinal study, including two apps with more than 1.5 million downloads, of how app stores behave with experiments. They show how the number of new downloads are affected by events like updates, changes in presentation, etc. In "My App is an Experiment: Experience from User Studies in Mobile App Stores," Niels Henze, Martin Pielot, Benjamin Poppinga, Torben Schinke, and Susanne Boll discuss five large-scale deployments that they have been involved in and discuss practical details of what has and hasn't worked in order to attract users and receive meaningful data when using an app as an experimental 'apparatus'. They discuss differences in use observed in an 'in the wild' evaluation versus a more controlled field experiment as well as the ethics of conducting research with unknown participants.

CLOSING COMMENTS

By bringing together the articles from Volume 3 of the IJMHCI which deliver exciting research and innovation, practical guidance, and research challenges in the field of Mobile HCI, this compendium serves as an essential publication for researchers, educators, students, and practitioners alike.

Continuing into a third year of publication of the IJMHCI would not have been possible without the ongoing efforts of the journal's amazing team of reviewers and associate editors. As the journal ages and becomes increasingly established as a key publication venue for research and innovation in the field of Mobile HCI, there has been an increasing growth in the stream of submissions, making the voluntary contribution of the IJMHCI's team even more laudable. I would, therefore, like to take this opportunity to thank all the members of the team who make the IJMHCI such a great publication – the invaluable contribution of the journal's advisory boards and board of reviewers and, of course, the essential contribution made by the authors of published articles and guest editors of special issues. These combined efforts have not only led to another year of an exciting and vibrant journal, but have culminated in this comprehensive publication.

All that remains now is for me to welcome you to this book which draws together all the research achievements and challenges presented in the third volume of the IJHMCI.

Joanna Lumsden
Aston University, UK

ENDNOTES

1. One workshop – *Ensembles of On-Body Devices* organised by Daniel Ashbrook (Nokia Research, USA) and Kent Lyons (Intel Labs, USA) – did not nominate a best paper. Information on this workshop can be found at http://burx.com/ensembles/.

2. J. Kjeldskov and C. Graham, "*A review of mobile HCI research methods,*" Lecture Notes in Computer Science, 2003.

Chapter 1
Evaluating the Visual Demand of In-Vehicle Information Systems:
The Development of a New Method

Ainojie Alexander Irune
University of Nottingham, UK

ABSTRACT

In-vehicle information systems (IVIS) provide a variety of driver support and infotainment functionality; however, there is a growing concern that the resulting engagement with IVIS could present significant sources of distraction to drivers. This paper summarises the PhD thesis of Dr Ainojie Alexander Irune, which was awarded at the University of Nottingham in December 2009. The primary aims of the research were to develop a framework to aid the selection of an appropriate HF/HCI method, for use at particular stages in the design process, and to develop a novel method (with a focus on glance duration) for assessing the visual demand afforded by IVIS. Five empirical studies are reported in the thesis. In the first study, interviews were conducted with subject experts and the results were combined with the literature to provide guidance regarding the appropriate use of human factors methods. The remaining four studies present an iterative development of a novel method capable of predicting the visual demand imposed by an IVIS.

INTRODUCTION

The last two decades have seen computer and communication technology become more prevalent in cars, enabling the incorporation of systems and devices for both driver support (e.g., navigation

aids) and infotainment (e.g., news and email). These systems are generally designed to improve the safety, efficiency and comfort of the driver. There is growing concern amongst HCI specialists and Human Factors researchers that new forms of technology, accessible to drivers, could become

DOI: 10.4018/978-1-4666-2068-1.ch001

significant sources of distraction and may have an impact on road accidents (Green, 1999a; Burnett et al., 2004; Pettitt, 2007). Although, distraction can occur as a result of events from either within or outside the vehicle, there is a real concern that poorly designed IVIS may introduce several dangerously distracting tasks into the driving environment (Pettitt, 2007).

In statistics from the UK, loss of control due to inattention was the most frequently reported contributory factor, involved in 35 percent of fatal accidents in 2006 (DfT, 2007). Although all forms of distraction could potentially negatively affect the driving task, visual distraction appeared to be a key contributing factor (Neal et al., 2005; Angell et al., 2006). It was reasonable therefore to infer that eye glance data contained important information for assessing the distraction effects of IVIS; in particular, glance duration. This has been noted by many authors to be an important ocular-based indicator of attention (Rockwell, 1988). Experimental research investigations into the degradation in primary driving tasks, as a result of sustained visual attention needed to achieve a secondary task, have generally attempted to highlight certain constraints, which secondary tasks must meet in order to be acceptable while driving (Gelau & Krems, 2004; Bhise, Forbes, & Farber, 1986; Green, 1999a). It is generally agreed that glances lasting longer than approximately 2.0 seconds are not acceptable while driving (Gelau & Krems, 2004).

Although a range of HF/HCI evaluation methods are available which allow researchers to describe, understand and predict relationships between variables in this area, practical concerns exist which limit their use in the real world. These include:

- **Deciding Which Method is Most Appropriate at a Particular Stage of Development:** Although some of the existing methods have been formalised e.g. occlusion (ISO, 2007), the majority of these methods could benefit greatly from further development (e.g., lane change test (LCT)) (ISO/WD, 2005), peripheral detection test (PDT) (Van Winsum et al., 1999). This lack of development often implies that there are no set procedures for implementing or assuring the reliability, and in some cases, the validity of the method. In this regard, designers and practitioners are often unsure as to which method would yield the best assessment at a particular stage in design.

- **Deciding Which Methods are Most Useful:** A further concern surrounds the measures obtained from the evaluation methods and their implications in relation to distraction. For example, methods such as field trials and simulator trials provide the opportunity to obtain direct measures of distraction. However, this involves frame-by-fame video analysis of IVIS interactions, which is a very expensive approach. On the other hand, methods such as occlusion (ISO, 2007), 15 second rule (SAE, 2000), PDT and LCT provide a less costly approach but predominantly focus on providing measures for visual allocation (glance frequency, total glance time, resumability (R)), vehicle control (lane deviation, deviation from a selected norm) and object and event detection (latency to detect, missing detection, situation awareness). These measures may imply degrees of visual and/or cognitive distraction but do not provide a direct measure of distraction. Practitioners are often unaware of which aspects of distraction are being measured.

Pettitt (2007) proposed a framework to group evaluation methods into three categories based on their economy, validity and reliability. To ensure the practical applicability of a method in the real world (i.e., amongst car and system manufacturers), a key additional concern is usability (how easy is it to use this method/technique in practice,

Figure 1. Categorisation framework for evaluation methods (Adapted from Pettitt, 2007)

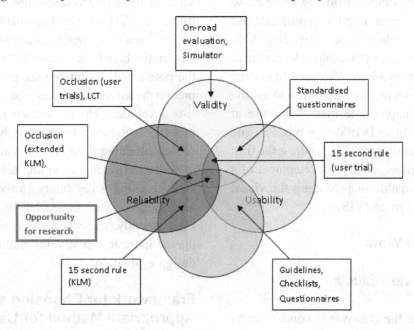

with real systems, in an iterative manner). Figure 1 updates Pettitt's original framework to include usability. The revised framework indicates that a method has greatest utility if it lies within the central zone bubble created by the intersections of all four factors (i.e., it is valid, reliable and economical, as well as easy to use).

Scope of the Thesis, Aims and Objectives

Two key aims of this research are:

1. To develop a framework to aid the selection of an appropriate human factors method(s) for use at particular stages in the design process of in-vehicle information systems.
2. To develop a valid, economical, reliable and easy to use evaluation method, focused on obtaining glance duration measures, for assessing the visual demand afforded by IVIS.

In order to meet these aims, five objectives were identified:

1. To investigate issues of distraction (definition, types, size of the problem and the role of technology in distraction related accident) and workload (specifically, the exploration of visual workload and its associated metrics).
2. To investigate the issue of IVIS design and evaluation requirements with a focus on the automobile industry and IVIS manufacturers.
3. To assess, through available literature, and interviews conducted with experts, the range of methods currently available for measuring distraction potential.
4. To aid the selection of an appropriate human factors method(s) for use at particular stages in IVIS design process.
5. To test the validity, reliability and usability of a method developed for measuring distraction potential.

Initially, a review of the available literature aimed to address the first objective. The first of five empirical studies sought to address the second objective through interviews conducted with ten Human Factors experts. The results were combined with information obtained from the literature to provide guidance regarding the appropriate use of human factors methods when evaluating IVIS. The information gathered alongside pertinent parts of the literature were also used to aid the development of a tentative framework to aid the selection of an appropriate method for use at particular stages in an IVIS design process. The remaining four studies sought to address the fifth objective through an iterative development of a novel method capable of predicting the visual demand imposed by an IVIS.

Experimental Work

Study 1: Interview Study

An initial step of the PhD was to conduct interviews to gain an insight into the working practices in industry and to understand the contribution of ergonomics and HCI to the current design and development of in-vehicle interfaces.

A series of telephone interviews were conducted with ten ergonomic and human factors experts, within relevant industrial research institutions (e.g., TNO, BAST, TRL, and MIRA), as well as car and equipment manufacturers (e.g., Chrysler, Ford, Visteon and Alpine). Questions asked aimed to understand the use of human factors methods in the assessment of IVIS, the context in which evaluation methods are used and the relationship between designer/stylist and human factors /HCI practitioners.

The results from the interviews conducted highlighted particular issues of interest pertaining to the general attitude of industry to human factors/HCI evaluation methods and their use, some key questions surrounding product development focus, integration of the design team, and some general human factors concerns.

Some recommendations on how human factors should integrate with design evaluation were presented by a number of interviewees. Although

some of these interventions are currently being used (e.g., PDT, LCT and simulation), others are not used and these include predictive human factors methods (e.g., extended KLM) which have the potential to estimate user performance and predict the safety implications of a system before it is developed. The motivation for using these methods in industry, tie in strongly with the cost of implementing them (e.g., resource and time), as well as providing clear and concise evidence which highlights key safety implications of system design choices from a HF/HCI perspective. Consequently, there is a drive for industry to pursue more low cost valid methods for use in design evaluation.

Framework for Choosing an Appropriate Method for Use at a Given Stage in Design

A tentative framework known as a T-Matrix was constructed from information available within the literature and obtained from the interviews conducted (see Figure 2). When choosing a method or combination of methods for use in the evaluation of IVIS, consideration must be given to three key components: the range of methods available (e.g., road and simulator trials, etc.); the nature of the evaluation (e.g., analytic/evaluative, formative/ summative); and key factors that can affect the use of the method (e.g., ethical constraints, resource issues, etc.).

Effective design evaluation must take these constraints into account when deciding on an appropriate method for use at a given stage of development. For this, the T-Matrix compares one list against two others in pairs, that is, a comparison of the methods available against a range of requirements and against the tasks demands they are capable of measuring.

A value of High, Medium and Low is given in each column to the right of the 'Method' column, for each method, denoting the level of confidence

Figure 2. T-matrix comparing methods against measures and metrics

STd	SDTsk	PVCtrl	PDTsk	Method	Stage in design	Environment	Validity	Reliability	Resource	Usability	Reproducibility	Efficacy
XXX	XXX	XX	X	Field Trials	Sum	Field	High	High	Low	Low	Low	Low
XXX	XXX	XX	X	Road Trials	Sum	Field	High	High	Low	Low	Low	Low
XXX	XXX	XXX	XXX	Simulator Trials	Form/Sum	Lab	Mid	Low	Mid	Mid	High	High
XX	XX			Occlusion	Form/Sum	Lab	High	High	High	High	High	Low
	XX			KLM	Form	Lab	High	High	Mid	Mid	High	Mid
XX	XX			EKLM	Form	Lab	High	High	High	Low	Mid	Mid
	XX			15 second Rule	Form/Sum	Lab	Mid	High	High	High	High	High
X	X	XXX	XXX	LCT	Form	Lab	High	Mid	High	High	High	High
XXX	XX	XX	XX	PDT	Sum	Lab/field	Mid	Mid	High	High	High	Mid
XX	XX			Heuristic Evaluation	Form/Sum	Lab	Low	Low	High	High	Mid	Low
X	X		X	Standardised Questionnaire	Form/Sum	Lab/Field	Mid	Mid	High	High	High	Low

noted for that method, for any of the given requirements. Although tests have not been carried out to support the method categorisation proposed (i.e., low, medium and high), it has been inferred, subjectively based on the literature explored thus far. More research work is required to further establish the validity of the matrix.

On the left side of the 'Method' column, the tasks of manipulation to be considered in the evaluation of in-vehicle systems are presented in the subsequent columns. Tasks are denoted as measurable by a particular method if there is an 'X' in the column with the task heading (e.g., PDTsk). A single 'X' (low) to three 'X's (high), express the degree to which the method in question is able to provide measures for that task.

The matrix allows a clear comparison between the methods available and the above listed variables, as well as a cross comparison against the tasks/interactions that can be measured. This diagram represents a starting point for the development of a framework for selecting a method suitable for use, at a particular stage in the design process.

Study 2: Feasibility of a Novel Method for Evaluating the Visual Demand of IVIS

As part of an iterative development process, study 2 explored a novel approach loosely based on the occlusion method (ISO, 2007). When engaging in a secondary task while driving, visual attention is shared between the primary driving task and the secondary in-vehicle task (see Figure 3). The occlusion method aims to mimic this sharing by allowing participants only brief periods of vision to simulate the glance behaviour of performing a secondary task in the driving environment. Participants in an occlusion study are required to carry out a range of in-vehicle tasks whilst wearing a pair of occlusion goggles. These are computer managed goggles that conceal the driver's vision at specific and controlled time intervals. By providing vision for a period of time, immediately followed by a period of occlusion, glancing behaviour is mimicked in a controlled fashion (ISO, 2007). Measures obtained using the occlusion method include, TSOT (Total Shutter Open Time - time the participant had vision whilst performing the

Figure 3. Illustration of the relationship between occlusion and PLT approach

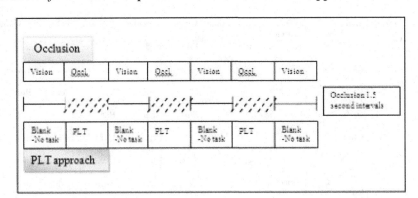

task) and R (Resumability ratio - the ease of resumption of a task following an interruption).

Occluded periods (where participants are without vision) within the occlusion method may be viewed as representative of glances to the road scene however, a driving related task is not provided in this period. It is argued that improving the method should involve the inclusion of a task in the occluded periods to prevent participants from rehearsing their suspended goals (Trafton et al., 2003). Suggestions for the nature of the task to replace the occluded period have prescribed one that closely resembles the driving task. In this regard, the occluded period should incorporate some other task, much like people attend to when engaged with the primary driving task, thereby creating an interruption that is contextually appropriate.

For study 2, an investigation was conducted into the feasibility of a primary loading task (PLT) approach, aimed at exploiting the underpinning concept of the occlusion technique. This involved presenting drivers with a time window for interacting with the secondary task at specific and controlled time intervals, achieved by employing a PLT which incorporated key components of driving (e.g., visual-spatial search and recognition). Subsequently, the occluded period within each occlusion cycle was substituted with a PLT.

Key reasons for exploiting the occlusion technique were due to its characteristics of:

- **Systematic Control:** Imposed periods of vision and non-vision whilst performing secondary tasks, ensure minimal variability in subjects' behaviour and consequently reduced variability in the data gathered.
- **Reliability:** The consistency and repeatability of a single measure; allows for the replicability of the method.

A key consideration of the PLT approach was the nature of the primary loading task employed (e.g., visual, auditory, cognitive etc.) and the issues surrounding its design. Driving is viewed as primarily a visual task, even though the motor demands are significant (Green, 1994; McGwin & Brown, 1999). A statement to the effect that 90% of the information required for driving is visual is common in the literature on driving (Sivak, 1996). From such a perspective, it seemed justified to explore a visual-spatial oriented primary loading task, as this contained the key components of real driving.

The PLT task required participant to alternate their vision between the PLT and the ST. The PLT occurred as part of a Microsoft PowerPoint slide show that appeared on a 17" monitor placed directly in front of the driver. The slide show progressed between variants of the PLT slide followed by a blank slide every 1.5 seconds. Participants were asked to search an array of four blocks of arrows on each PLT slide and indicate verbally,

Figure 4. Primary loading task

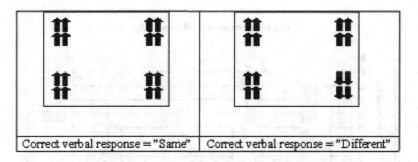

whether one was different (i.e., all arrows are not facing the same direction) or the same (i.e. all arrows facing the same direction) (see Figure 4).

To summarise, a total of 10 people took part in a user trial and they performed 4 tasks using two IVIS. Mean values for Total Task Time (TTT), Total Interrupted Vision time (TIVT) (equivalent of TSOT in occlusion) and R(PLT), equivalent of Resumability ratios (R) in occlusion were compared.

Results from the exploratory study demonstrated promising aspects to the PLT approach. A key observation was the potential of the approach to discriminate between tasks. Task 1 - (POI entry task) required users to assess a stored list, in contrast to task task 2 – (Destination entry task) which required users to navigate a menu system were repeated manual entries were required. A similar trend was found across tasks 3 and 4 (see Figure 5). It was evident from observing the trials, that task 4 was more complex as it possessed a

scrolling list with three times more columns than task 3 for the participant to search through.

Figure 6 shows a graphical representation of the TIVT values illustrated in the Figure 5 across all four tasks against time.

By introducing a primary loading task in the occluded cycle, it is anticipated that participants may be prevented from rehearsing their suspended goals. Observations of the trials and post-trial interviews indicated that there was a reduced effect of blind operation. Another potential merit of this approach is in its cost of implementation, the PLT method only requires mundane computing equipment (e.g., computers, monitor screens, keyboard and mouse) which are cost effective and more widely available in comparison to other methods (e.g., occlusion and simulation).

It is evident that efforts in developing the PLT method further, may potentially lead to a novel, economic, usable and valid evaluation method. In light of the above, further investigation is needed

Figure 5. PLT study main results

	TTT (Seconds)	TIVT (Seconds)	R(PLT) (Seconds)
Task 1 – POI	10.1(4.08)	10.0 (3.77)	1.07 (0.274)
Task 2 – Address	15.2(5.30)	11.6 (2.93)	0.82 (0.189)
Task 3 – Short scrolling	8.0(2.57)	5.0 (2.48)	0.77 (0.430)
Task 4 – Long scrolling	9.3(5.16)	6.4 (2.92)	0.96 (0.507)

STd = Secondary task demands, SDTsk = Secondary driving task, PVCtrl = Primary vehicle control task, PDTsk = Primary driving task Sum = Summative development stage, Form = Formative development stage

Figure 6. PLT study main results

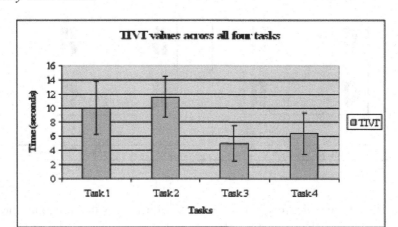

to better understand and improve aspects of the primary loading task (e.g., experimental protocol in relation to task prioritisation, PLT response protocol and PLT design, exploration of other PLT measures (e.g., number of PLT missed) in relation to visual demand).

Study 3: Exploring PLT Measures

This study presented an empirical evaluation of the PLT method, focused on addressing issues raised in the previous study regarding the viability of the novel approach. In addition, study 3 aimed to establish the logic underpinning the method and explored a range of PLT measures that had the potential to link to measures of visual demand (e.g., the number of PLT missed could potentially highlight visual inattention periods to the PLT task). As part of an iterative design process, a variety of improvements to the PLT method were addressed. These included issues concerning the experimental protocol, the nature of the PLT response and the length of the PLT task (see Figure 7).

Figure 8 illustrates the visual interaction of participants in a PLT study. They will begin the PLT at the start of the trial and after a short period they engage with the secondary task. The PLT continues to run simultaneously while the participant attempts to perform the secondary task. Shortly after this is completed, the PLT is terminated.

A full description of this study and its results can be found in Irune (2009). To summarise, a total of 10 people took part in a user trial performing a range of waiting task in which a series of squares were presented to participants (Participants had to wait as sequences were displayed in 0.5 second intervals, until key information was made available), in an attempt to manipulate drivers' visual attention in a controlled way by encouraging a wide range of glance durations. This was designed to explore the relationship between real glance duration (i.e., measured through frame-by-frame analysis of the videos from the trials) and expected glance duration (i.e., secondary task waiting duration 1, 1.5, 2, 2.5 and 3 sec).

Figure 7. Primary loading task

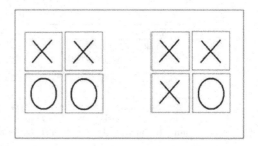

Figure 8. Dual task interaction with PLT method

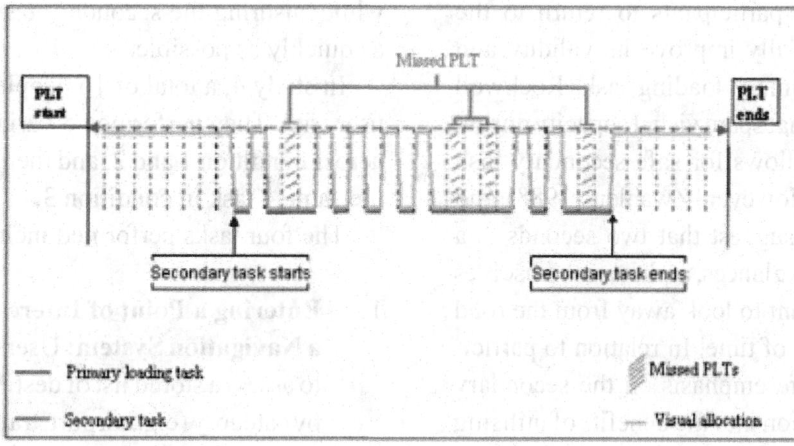

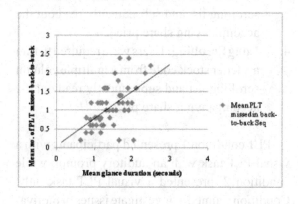

Figure 9 highlights the relationship between mean glance duration and number of PLT missed in a back-to-back sequence.

Figure 10 summarises the Pearson correlation values for mean glance duration and mean glance frequency, against PLT. The highest correlation value was found between the mean values for

Figure 9. Glance duration against PLT missed in a back-to-back sequence

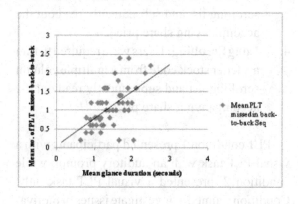

glance duration and the number of PLT missed in a back-to-back sequence (r=0.563, p < 0.01).

Analysis of the data showed that the highest correlation across all PLT measures in relation to real glance measures, was between the PLT missed in a back-to-back sequence, and the mean secondary task glance duration (r=0.563, p < 0.01). The results clearly highlight the degradation in primary task performance across this specific PLT measure, as a result of sustained visual attention needed to achieve a secondary task.

The variability in the results obtained from the study can be attributed to the different strategies employed by participants. It is important to note that participants had three attempts to perform the secondary task, whilst also engaging with the PLT. Although the results obtained were desirable, minimising the strategies adopted by participants would have produced less data variability. This potentially could have been achieved by simply reducing the number of attempts at achieving the secondary task, from three to a single attempt.

Figure 10. Correlation between real measured values of visual demand and PLT values

Independent variables	Mean no. of PLT missed in back-to-back sequence	Mean no. of PLT missed
Mean glance duration	0.563 (p < 0.01)	0.528 (p < 0.01)
Mean glance frequency	0.167	0.205

Furthermore, providing the necessary degree of motivation for participants to return to the PLT could potentially improve its validity and relevance as a primary loading task. Rockwell (1988) indicated that spare visual capacity during the driving task allows for safe secondary task related glances. However, Zwahlen (1988) and Rockwell (1988) suggest that two seconds is a maximum for such glances, as drivers themselves are usually reluctant to look away from the road for longer periods of time. In relation to participants placing more emphasis on the secondary task, an investigation into the benefits of utilising an auditory signal to alert participants about PLT changes, could potentially serve as a reminder to participants of the continuity of the PLT, while their attention is diverted away from it.

Study 4: Real IVIS Task Study and Comparing PLT Against Driving Simulator Method

The previous study investigated the feasibility of the approach and improvements required to further develop aspects of the PLT method. The promising results obtained demonstrated that there was merit in the PLT approach and highlighted the potential for the method to be utilised as an evaluation tool, capable of providing a surrogate measure for glance duration.

Nevertheless, it was concluded that in order to present the new method as both internally and externally valid. A study was conducted to examine a wide range of real-world, in-vehicle secondary tasks and to evaluate the method alongside an established user trial technique (i.e., simulation).

The PLT task used was identical to that used in the previous study with the only key change made to the verbal instructions given to the participants before the study. Participants were requested to alternate their vision between the PLT and the secondary task (i.e., IVIS task) with the following aim: To give the primary task the highest priority for the duration of the trial, with

the aim of maintaining error free performance, while ensuring the secondary task is completed as quickly as possible.

In study 4, a total of 16 people took part in a user trial. They performed 5 variations of 4 task across condition 1 and 2, and then 2 variations of the same 4 task in condition 3.

The four tasks performed included:

1. **Entering a Point of Interest (POI) Using a Navigation System:** Users were required to assess a stored list of destinations ordered by category (e.g., airport, train station, shopping centres). This list was accessed through the menu system and was ordered, based on proximity.

2. **Entering a Street Address Using a Navigation System:** Users were required to navigate a menu system in order to reach the data entry screen. The data was structured by town, street and house/building number. The data entry method was touch screen and the input keyboard followed an ABC layout, as opposed to a QWERTY layout (common with standard desktop computer keyboards).

3. **Short Scrolling:** Users were required to find a 3 letter stock code from a single column scrolling list and subsequently read out the accompanying share price.

4. **Long Scrolling:** Users were required to find a 3 letter stock code from a multiple column scrolling list and subsequently read out the accompanying share price.

PLT condition 1 presented participants with a visual PLT task with an auditory prompt, while condition 2, presented a visual PLT task only. Condition 1 aimed to investigate issues of motivation - it was envisaged that the auditory prompt would act as a reminder to participants engaged in the secondary task that the PLT was progressing ahead, even whilst their attention was being diverted away to the secondary in-vehicle task.

Figure 11. Correlation of measures from PLT condition 1 (auditory prompt)

Condition 1	MGD	MGF	MTTT	MissPLT
MGF	-.477**			
MTTT	.112	.728**		
MissPLT	.668**	.038	.673**	
MissPLT_seq	.928**	-.327**	.210	.710**
**. Correlation is significant at the 0.01 level (2-tailed).				

MGD = Mean glance duration, MGF = Mean glance frequency, MTTT = Mean total task time, MissPLT = Mean no. of PLT missed, MissPLT_seq = Mean no. of PLT missed in sequence.

Condition 3 was a simulator trial conducted in a fixed based, low-cost simulator, interactive driving simulator at the University of Nottingham. The trials aimed to assess the validity of the measures obtained via the PLT method.

Pearson's two-tailed correlation analysis was performed on the data. Figure 11 and Figure 12, present a correlation analysis of the values of each measure from the PLT trials, compared against each other, for conditions 1 and 2 respectively.

As established in the previous study, there was a significant correlation between glance duration and the number of PLT missed in a back-to-back sequence for both conditions 1 and 2.

Both PLT conditions show a significant correlation between MissPLT and MTTT, i.e., condition 1 (r=0.673, p<0.01) and condition 2 (r=0.781, p<0.01). This correlation highlights a link between the numbers of PLT missed (MissPLT) and total task time (MTTT).

Overall, both PLT conditions appear to show minimal differences in the data obtained, subjective data suggests that condition 2 (no-auditory prompt) was less stressful than condition 1 (auditory prompt).

Comparing PLT Against Simulator Trials

Figure 13 and Figure 14 show the correlation of mean values for all measures obtained from the PLT trials under condition 1 and 2 respectively, against those obtained from the simulator trials. Firstly, mean glance duration values with standard deviation in brackets, are presented for the simulator trials: Task 1: 1.47 (0.32), Task 2: 2.22 (0.5), Task 3: 1.63 (0.26) and Task 4: 2.06 (0.44). Figure 13 and Figure 14 highlight the correlation of identical measures (i.e., glance duration, glance frequency and total task time) across both PLT and simulator trials.

Figure 12. Correlation of measures from PLT condition 2

Condition 2	MGD	MGF	MTTT	MissPLT
MGF	-.420**			
MTTT	.064	.842**		
MissPLT	.559**	.393**	.781**	
MissPLT_seq	.930**	-.250*	.210	.661**
*. Correlation is significant at the 0.05 level (2-tailed).				
**. Correlation is significant at the 0.01 level (2-tailed).				

MGD = Mean glance duration, MGF = Mean glance frequency, MTTT = Mean total task time, MissPLT = Mean no. of PLT missed, MissPLT_seq = Mean no. of PLT missed in sequence.

Figure 13. PLT measures (condition 1) against simulator trial measures

PLT cond. 1 vs. Simulator trials	SMGD	SMGF	SMTTT
PMGD	.537**	-.116	.131
PMGF	-.500**	.699**	.513**
PMTTT	-.060	.606**	.690**
MissPLT	.350**	.156	.417**
MissPLT_seq	.430**	-.027	.184
**. Correlation is significant at the 0.01 level (2-tailed).			

SMGD = Simulator mean glance duration, SMGF = Simulator mean glance frequency, SMTTT = Simulator total task time, PMGD = PLT Mean glance duration, PMGF = PLT Mean glance frequency, PMTTT = PLT Mean total task time, MissPLT = Mean no. of PLT missed, MissPLT_seq = Mean no. of PLT missed in sequence.

To illustrate glance behaviour exhibited in each trial (PLT condition 1, 2, and simulator), Figure 15 presents glance duration values for all three condition. Although values for the simulator trials generally appear to be lower, a similar pattern is highlighted across the four tasks.

Values for MGD for the simulator show differences between task 1 and 2 (navigation task), and task 3 and 4 (visual search tasks). A within subjects paired sample t-test was conducted to determine the significance of this differences. MGD (secs): For task 1 (M = 1.47, SD = 0.32) versus task 2 (M = 2.22, SD = 0.5), and task 3 (M = 1.63, SD = 0.26) versus task 4 (M = 2.06, SD = 0.44). The results showed that the differences were significant for task 1 versus 2 ($t(15) = -7.608$,

$p < 0.001$) and task 3 versus 4 ($t(15) = -4.665$, $p < 0.001$). Values for MGF and TTT also show a relatively similar pattern for all four tasks, across the three conditions.

The results of the study conducted were encouraging. They highlighted a significant correlation between the number of PLT missed in a back-to-back sequence and glance duration, for PLT condition 1 ($r=0.928$, $p<0.01$) and condition 2 ($r=0.930$, $p<0.01$). In addition to studies presented earlier, these correlations provided more evidence to support the PLT method's ability to provide a surrogate for glance duration.

A key outcome from this trial was the effect of the auditory prompt versus the no-auditory prompt conditions, on participants' motivation to switch

Figure 14. PLT measures (condition 2) against simulator trial measures

PLT cond. 2 vs. Simulator trials	SMGD	SMGF	SMTTT
PMGD	.441**	-.226	.033
PMGF	-.192	.621**	.587**
PMTTT	.091	.486**	.626**
MissPLT	.387**	.134	.428**
MissPLT_seq	.385**	-.072	.140
**. Correlation is significant at the 0.01 level (2-tailed).			

MGD = Simulator mean glance duration, SMGF = Simulator mean glance frequency, SMTTT = Simulator total task time, PMGD = PLT Mean glance duration, PMGF = PLT Mean glance frequency, PMTTT = PLT Mean total task time, MissPLT = Mean no. of PLT missed, MissPLT_seq = Mean no. of PLT missed in sequence.

Figure 15. PLT (condition 1 and 2) and simulator mean glance duration

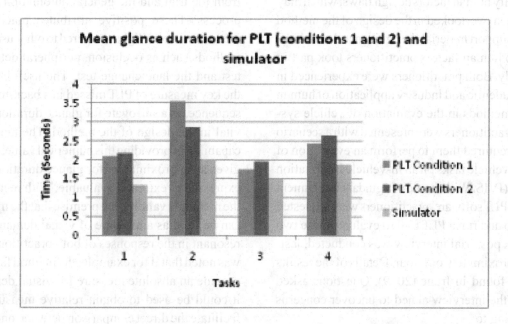

between tasks. Although it was observed that the beep sound encouraged participants to return to the PLT, there was an adverse effect caused by the irritation and anxiety induced by the constant, repetitive nature of the sound. Subsequently, participants expressed a preference for condition 2 (no-auditory prompt), as it was perceived to be less unsettling.

Unexpectedly, significant correlations were found between the number of PLTs missed and total task time for both PLT conditions. Although not pertinent to the central issue of obtaining glance duration measures, such significant correlations highlight a further benefit of the PLT method. Measures of total task time have been used to imply acceptable and unacceptable levels workload afforded by an IVIS (e.g., 15 second rule) (Green, 1999b; SAE, 2000). Therefore, the PLT method could potentially be useful for this measure. Further research is required however to validate the finding.

In relation to the external validity, the results obtained from both PLT conditions appeared to be in line with those obtained from the simulator

condition. Although measures differed across experimental conditions on an absolute basis, comparisons of performance differences between trials, show the differences to be in the same direction and of similar magnitude. In this regard, relative validity was established. In addition, the correlation between the number of PLT missed in a back-to-back sequence and glance duration in the simulator (r=0.385, p< 0.01), further establishes the external validity of the PLT measure and the method as a whole.

Study 5: Preliminary Assessment of PLT Method Usability

With 'usability' being one of the key considerations, highlighted in the development of a novel IVIS evaluation method, it was imperative to conduct an evaluation to explore the 'ease of use' of the PLT method.

With the aid of the PLT guidance document and software program designed to automate the PLT method process (including the analysis of the data), a preliminary investigation was conducted to identify human factors design flaws, which may have been overlooked in the design of the method and its support material (i.e., guidance document).

Two human factors practitioners took part in the study. Both practitioners were experienced in both academic and industry application of human factor methods in the evaluation of vehicle systems. Practitioners were presented with a scenario which required them to perform an evaluation of two in-vehicle tasks on an in-vehicle information system (IVIS. With the aid of a guidance document and the PLT software, practitioners were requested to setup and run a PLT trial to evaluate these two tasks. A post trial interview was conducted, lasting approximately one hour. Details of the results can be found in Irune (2009). Questions asked during the interview aimed to uncover concerns pertaining to:

- The structure of the guidance document and the method.
- The learnability of the procedure.
- Time and cost.
- The usefulness of the key measures obtained by the method.
- The validity of the method.
- The reliability of the method.
- Overall impression of the method.

For the most part, both practitioners were able to follow the instructions provided in the guidance document. Concerns were however expressed about the layout of the instructions contained in the guidance document. These concerns highlighted a need for a more chronologically ordered sequence of events across the entire process. There were no concerns raised regarding the method's learnability. Both practitioners found the method easy to learn; the equipment was simple to setup and instructions for running the trial were easy to remember. Generally, the method was seen not to be time consuming or require excessive effort (laborious) in its administration. Practitioners made reference to the ease of obtaining measures

from the trial and the general automation of the process. These positive attributes placed the method in a better light compared to other user trial methods such as occlusion, peripheral detection test and the lane change test. The usefulness of the key measure of PLT missed in a back-to-back sequence, as a surrogate for glance duration, was vital in the design of the method. The method's capability of providing this numerical value, which gives an approximation for glance duration, was expressed as extremely valuable. With regards to the method's validity, the premise that the method can be used as a measure of visual demand was resonant in the responses of both practitioners. It was noted that if for example, the method failed to provide an absolute measure for visual demand; it could be used to obtain relative measures to facilitate the direct comparison between one task/system, against another. A concern raised however, in relation to the measures obtained was that it was unrealistic that participants would correctly respond to each PLT. In this regard, it was suggested that a baseline performance measure of PLT alone, compared with PLT and secondary task interaction, could potentially yield a more precise measure in relation to visual demand (more specifically, glance duration). However, it was acknowledged by both practitioners that not considering a baseline measure could be the reason why statistically, the number of PLT missed in a back-to-back sequence, is a more appropriate measure than say, the number of PLT missed. A general view held by both practitioners was that the method appeared to be reliable, especially due to the fact that the process was automated. As with newly developed methods, it was acknowledged that several trials would need to be conducted to establish this in principle.

The PLT method was acknowledged not to require any form of simulation; it was simple to setup; proprietary hardware and software were not required; and the automation of the process was of considerable benefit. It was however emphasized that the method's key strength, was in the ease of use and its ability to provide a surrogate for glance duration in a simple and cost effective way. Nevertheless, a more comprehensive usability evaluation is required (e.g. observing designers applying the method in situ). It is envisaged that this will aid further development of the PLT method.

General Discussion and Conclusion

The two key aims of this research were to: 1) develop a framework to aid the selection of an appropriate human factors method(s) for use at particular stages in the design process of in-vehicle information systems and; and 2) develop a valid, economical, reliable and easy to use evaluation method, focused on obtaining glance duration measures, for assessing the visual demand afforded by IVIS.

1. Framework to aid selection of an appropriate method Glance duration prediction capability.
    Based on the review of available literature and interviews conducted, a tentative framework was realised. Within this framework, a pool of methods was compared against a range of criteria for their suitability, in relation to the nature of the intervention proposed. Based on the requirements of the evaluation, designers and human factors practitioners can utilise this matrix to determine the most suitable method to apply.

2. Development of a valid, economical, reliable and easy to use evaluation method, focused on obtaining glance duration measures, for assessing the visual demand afforded by IVIS.

The following section reviews the PLT method with respect to its economy, usability, reliability and validity.

Economy

Analysis

A cost benefit of the PLT method is in its ability to provide measures for glance duration in a cost effect manner. Comparing a frame-by-frame video analysis with the PLT method, based on the average time taken for an experimenter to analyse one hour of video footage a comparison is presented in Figure 16.

The implications for such cost savings are immense for any manufacturer. They can divert funds to other areas of research and development, assessment cost is reduced greatly and product time-to-market is also reduced. In a human-centred design approach there is also potential for more iteration in the design evaluation cycle.

Equipment

Compared to existing predictive methods which all require proprietary or specialist equipment,

Figure 16. PLT method versus frame-by-frame video analysis

Method	Video analysis method	PLT method
Study time per participant	1 hour	1 hour
Number of participants	16 participants	16 participants
Video length	16 hours	-
Data analysis time	44 days	Less than 1 hour

the PLT method requires conventional computing equipment (PC, mouse and keyboard) to ensure its implementation.

Design

Another cost saving property of the PLT method is its design as either a desktop or car buck, or real car evaluation procedure. Manufacturers can employ either test environment, depending on the resources available to them. Utilising the desktop procedure for example reduces costs considerably.

Usability (Ease of Use)

A key benefit of automating the PLT method's process was to ensure ease of use through a series of step-by-step instructions, imposed by the software and the accompanying guidance document. Practitioners involved in study 5 underlined a range of benefits which contributed to the overall usability of the method which concluded: it was simple to setup; proprietary hardware and software were not required; the measures obtained were simple to interpret; and the automation of the data collection, collation and analysis process was extremely beneficial. It is also important to note that none of the low-cost methods available provide the automated approach to analysis which is available with the PLT method.

Reliability

One way of improving this reliability is through standardising the method. Consequently, it was vital to the reliability of the PLT method to develop a set of guidelines that specify how the assessment should be conducted. It was apparent that the inclusion of the guidance document, coupled with the automation of the process, ensured a systematic progression through the evaluation process. Similar to the occlusion method, it was observed in study 5 that the standardisation approach greatly reduces issues of variability,

caused during independent application of user-trial methods and improves the reliability of the results obtained.

Validity

Two studies (study 3 and 4) were conducted to assess the validity of the PLT method. The approach in the first study was to compare measures generated by the PLT method and observed participant's glance behaviour, to assess the internal validity of the method (i.e., if the measures obtained were in line with observed measures). The results showed a significant correlation between the number of PLT missed in a back-to-back sequence, and secondary task glance duration, thus highlighting the degradation in primary task performance as a result of the sustained visual attention needed to achieve a secondary task. The second study compared the measures generated by the PLT method, against measures obtained from a simulator study. A significant correlation was found between the number of PLT missed in a back-to-back sequence and mean glance duration and also the number of PLT missed and total task time. Overall, the results from the PLT study were in line with those obtained from the simulator.

Surrogate for Glance Duration

Aside from the PLT method's attributes (i.e., low cost, easy to use, reliable and valid), one key benefit of this method is its ability to provide a surrogate measure for glance duration. Other methods such as the occlusion method, 15 second rule, peripheral detection test and lane change test, provide some indicative measures of the potential visual and/or cognitive distraction of an interface in a cost effective way. However, these methods predominantly focus on providing measures either for visual allocation (glance frequency, Resumability); vehicle control (lane deviation, deviation from a selected norm); and/or object and event detection (latency to detect, missing detection and situation awareness). All of these measures imply 'degrees' of visual and/or cognitive distraction. The PLT method however, provides a surrogate measure for glance duration (an ocular indicator of attention and exposure to risk).

There are certain constraints identified in previous experimental research that secondary tasks have to meet in order to be acceptable while driving. With respect to the glance duration measure, Bhise, Forbes, and Farber (1986) state that based on speed and travel distances of a moving vehicle, a single display glance greater than 2.5 seconds is inherently dangerous. The British Standards Institution Guidelines, published in 1996, recommend that a single glance should be no more than 2 seconds so that it does not affect driving (Green, 1999a). The Battelle Guidelines suggest that the average glance time should not exceed 1.6 seconds, for a task to be considered safe for on-road use (Green, 1999a). It is generally agreed that glances lasting longer than approximately 2.0 seconds, are not acceptable while driving (Gelau & Krems, 2004). The PLT method could potentially provide upper limits for glance duration based on PLT performance, thereby highlighting interac-

tions which promote glance behaviours that are unacceptable.

Results from the research support the relationship between PLT performance and glance duration. One key benefit from this concerns the presentation of glance duration limits (which directly relate to specific PLT measures) for system interaction. In Figure 17, mean glance duration of 2 seconds has been highlighted on the x-axis. When this is further extended to the y-axis, it indicates the number of PLT missed in a back-to-back sequence (i.e., 1.8) is equivalent to the 2 second glance duration value (with the safe area highlighted). Based on guidelines in the literature, human factors practitioners in industry can now provide a 'figure' which highlights key safety implications of system design choices, thus strengthening the case from an ergonomic/safety perspective. This lack of explicit implication of design choices was highlighted as a major drawback for human factors contribution within industry (study 1).

Figure 17. Mean glance durations against PLT missed in sequence

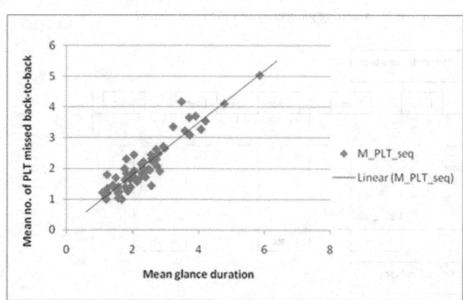

Limitations of the PLT Method

The method can only be used for assessing tasks that are likely to have single glances greater than one second. Due to the duration and frequency of the PLT, it is possible for participants to attend to the PLT (which is achievable in under a second), attend to a secondary task (which may require a glance of less than one second) and then return to the next PLT slide, in time to give the appropriate response (see Figure 18). This implies that there will be no degradation in PLT performance over a series of such glances. Therefore, secondary task demands will not be highlighted as expected, by PLT performance.

Although this highlights the method's insensitivity to assessing tasks with single glances of less than a second, tasks of this nature are assumed to be less likely to cause serious distraction problems. They are therefore seen as safer than tasks requiring single glance durations of longer periods (e.g., 1.6 to 2 seconds and above) (Green, 1999a; Tijerina, 1998, 2000; Farber, 2000; Zwahlen, 1988). A need to assess interactions of shorter glance durations which relate to specific driving situations will require modifications to be made

to the PLT method. For example, implementing a PLT with a shorter task duration of 0.5 seconds (analogous to driving in densely populated vehicle and pedestrian traffic) will force drivers to make shorter but more frequent glances to the PLT screen. This in turn should highlight secondary task demands of shorter durations, due to the intensity of glances required to achieve the PLT.

Future Work

Scope

Distraction caused by the use of IVIS is perceived to have an unfavourable effect on driving performance. Although the PLT method was designed to address issues of visual distraction and appears to be valid predominantly for visually oriented tasks, other forms of distraction can occur due to the use of IVIS. These include cognitive, biomechanical and auditory.

Presently, IVIS represent a wide range of multi-modal devices found within, or brought into, vehicles for driver support and infotainment (Irune & Burnett, 2007). Although these devices have various interaction styles (voice control,

Figure 18. Implications for assessing glances of less than one second

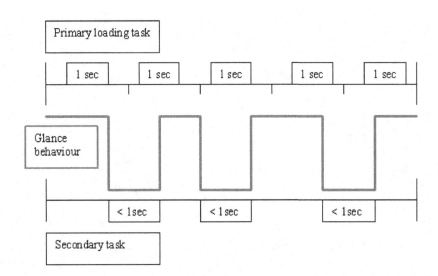

button control etc.), those that call upon the visual resources of the driver are of the greatest importance, simply because driving is predominantly a visual task. Nevertheless, further work may wish to extend the scope of the PLT method to cater for other forms of interaction (e.g., voice interaction).

Enrichment

As previously noted, the PLT aimed to mimic the visual/spatial and cognitive demands imposed by the primary task of driving. However, evaluating every IVIS from the perspective of a single primary loading task may not provide a range of scenarios representative of the real driving context. It is imperative that future improvements of the method address the implementation of a range of PLT activities, which simulate a variety of driving scenarios (e.g., motorway driving with light traffic, urban town driving with dense vehicle and pedestrian traffic).

Usability

Further assessments of the methods usability with a larger sample in an industrial setting as well as for academic research will be beneficial to improving the usability of the method. It is anticipated that evaluation of the method in an IVIS design context may highlight issues that have been missed by preliminary investigations conducted through its development.

Reliability

Future work may focus on assessing the overall reliability of the measures obtained from the method, especially the correlation between real glance duration and PLT missed in a back-to-back sequence. It may also be beneficial to evaluate the method in a realistic design cycle, where it is used to assess iteration of prototypes being developed.

Implementation

Although the PLT method has been developed into a software program to facilitate its ease of use and reliability, future research may improve aspects of this software program to include graphical representation tools to the software. This will facilitate more comprehensive analysis of the data obtained from the process and reduce the dependency on other analysis tools, such as MS Excel and SPSS.

Contribution to Research

It is evident from the literature that the academic research community have continued to develop methods for evaluating user interfaces. More recently, with minimal participation from industry, newer methods and techniques have been developed for use in the driving context, for evaluating IVIS (occlusion, simulator trials etc). However, these methods are rarely used in practical development, despite the significant research interest (John, 2003). This is an undesirable situation, especially since the proven benefits are well grounded and substantial (Dix et al., 2004; John, 2003). Broadly, two key reasons were identified by this research for the lack of practical use of available methods: firstly the lack of a clear understanding of the benefits of each method to the development process; secondly and more importantly, in comparison to other methods, the unavailability of an easy to use, economic, valid and reliable method. This research presents two key contributions in this respect: a framework that facilitates the selection of an appropriate method for use at a particular stage in IVIS development; and the development of a method that is easy to use, economic, valid and reliable. It is also important to note the uniqueness of this research in relation to the contribution from industry. This contribution has been key in realising the requirements considered in the development of both, the framework and the novel method previously mentioned. As such, the thesis has made an important contribution to human factors/HCI evaluation research, specifically in the context of IVIS design, by facilitating their development in the real world.

ACKNOWLEDGMENT

Special thanks to Honda R&D Europe Ltd., especially Catherine Prynne, for the tremendous support and for partly funding the research.

REFERENCES

Angell, L., Auflick, J., Austria, P. A., Kochhar, D., Tijerina, L., & Biever, W. (2006). *Driver workload metrics task 2 final report DOT HS 810 635*. Washington, DC: National Highway Traffic Safety Administration.

Bhise, V. D., Forbes, L. M., & Farber, E. I. (1986). *Driver Behavioral Data and Considerations in Evaluating In-vehicle Controls and Displays*. Paper presented at the Transportation Research Board, National Academy of Sciences, 65th Annual Meeting, Washington, DC.

Department for Transport. (2007). *Road Casualties Great Britain*. Retrieved from http://www.dft.gov.uk/pgr/statistics/datatablespublications/accidents/casualtiesmr/rcgbmainresults2007

Dix, A., Finlay, J., Abowd, G. D., & Beale, R. (2004). *Human-Computer Interaction* (3rd ed.). Upper Saddle River, NJ: Pearson Education Limited, Prentice Hall.

Gelau, C., & Krems, J. F. (2004). The occlusion technique: A procedure to assess the HMI of in-vehicle information and communication systems. *Applied Ergonomics*, *35*(3), 185–187. doi:10.1016/j.apergo.2003.11.009

Green, P. (1994). *Measures and Methods Used to Assess the Safety and Usability of Driver Information Systems* (Tech. Rep. No. UMTRI-93-12). Ann Arbor, MI: The University of Michigan.

Green, P. (1999a). *Visual and Task Demands of Driver Information Systems (Tech. Rep. No. UMTRI- 98-16)*. Ann Arbor, MI: The University of Michigan.

Green, P. (1999b). The 15-Second Rule for Driver Information Systems. In *Proceedings of the Intelligent Transportation Society of America Conference*.

Irune, A. (2009). Evaluating the visual demand of in-vehicle information systems: The development of a novel method.

Irune, A., & Burnett, G. E. (2007). Locating in-car controls: Predicting the effects of varying design layout. In *Proceedings of Road Safety and Simulation conference (RSS2007)*, Rome, Italy.

ISO. (2007). *Road vehicles – Ergonomic aspects of transport information and control systems – Occlusion method to assess visual demand due to the use of in-vehicle systems*. Geneva, Switzerland: ISO International Standard.

John, B. E. (2003). Information processing and skilled behaviour . In Carroll, J. M. (Ed.), *HCI Models, Theories and Frameworks: Toward a Multidisciplinary Science*. London: Morgan-Kaufmann. doi:10.1016/B978-155860808-5/50004-6

Kanis, H., & Wendel, I. E. M. (1990). Redesigned use, a designer's dilemma. *Ergonomics*, *33*(4), 459–464. doi:10.1080/00140139008927151

McGwin, G., & Brown, D. B. (1999). Characteristics of traffic crashes among young, middle-aged, and older drivers. *Accident; Analysis and Prevention*, *31*, 181–198. doi:10.1016/S0001-4575(98)00061-X

Neal, V. L., Dingus, T. A., Klauer, S. G., Sudweeks, J., & Goodman, M. J. (2005). *An overview of the 100-car naturalistic study and findings* (Paper No. 05-0400). Washington, DC: National Highway Traffic Safety Administration.

Pettitt, M. A. (2007). *Visual demand evaluation methods for in-vehicle interfaces*. Unpublished doctoral dissertation, University of Nottingham, UK.

Rockwell, T. H. (1988). Spare visual capacity in driving revisited: New empirical results for an old idea. In Gale, A. G., Freeman, M. H., Haslegrave, C. M., Smith, P., & Taylor, S. P. (Eds.), *Vision in Vehicles II*. London: Elsevier Science.

Sivak, M. (1996). The information that drivers use; is it indeed 90% visual? *Perception, 25*(9), 1081–1089. doi:10.1068/p251081

Society of Automotive Engineers. (2000). *SAE J2364 Recommended practice: Navigation and route guidance function accessibility while driving*.

Stephen, P. (1996). *Bodyspace: Anthropometry, Ergonomics and the Design of Work*. London: Taylor & Francis.

Stutts, J. C., Reinfurt, D. W., Staplin, L., & Rodgman, E. A. (2001). *The role of driver distraction in traffic crashes*. Washington, DC: AAA Foundation for Traffic Safety.

Tijerina, L., Palmer, E., & Goodman, M. J. (1998). *Driver workload assessment of route guidance system destination entry while driving* (Tech. Rep. No. UMTRI-96-30). Ann Arbor, MI: The University of Michigan Transportation Research Institute.

Trafton, J. G., Altmann, E. M., Brock, D. P., & Mintz, F. (2003). Preparing to resume an interrupted task: Effects of prospective goal encoding and retrospective rehearsal. *International Journal of Human-Computer Studies, 58*, 583–603. doi:10.1016/S1071-5819(03)00023-5

Trafton, J. G., Altmann, E. M., Brock, D. P., & Mintz, F. (2003). Preparing to resume an interrupted task: Effects of prospective goal encoding and retrospective rehearsal. *International Journal of Human-Computer Studies, 58*, 583–603. doi:10.1016/S1071-5819(03)00023-5

Zwahlen, H. T. (1988). Safety aspects of cellular telephones in automobiles. In *Proceedings of the International Symposium on Automotive Technology and Automation (ISATA)*, Florence, Italy.

This work was previously published in the International Journal of Mobile Human Computer Interaction, Volume 3, Issue 1, edited by Joanna Lumsden, pp.1-21, copyright 2011 by IGI Publishing (an imprint of IGI Global).

Chapter 2
Classic and Alternative Mobile Search:
A Review and Agenda

Matt Jones
Swansea University, UK

ABSTRACT

As mobile search turns into a mainstream activity, the author reflects on research that provides insights into the impact of current interfaces and pointers to yet unmet needs. Classic text dominated interface and interaction techniques are reviewed, showing how they can enhance the user experience. While today's interfaces emphasise direct, query-result approaches, serving up discrete chunks of content, the author suggests an alternative set of features for future mobile search. With reference to example systems, the paper argues for indirect, continuous and multimodal approaches. Further, while almost all mobile search research has focused on the 'developed' world, the paper outlines challenges and impact of work targeted at 'developing' world contexts.

INTRODUCTION

Mobile search is becoming an everyday activity for millions of people, with a recent study reporting that forty percent of US mobile users already use a search engine to navigate to mobile sites (Kamvar et al., 2009). In addition to web-wide searches, device-based tools are now needed to manage the quantities of on-board content that ranges from apps to books to music and more

The significance of searching as a key to effective mobile use cannot be overstated; and, this is before the billions of relatively new mobile users – in developing or 'emerging' markets – are taken

DOI: 10.4018/978-1-4666-2068-1.ch002

into account. For people in many parts of India, China, Africa or Latin America, the mobile will be their primary experience of the web and computers. Better search engines that accommodate these users' particular needs - their lack of textual literacy and limited exposure to conventional computers, to name just two – could have a hugely significant impact on the quality of life of the majority of the planet's inhabitants.

In this article, we will begin by considering the nature of mobile search. How, if at all, does it, or will it, differ from conventional desktop search? To do this, we will look at what the data seems to be saying about the trends in things people use mobiles to search for; how they do mobile search; and, what drives them to use a mobile in the first place.

Classically, mobile search engines have been textually driven – you pose a query by typing and result lists with various forms of artefact surrogates, such as the web page titles or keywords, are returned. We will explore a number of interaction methods and interfaces that have been proposed to make this process more efficient, effective and satisfying. Search engines have always been important in helping people make sense of the otherwise overwhelming complexity of vast information spaces. Without such innovative interaction designs, these spaces will feel like black holes when traversed through the relatively tiny screens and input mechanisms found on mobiles.

While conventional mobile search engines will remain important, in the second half of the article we will lay out the case for alternative approaches. Ones that take account of social, real-time activity; others that allow us to reflect on our search activities, seeing them as processes that create persistent, interogatable archives of information seeking behaviours. Then, there are the research prototypes that move away from simple text input and output, engaging a wider set of modalities. There has been a great deal of work on image-based and speech-based mobile search often combining these forms of interaction with others (e.g.,

Edwards et al., 2008; Paek et al., 2008). We will not be focussing on these; rather we will explore less explicit forms of query specification and result presentation through the use of gesture, audio adaption and haptics. Straight-forward speech and visual search, though, may well prove critical in developing world contexts, contexts that we will return to at the end of the article.

THE NATURE OF MOBILE SEARCH

Consider the last time you used a mobile search engine. What did you need to know? How urgent was your information need? What other resources did you have to hand to help you satisfy this need? Did the mobile help or were there some interaction or content issues?

A number of studies have been published in an attempt to understand the form and changes in mobile search behaviour. Two complementary types of analysis have been performed. First, log-files, held by mobile operators and search engine providers, containing millions of individual queries have been sifted. Typically, these studies determine mobile query lengths; click-through rates (where a click-through is defined as a search leading to a click to a result site); and, popular topics.

Log-files are only available to a very limited number of researchers, typically those working with the companies that create them. They also fail to capture important elements such as the intentions or needs behind the stark search terms or the context the query was performed within. Resources available to a user like other people, street signs, notebooks along with hurdles to entering a query and reviewing results – the busyness of a commuter street; the immersive chatter of a social gathering - need to be understood to gain a rich picture of mobile information needs. To overcome these log-file limitations, then, researchers have deployed *human* logging techniques: diary approaches and experience sample methodologies.

Figure 1. Mobile log-file analysis studies

Study	Region	Average Query Length	Click Thrus	Popular Topics	Trends
2006 Kamvar & Baluja (Google)	Global	2.3-2.7	<10%	Adult Internet & Telecom Entertainment	
2007 Baeza-Yates et al (Yahoo!)	Japan	2.29	Not given	Online shopping Sports Health	
2007a Kamvar & Baluja (Google)	Global	2.56	>50%	Adult Entertainment Internet & Telecom	'adult' and click-throughs increasing compared to 2006 study
2008a Church et al	Europe	2.2	12%	Adult Email, messaging , chat Resources for finding things Entertainment	'adult' increasing compared to earlier study by these researchers

Machine Logs

Turning first to the log-file analysis work, Figure 1 summarises four such studies published between 2006-2008. Three are from researchers working at search engine companies; the other is by a team who had access to a European mobile operator. Kamvar and Baluja (2009) provide a good summary of these articles.

What do we learn from these studies? The first task for a user is to enter a query consisting of one or more words. Users it seems are entering queries of a similar length to those seen on desktop search, despite the limitations of mobile keypads. Techniques that reduce the time and frustration to complete this search step will obviously improve the overall user experience. Harder to use interfaces also reduce the average length of a query (Kamvar et al., 2009).

There is a lower level of click-through from search results to web sites than seen with desktop search. Users are reviewing result lists but often not navigating away from the service to third-

party pages. This could be due to factors that are outside the remit of interface design. Uncertainty about data costs incurred by jumping from the sparse search result page to a rich web page can significantly deter a user from the sorts of freeform navigation seen with unlimited data packages on home broadband connections. There could, though, be reticence due to the perceived high *interaction* costs in clicking through to a hard to navigate, mobile unfriendly website (Jones et al., 1999). Indeed, one of the explanations given by Kamvar and Baluja (2007a) for the increase in click-throughs is the better transcoders that make web pages usable on the small screen. Whatever the reasons, more expressive search results might help users judge which ones are worth exploring.

Certain topics are found to represent high proportions of queries. It is not surprising that "adult" searches are seen to be popular. The discreetness of the mobile platform perhaps is well suited to such private content. It is also well known that pornography is often in the vanguard of new technologies. In the long-term, however, as was the trend with desktop searching, this sort of content will likely decrease in relative prominence. Even so, mobile search services tailored for the significant topic areas including shopping and entertainment might prove popular with users.

All of these studies came before the mobile landscape changed dramatically, a couple of years ago, with the arrival of sophisticated touch-screen mobile phones such as the iPhone or the Nokia N900. Kamvar et al. (2009) have explored the impact of these devices on search behavior by comparing query log entries from three platforms: the desktop; conventional internet-enabled mobile phones (that have a smaller screen and fixed, physical keypad); and, the iPhone.

The key finding, in short, was fascinating. The iPhone and desktop searches had much in common while standard mobile phone search was, in many respects, distinct, as the earlier studies suggested they would be. It seems that people are using iPhones – and, we can hypothesize, other fully featured touch screen mobiles – to find information on a broad range of content. They are treating their devices as handheld versions of their desktops. Mobile search on these platforms is not just about finding urgent, 'being-mobile' relevant content. There were differences though such as the degree of interactivity measured in terms of query-result iterations. This was lower on iPhones than the desktop, and seen to be lower still on the standard mobile.

The authors present some useful pointers from their work to guide future search interface design. For conventional mobile phone users where the set of queries is less diverse it could help to anticipate the queries and results and, perhaps, theme the search home page with content based on the user's primary interest.

Meanwhile for the high-end phones, phones that are going to become predominant, at least in developed markets, the search experience should be better integrated to the desktop search. What you look for on your desktop should be available on your mobile and vice versa. Searching on these

sorts of mobile is not all about being mobile – we should not, then, get fixated with location-based tweaks to the multi-purpose, standard search interfaces we provide. Tailored interfaces – such as the Google Maps App - might be better for when we want results relevant to our mobility.

When we are truly mobile, though, say when walking or being jostled on a commuter train there are a number of constraints ranging from the user's attention to the network availability that will limit the desire for interactive search (Sohn et al., 2008). Kamvar et al suggest accommodating these limitations by consider schemes that provide better snippets of content in the search result; or more sophisticated content summaries. Alternatively, users could be helped to follow up their search needs when they have more time later.

Human Logs

The studies reviewed so far analysed vast slews of log data in an automated fashion. Alternative approaches have been used by researchers including Sohn et al. (2008) and Church and Smyth (2009). Their interest is to delve into the needs and intentions behind the millions of pieces of search data generated every day. They wish to go beyond the *effects* found in the log-file studies and to determine the *causes*, exploring when, where and why information needs arise and how people act on them.

Consider, for example, the work of Sohn et al. (2008). Here, the researchers followed twenty people over a two-week period. These participants sent a text message to the investigators' server whenever they had an information need away from their home or office. At the end of each study day, participants logged onto a web site and answered questions about each of the information needs they had submitted.

Amongst the findings is a series data on how people satisfy an information need when mobile. Although 45% of needs were met immediately, 25% were satisfied later and 30% never met. Of the needs that were met immediately, the web was used in a significant number of times (30%) but many needs were also met by calling someone (39%); interestingly, 7% of needs were met in a less than web-speed way by referring to content previously printed out.

The reasons why people did not act on information needs immediately but at a later time included being occupied by another activity – such as driving or being in a meeting (54% of time) – or due to a lack of web access (32%). Reasons for not following up on an information need included the need being unimportant (35%); not knowing how to address it (23%); or no access to the web (23%).

Context is often considered as a key driver for mobile information needs. This study certainly provides evidence to support this: some 72% of needs were classified as context related. However, location - where a user is – was not the only form of context trigger, accounting for 35% of needs. Other factors included the time (28%), conversation (27%) and activity (24%).

As with the log-studies, this work provides spurs for future mobile search thinking. First, we are reminded that people are ecological and not technological. Too often mobile designers and researchers (this author included) view the device as a panacea to information needs. The reality is that people will use it in conjunction with other useful resources – including people – to hand. In this respect, a simple interface refinement suggested by Sohn's study is for results to highlight telephone numbers of useful contacts if appropriate.

The work further reinforces the need to support the stretched-out information seeking highlighted in the iPhone log-study. Search does not always have to be fast-paced and immediate. As people cannot or do not want to always act immediately on an information need, ways of allowing them to delay their information seeking activities, satisfying a need later might prove important. Mobile, lightweight, low-cost interactions to capture the essence of the 'need' will also be needed. Finally,

more work on context capture is required, particular to address the subtler, dynamic elements of what makes for 'context'.

CLASSIC MOBILE SEARCH INTERACTION DESIGNS

Standing between the user and the endless information possibilities of the web on mobiles is a relatively impoverished interface: small keypads, both physical and on screen; and, limited display real estate. This interface provides two sets of challenges to what we can call 'classic' or conventional mobile search. Here, the interaction process consists of entering textual queries and reviewing text dominated result lists. The aim of an interface designer in this context is to reduce the query entry burden and to allow users to quickly review potentially long result lists, making good choices of links to follow.

Improving Query Term Entry

Ways of speeding up text entry for mobiles has been the subject of much research for sometime (see, for example, Mackenzie et al., 2001) and predictive schemes are available on most mobile phones.

For mobile search queries, researchers have looked at extending dictionary-based predictions. Some methods take account of query patterns seen in logs; others adjust completions based on contextual factors; and a further class promote exploration of the potential search space by suggesting a number of alternatives rather than simply completing one query.

Let's begin by considering how context might be used to provide clever query completion. This was the question addressed by Kamvar and Balja (2007b) as they explored several potential query adaption factors including the user's location; the time of day a query is entered; and, the mobile phone network carrier.

Figure 2. Suggestion interface helps user explore the information space (from Marsden et al., 2002)

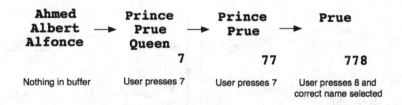

As an example of location query completion, they give the case of a user entering 'al' while in San Francisco, CA and the most likely completion 'alcatraz' is presented. Such location completions were effective in reducing the key presses but temporal and carrier based adaptations did not bring benefits. The authors remind us, though, that algorithms proven in the lab might lead to frustrations in the wild. People like interfaces to be consistent and intelligent query interfaces could confuse. So, our San Francisco user may be expecting "al" to always be completed with "alcatraz" and stumble when, in Chicago, "Al Capone" appears in the query box.

Interfaces that attempt to pre-empt a user by, for example, providing just one query completion can be less pleasing than ones that offer a wider choice, letting them feel more in control of the dialogue. In an early example of suggestion rather than completion, Marsden and colleagues (Marsden et al 2002) describe a search interface for mobiles that iteratively prunes and displays the search space, key press by key press. At first all of the phones functions, address book contacts and network services are shown in one long scrollable list. However, with each press of the numeric keypad the list is reduced.

Figure 2 illustrates the process – as the user presses '7' all of the entries in the system starting with 'pqrs' (the letters associated with key 7 in the ISO keypad standard) are displayed; pressing '7' again further prunes the list to show entries starting with 'pq', 'pr', 'ps', 'qp', 'qq', 'qr' and so on. When the user feels they have pruned the list to a small enough they can scroll down and select the option they want to activate or simply press the select button when only one option is left (in the example, to ring "prue's" number).

The positive impact on key-press efficiency and user satisfaction of suggestion schemes has been demonstrated on the Google interface (Kamvar & Baluja, 2008). Their suggestion interface dynamically offered a number of potential terms as the user typed. In comparison with an interface without the suggestions the researchers found key presses halved and users' perceptions of workload was lower and enjoyment higher.

Queries are excellent if you know what you are looking for. So, if you want to know what the weather in Swansea is, then formulating a term is not hard. But, if you only know the dimensions – or facets - of what you are looking for then query interfaces can be frustratingly tedious. Imagine, then, being in a new city and wanting to find a cheap, nearby, child-friendly, Asian restaurant with good reviews. The likely high number of query-result iterations and scrolling through web pages is, as the earlier discussed studies show, not best suited to the mobile platform.

An alternative is to allow users to explore the possibilities by filtering results with regard to the facets that define the information space. FaThumb (Karlson et al., 2006) demonstrates the effectiveness of the approach in searching a Yellow Pages directory. At the top level, users are presented with six facets such as 'price', 'location' and 'rating'. Selecting from this group, they are then shown any sub-facets that may similarly be added to the set that filters the information space.

Figure 3. Categorising search results to reduce scrolling and to help users decide which links to select. From Jones et al. (2003)

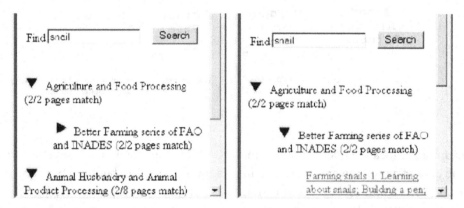

Presenting Search Results: Tiny Screen, World of Possibilities

When a search engine returns a list of results, users need to be quickly assess the set and decide whether to issue another query. If the list seems like a good one, they also need to be supported to decide which of the results is the best to click on (Hearst, 1999).

On a small screen, the first requirement can be hampered by the need to scroll to view more than just a few of the results. Result categorisation and clustering has been proposed to reduce this scrolling, helping the user get a better sense of the content the results represent.

LibTwig groups results with respect to the topics and sub-topics they related to (Jones et al., 2003) - Figure 3 illustrates the interaction. The user has entered the term 'snail' and system shows there are results in several categories. By selecting the first category (left hand image) the user is able to see that the results relate to the sub-category of 'Better farming...' If this is the sort of information they are interested in they can further select that category and see the titles of the documents themselves (right hand image). The category presentations help give the user an overview of the range of content available and, as categories are expanded, their labels help provide context

for the results they contain. Related cluster-based approaches are discussed in Heimonen and Kaki (2007) and Carpineto et al. (2009)

The categories in Figure 3 do provide additional data to help judge whether a search result is worth exploring. It is possible too, of course, to extract descriptive information from the documents themselves; a technique now commonly deployed in commercial search services. In Jones et al. (2004), we considered whether automatically extracted key phrases as document surrogates could help in a mobile context. Figure 4 shows PDA screens to illustrate our experimental apparatus. Each PDA displays a result list of documents. The left-hand image gives the titles while the right-hand list gives key phrases mined from those documents. Our studies indicatedthis sort of automatically extracted data could help users make judgements about the content of the document they represent. Keyphrases are just one sort of summary information that might be helpful in the small screen context; Sweeny and Crestani (2006) have looked at varying the size of summaries and conclude that smaller summaries may be the most helpful for a range of platforms.

Key phrases picked out of documents and other forms of summary presented in the search result lists as 'snippets' can often help users meet their information need without having to click on

Figure 4. Document titles (Left hand side image); and, key phrases (Right hand side). From Jones et al. (2004)

the link and explore the third-party site. Certainly, if a mobile user wants a telephone number, an address, the weather, traffic conditions and the like, then conventional text-result interfaces should – and many do - endeavour to highlight potential answers in result lists. Google has a mobile version that aims just to give the user the answer: send a text message to their server (e.g., "weather boston") and the information, not the web page are returned as results (Schusteritsch et al., 2005).

SEEKING ALTERNATIVES

The sorts of scheme discussed, above, are direct, query-result orientated. They aim at providing immediate answers in discrete chunks of content. They are primarily textually and visually orientated and their target users are the literate, educated 'developed' world.

In the second half of this article, we will outline an alternative set of characteristics for future mo-

bile search. Instead of directly looking for something, what about letting the user take advantages of other peoples' interactions; that is search as an incidental activity? Instead of always providing the results immediately, what about designing for delayed, reflective interactions? Instead of discrete chunks of content, what about providing results through continuous media? Instead of text, how can audio and touch assist in mobile search? And, instead of continuing to focus on the developed world, what can researchers do to provide effective mobile search interfaces for the billions of new users in the developing world?

Incidental Search

When Peter Day, a reporter for the UK's BBC Radio 4, visited Google's headquarters a few years ago, he was intrigued by the scrolling display of words displayed above the receptionist's desk. They were search terms being entered by users around the world just at that moment: "We are looking into the mind of the world", he remarked.

Inspired by this visualisation of other people's search interactions, we developed the *Questions Not Answers* mobile system (Jones et al., 2007; Arter et al., 2007). Users are shown the queries that happen in a location overlaid on a map (see Figure 5). The visualisation updates automatically, queries appearing and disappearing as remote users enter new search terms.

We hypothesised that while queries are typically very short, they can communicate the essence of aspiration, needs, values and give insights into the character of a place. The dynamic view further captures the changing dialogue people have with a place. The types of thing searched for in Trafalgar Square on July 6th 2005, for instance, (the day London won the right to host the Olympics in 2012) would be very different from the queries the day later (after the 7/7 bombings). Users benefit from other people's search activity for 'free' and the glance-based interaction allows for minimal attention if needed. However, the interface also has the potential to promote interactions, with users clicking on terms to explore results or generating their own terms prompted by other people's examples.

Studies of the approach showed that it can indeed give a sense of people and place and could be a powerful complementary approach to direct search. There are issues, though, such as how to deal with large volumes of queries in a place – should they for instance be aggregated or summarised in some way or is it better to present a fast moving, chatter of terms? Similar interfaces have been explored by Church and Smyth (2008b) who also provide ways of filtering terms – allowing the user to see queries earlier than current time, for instance.

Of course, since this work was completed, social information systems such as Twitter have become prominent. Search systems of all varieties, especially mobile, will increasingly draw on these real time pieces of information.

Figure 5. PDA version of the Questions Not Answers system (Arter et al., 2007). Queries (e.g., "transport, "welsh dept") are collaged dynamically over a map

Delayed, Reflective Search

As we saw earlier, in many situations, people will not want or be able to act on an information need immediately, when mobile. The *Laid Back Search* system (Jones et al., 2003, 2006c) was designed for people to capture written queries quickly and to later explore the result sets associated with them. Queries could be jotted into a PDA and when the user returned to their computer, the previously collected search terms were sent to their computer. These were then submitted as queries to a search engine and for each of them a series of results and associated web pages were downloaded to the PDA. These packages of content could be reviewed later at will without the PDA having to be connected to the internet. The content was also made available on the user's desktop computer as well as a shared, large screen display. Figure 6 illustrates the system.

We applied these stretched-out search ideas in a later prototype, using gestures rather than text to specify queries. In the *Point-to-Geoblog* system

Figure 6. Laid Back Search System (Jones et al., 2003, 2006c). Query terms jotted during a meeting (top left); later transferred to a PC with results returned to the PDA for offline review (top right) or access via the PC and large screen display. The mobile, then, is one element of the ecology of devices in an extended search process

Figure 7. The Point-to-Geoblog system (Robinson et al., 2010). Using a small sensor pack (top), the user points at a location and tilts (to distance). Later, the user can see their route on a map and review their previously targeted locations (bottom left). Clicking on target markers shows content related to that target (bottom right).

(Robinson et al., 2010) the user points and tilts the mobile device towards an area they want to find out about. This direction and distance marker is later used by the system to retrieve web pages, Wikipedia articles and social network information associated with that location. Users can then review their routes, seeing where they were when they requested information and what they were pointing at (see Figure 7).

Continuous, Multimodal Searching

An increasingly important form of mobile searching is pedestrian navigation. There are many commercially available systems that provide on-screen, turn-by-turn directions allayed, in some cases, with spoken instructions. Before such systems became widely available we were interested in using the mobile user's music to help them navigate from one location to another. Unlike the turn-by-turn systems, the approach did not use discrete, navigation commands ('turn left', 'go on a hundred metres' and the like). Instead, it simply panned the playback of the music through the listener's headphones to indicate the direction they should head (Warren et al., 2005; Jones et al., 2008). A similar technique is described in Stachan et al. (2005) with their *gpsTunes* system.

Music playback could also be adapted to alert the user when previously specified information needs could be met enroute (Jones & Jones, 2006d). Consider then, a user planning a visit to a new city and wanting to find out more about key architectural highlights. They could express this need before leaving on their travels. Then, while wandering through the city, listening to their music, as they pass buildings of note, the music could pan to prompt them to look around, pause and learn more. This scheme, then, would be the counterpoint to *Point-to-Geoblog*.

Turning from audio and music as a modality to present search 'results', let us consider another continuous form of seeking and finding through

the use of gestures and haptics. In the *Sweep-Shake* system (Robinson et al., 2009), the user holds a device in their hand and as they pan it front of them, the device vibrates to indicate the presence of digital content (e.g., web pages, images, audio or video) associated with those locations. Instead of immediately viewing the results of these gestural queries, the user can explore and filter them with further movements and haptic feedback (see Figure 8).

Figure 8. The Sweep-Shake System (Robinson et al., 2009). The user receives haptic feedback as they point toward content hot-spots. Gesturing in different directions provides further haptic feedback to indicate the availability of different types of content at that location.

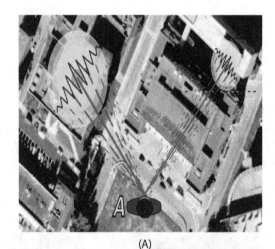

(A)

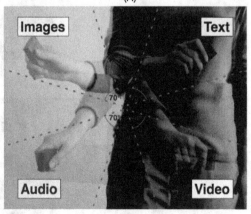

(B)

Developing World Needs

Finally, we turn to the real future of mobile search – the billions of users and their needs in what has variously been called the 'developing world' or 'emerging markets', places such as China, India and the continent of Africa. The challenges and opportunities are great; there is much research to be done to define appropriate and effective mobile search and browsing schemes for these contexts.

Non-textual interfaces will certainly be important due to the low levels of textual literacy in many of these who live in these regions. In the *StoryBank* system, for instance (Frohlich et al., 2009) we developed simple, visual icon based ways for users to tag audio-visual content created on the phone to aid later retrieval. Another form of non-text information resource is being developed by IBM Research India. Their Spoken Web project enables anyone with access to a phone (both fixed line and mobile) to create voice sites simply through speech and touch tone selections (Ararwal et al., 2008). Interesting information access challenges, particularly for voice driven interactions, will need to be met as the amount of content grows and as it becomes more diverse (Patel et al., 2009).

Current mobile search interfaces can assume their users have experience in using search on the desktop; indeed, as we've noted, some research suggests more closely integrating search experiences across the different platforms these 'advanced' users have at their disposal. In the new developing world contexts, however, most users will have very little previous exposure to using a computer to find content (Gitau et al., 2010). Furthermore, their view of how information can be organised and accessed may be very different from those in the developed world (Walton et al., 2002). These factors, then, will mean not only the need to accommodate non-textual content but perhaps completely different ways of allowing people to formulate queries and to visualise results.

CONCLUSION

Mobile search was an exciting area to get into when I started handheld research fifteen years ago. Research from many groups and individuals has provided good foundations for anyone considering implementing a search tool on a handheld device.

Mobile search is still a fascinating topic with many challenges left to address. There are many interesting opportunities to further explore the role of context; to extend the input and output modalities deployed; and, to integrate social, real-time information whilst not overlooking desires to reflect and consider. Crucially, though, I would urge anyone thinking of beginning a programme of research in this area to work on the hugely important area of mobile interaction design for developing world contexts. There are hard problems to address with a huge potential for impact.

ACKNOWLEDGMENT

I am indebted to the many colleagues whose work I have drawn on. Part of the review of classic mobile search interaction designs is based on a section in *Mobile Interaction Design* (Wiley and Sons 2006) co-authored with Gary Marsden. The *Questions Not Answers* work was funded by Microsoft Research and other systems were developed through the support of the EPSRC (EP/E042740/1). The article began life as an invited tutorial for Mobile HCI 2009; slides associated with that presentation may be found at: http://is.gd/dhu8i.

REFERENCES

Agarwal, S., Kumar, A., Nanavati, A. A., & Rajput, N. (2008, April 21-25). The world wide telecom web browser. In *Proceedings of the 17th international Conference on World Wide Web (WWW '08)*, Beijing, China (pp. 1121-1122). New York: ACM. Retrieved from http://doi.acm. org/10.1145/1367497.1367686

Arter, D., Buchanan, G., Jones, M., & Harper, R. (2007, September 9-12). Incidental information and mobile search. In *Proceedings of the 9th international Conference on Human Computer interaction with Mobile Devices and Services (MobileHCI '07)*, Singapore (Vol. 309, pp. 413-420). New York: ACM. Retrieved from http://doi.acm.org/10.1145/1377999.1378047

Baeza-Yates, R., Dupret, G., & Velasco, J. (2007). A study of mobile search queries in Japan. In *Proceedings of the WWW 2007 Workshop on Query Log Analysis: Social and Technological Analysis*. Retrieved July 8, 2010, from http://querylogs2007.webir.org/program.htm

Carpineto, C., Mizzaro, S., Romano, G., & Snidero, M. (2009). Mobile information retrieval with search results clustering: Prototypes and evaluations. *Journal of the American Society for Information Science and Technology, 60*(5), 877–895. Retrieved from http://dx.doi.org/10.1002/asi.v60:5. doi:10.1002/asi.21036

Church, K., & Smyth, B. (2008b). Who, what, where & when: a new approach to mobile search. In *Proceedings of the 13th international Conference on intelligent User interfaces*, Gran Canaria, Spain (pp. 309-312). New York: ACM. Retrieved from http://doi.acm.org/10.1145/1378773.1378817

Church, K., & Smyth, B. (2009). Understanding the intent behind mobile information needs. In *Proceedings of the 13th international Conference on intelligent User interfaces*, Sanibel Island, FL (pp. 247-256). New York: ACM. Retrieved from http://doi.acm.org/10.1145/1502650.1502686

Church, K., Smyth, B., Bradley, K., & Cotter, P. (2008a, September 2-5). A large scale study of European mobile search behaviour. In *Proceedings of the 10th international Conference on Human Computer interaction with Mobile Devices and Services (MobileHCI '08)*, Amsterdam, The Netherlands (pp. 13-22). New York: ACM. Retrieved from http://doi.acm.org/10.1145/1409240.1409243

Edwards, G. T., Liu, L. S., Moulic, R., & Shea, D. G. (2008). Proxima: a mobile augmented-image search system. In *Proceeding of the 16th ACM international Conference on Multimedia*, Vancouver, BC (pp. 921-924). New York: ACM. Retrieved from http://doi.acm.org/10.1145/1459359.1459522

Frohlich, D. M., Rachovides, D., Riga, K., Bhat, R., Frank, M., Edirisinghe, E., et al. (2009). StoryBank: mobile digital storytelling in a development context. In *Proceedings of the 27th international Conference on Human Factors in Computing Systems*, Boston (pp. 1761-1770). New York: ACM. Retrieved from http://doi.acm.org/10.1145/1518701.1518972

Gitau, S., Marsden, G., & Donner, J. (2010). After access: challenges facing mobile-only internet users in the developing world. In *Proceedings of the 28th international Conference on Human Factors in Computing Systems*, Atlanta, GA (pp. 2603-2606). New York: ACM. Retrieved from http://doi.acm.org/10.1145/1753326.1753720

Hearst, M. (1999). User Interfaces and visualisation . In Baeza-Yates, R., & Ribeiro-Neto, B. (Eds.), *Modern Information Retrieval*. New York: ACM Press.

Heimonen, T., & Käki, M. (2007). Mobile findex: supporting mobile web search with automatic result categories. In *Proceedings of the 9th international Conference on Human Computer interaction with Mobile Devices and Services*, Singapore (Vol. 309, pp. 397-404). New York: ACM. Retrieved from http://doi.acm.org/10.1145/1377999.1378045

Jones, M., Buchanan, G., Cheng, T., & Jain, P. (2006c). Changing the pace of search: Supporting "background" information seeking. *Journal of the American Society for Information Science and Technology, 57*(6), 838–842. Retrieved from http://dx.doi.org/10.1002/asi.v57:6. doi:10.1002/asi.20304

Jones, M., Buchanan, G., Harper, R., & Xech, P. (2007). Questions not answers: a novel mobile search technique. In *Proceedings of the SIGCHI Conference on Human Factors in Computing Systems*, San Jose, CA (pp. 155-158). New York: ACM. Retrieved from http://doi.acm.org/10.1145/1240624.1240648

Jones, M., Buchanan, G., & Thimbleby, H. (2003). Improving web search on small screen devices. *Interacting with Computers, 15*(4), 479–495. doi:10.1016/S0953-5438(03)00036-5

Jones, M., Jain, P., Buchanan, G., & Marsden, G. (2003). Using a mobile device to vary the pace of search. In . *Proceedings of the Mobile HCI, 2003*, 90–94.

Jones, M., & Jones, S. (2006a). The music is the message. *Interaction, 13*(4), 24–27. Retrieved from http://doi.acm.org/10.1145/1142169.1142190. doi:10.1145/1142169.1142190

Jones, M., & Jones, S. (2006d). The music is the message. *Interaction, 13*(4), 24–27. Retrieved from http://doi.acm.org/10.1145/1142169.1142190. doi:10.1145/1142169.1142190

Jones, M., Jones, S., Bradley, G., Warren, N., Bainbridge, D., & Holmes, G. (2008). ONTRACK: Dynamically adapting music playback to support navigation. *Personal and Ubiquitous Computing, 12*(7), 513–525. Retrieved from http://dx.doi.org/10.1007/s00779-007-0155-2. doi:10.1007/s00779-007-0155-2

Jones, M., & Marsden, G. (2006b). *Mobile Interaction Design*. New York: John Wiley & Sons.

Jones, M., Marsden, G., Mohd-Nasir, N., Boone, K., & Buchanan, G. (1999). Improving Web interaction on small displays. In P. H. Enslow (Ed.), *Proceedings of the Eighth international Conference on World Wide Web*, Toronto, Canada (pp. 1129-1137). New York: Elsevier.

Jones, S., Jones, M., & Deo, S. (2004). Using keyphrases as search result surrogates on small screen devices. *Personal and Ubiquitous Computing, 8*(1), 55–68. Retrieved from http://dx.doi.org/10.1007/s00779-004-0258-y. doi:10.1007/s00779-004-0258-y

Kamvar, M., & Baluja, S. (2006). A large scale study of wireless search behavior: Google mobile search. In R. Grinter, T. Rodden, P. Aoki, E. Cutrell, R. Jeffries, & G. Olson (Eds.), *Proceedings of the SIGCHI Conference on Human Factors in Computing Systems*, Montréal, Québec, Canada (pp. 701-709). New York: ACM. Retrieved from http://doi.acm.org/10.1145/1124772.1124877

Kamvar, M., & Baluja, S. (2007a). Deciphering Trends in Mobile Search. *Computer, 40*(8), 58–62. doi:10.1109/MC.2007.270

Kamvar, M., & Baluja, S. (2007b). The role of context in query input: using contextual signals to complete queries on mobile devices. In *Proceedings of the 9th international Conference on Human Computer interaction with Mobile Devices and Services*, Singapore (Vol. 309, pp. 405-412). New York: ACM. Retrieved from http://doi.acm.org/10.1145/1377999.1378046

Kamvar, M., & Baluja, S. (2008). Query suggestions for mobile search: understanding usage patterns. In *Proceeding of the Twenty-Sixth Annual SIGCHI Conference on Human Factors in Computing Systems*, Florence, Italy (pp. 1013-1016). New York: ACM. Retrieved from http://doi.acm.org/10.1145/1357054.1357210

Kamvar, M., Kellar, M., Patel, R., & Xu, Y. (2009). Computers and iphones and mobile phones, oh my!: a logs-based comparison of search users on different devices. In *Proceedings of the 18th international Conference on World Wide Web*, Madrid, Spain (pp. 801-810). New York: ACM. Retrieved from http://doi.acm.org/10.1145/1526709.1526817

Karlson, A. K., Robertson, G. G., Robbins, D. C., Czerwinski, M. P., & Smith, G. R. (2006). FaThumb: a facet-based interface for mobile search. In R. Grinter, T. Rodden, P. Aoki, E. Cutrell, R. Jeffries, & G. Olson (Eds.), *Proceedings of the SIGCHI Conference on Human Factors in Computing Systems*, Montréal, Québec, Canada (pp. 711-720). New York: ACM. Retrieved from http://doi.acm.org/10.1145/1124772.1124878

MacKenzie, I. S., Kober, H., Smith, D., Jones, T., & Skepner, E. (2001). LetterWise: prefix-based disambiguation for mobile text input. In *Proceedings of the 14th Annual ACM Symposium on User interface Software and Technology*, Orlando, FL (pp. 111-120). New York: ACM. Retrieved from http://doi.acm.org/10.1145/502348.502365

Marsden, G., Gillary, P., Thimbleby, H., & Jones, M. (2002). The Use of Algorithms in Interface Design. *International Journal of Personal and Ubiquitous Technologies*, 6(2), 132–140. doi:10.1007/s007790200012

Paek, T., Thiesson, B., Ju, Y., & Lee, B. (2008). Search Vox: leveraging multimodal refinement and partial knowledge for mobile voice search. In *Proceedings of the 21st Annual ACM Symposium on User interface Software and Technology,* Monterey, CA (pp. 141-150). New York: ACM. Retrieved from http://doi.acm.org/10.1145/1449715.1449738

Patel, N., Agarwal, S., Rajput, N., Nanavati, A., Dave, P., & Parikh, T. S. (2009). A comparative study of speech and dialed input voice interfaces in rural India. In *Proceedings of the 27th international Conference on Human Factors in Computing Systems*, Boston (pp. 51-54). New York: ACM. Retrieved from http://doi.acm.org/10.1145/1518701.1518709

Robinson, S., Eslambolchilar, P., & Jones, M. (2009). Sweep-Shake: finding digital resources in physical environments. In *Proceedings of the 11th international Conference on Human-Computer interaction with Mobile Devices and Services,* Bonn, Germany (pp. 1-10). New York: ACM. Retrieved from http://doi.acm.org/10.1145/1613858.1613874

Robinson, S., Eslambolchilar, P., & Jones, M. (2010). Exploring Casual Point-and-Tilt Interactions for Mobile Geo-Blogging. *Personal and Ubiquitous Computing,* 4(14), 363–379. doi:10.1007/s00779-009-0236-5

Schusteritsch, R., Rao, S., & Rodden, K. (2005). Mobile search with text messages: designing the user experience for google SMS. In *Proceedings of CHI '05 Extended Abstracts on Human Factors in Computing Systems*, Portland, OR (pp. 1777-1780). New York: ACM. Retrieved from http://doi.acm.org/10.1145/1056808.1057020

Sohn, T., Li, K. A., Griswold, W. G., & Hollan, J. D. (2008). A diary study of mobile information needs. In *Proceeding of the Twenty-Sixth Annual SIGCHI Conference on Human Factors in Computing Systems*, Florence, Italy (pp. 433-442). New York: ACM. Retrieved from http://doi.acm.org/10.1145/1357054.1357125

Strachan, S., Eslambolchilar, P., Murray-Smith, R., Hughes, S., & O'Modhrain, S. (2005). GpsTunes: controlling navigation via audio feedback. In *Proceedings of the 7th international Conference on Human Computer interaction with Mobile Devices &Amp; Services*, Salzburg, Austria (Vol. 111, pp. 275-278). New York: ACM. Retrieved from http://doi.acm.org/10.1145/1085777.1085831

Sweeney, S., & Crestani, F. (2006). Effective search results summary size and device screen size: is there a relationship? *Information Processing & Management,* 42(4), 1056–1074. doi:10.1016/j.ipm.2005.06.007

Walton, M. Vukovic', V., & Marsden, G. (2002). 'Visual literacy' as challenge to the internationalisation of interfaces: a study of South African student web users. In *CHI '02 Extended Abstracts on Human Factors in Computing Systems*, Minneapolis, MN (pp. 530-531). New York: ACM. Retrieved from http://doi.acm.org/10.1145/506443.506465

Warren, N., Jones, M., Jones, S., & Bainbridge, D. (2005). Navigation via continuously adapted music. In *CHI '05 Extended Abstracts on Human Factors in Computing Systems*, Portland, OR (pp. 1849-1852). New York: ACM. Retrieved from http://doi.acm.org/10.1145/1056808.1057038

This work was previously published in the International Journal of Mobile Human Computer Interaction, Volume 3, Issue 1, edited by Joanna Lumsden, pp.22-36, copyright 2011 by IGI Publishing (an imprint of IGI Global).

Chapter 3
How Do People Use
Their Mobile Phones?
A Field Study of Small Device Users

Tianyi Chen
University of Manchester, UK

Simon Harper
University of Manchester, UK

Yeliz Yesilada
Middle East Technical University, Northern Cyprus Campus, Turkey

ABSTRACT

The usability evaluation of small devices (i.e., mobile phones and PDAs) is an emerging area of research. Compared with desktop computers, designing a usability evaluation for small devices is more challenging. Context of use, such as environmental disturbance and a user's physical activities affect the evaluation results. However, these parameters are usually ignored or excluded from simple and unnatural evaluation settings; therefore generating unrealistic results. This paper presents a field study that investigates the behaviour of small device users in naturalistic settings. The study consists of a series of unobtrusive remote observations and interviews. Results show that small device users normally use the device with just one hand, press the keys with thumb and make phone calls and send text messages while walking. They normally correct typing errors and use abbreviations. On average, small device users switch their attention between the device screen and the surrounding environment 3 times every 20 seconds, and this increases when they are walking.

DOI: 10.4018/978-1-4666-2068-1.ch003

INTRODUCTION

Small devices, such as mobile phones and PDAs, are widely used and are becoming increasingly important in our daily life. However, the study of human computer interaction with small devices is still a young field which is highly technology driven with less reflection on research methodologies. Results of a recent survey on mobile HCI research methods indicate a clear trend on building systems and doing evaluation in laboratory settings, whereas understanding the real problems that small device users are facing is less prioritized (Kjeldskov & Graham, 2003).

Small devices, by their nature, are intended to be used in mobile settings. Different from using a desktop computer, using a small device in motion means that a user often cannot devote all their visual attention to interacting with the device. The attention of a small device user is normally switched between a primary task, such as walking and navigating, and a secondary task, for instance the interaction with a small device (Lumsden & Brewster, 2003). Unfortunately, most user evaluations of small devices are conducted in simple laboratory settings where small device users get little distraction from the environment and thus can fully devote their attention resources to the tasks in hand. For example, some artificial settings used include simulated motion using treadmills (Bernard et al., 2005; Lin et al., 2007), walking down a quite corridor with a subjectively natural speed (Mustonen et al., 2004; Mizobuchi et al., 2005) or walking down a defined track in a laboratory room (Pirhonen et al., 2002; Lin et al., 2007). Such artificial settings argued to have the advantages of strong control of variables, easy to set up and highly replicable, but the problem is that they are less realistic and thus the results are hard to generalize (Kjeldskov & Stage, 2004). There have been also several field-based evaluations conducted outdoor (Brewster, 2002; Kane et al., 2008). However, the settings were still quite artificial as participants either walked

along a short and reasonably quiet path or had to repeat the same path several times continuously. Furthermore, field-based evaluations also suffer the disadvantages of complicated data collection and limited experimental control (Petrie et al., 1998) (see Background section).

Clearly, there is a need of conducting usability evaluation in a more realistic setting, and still maintains efficient experimental control and sufficient data collection. The first step though, is to understand how people use their small devices in real-world situations. Patterns of small device users' behaviours can then be feed back to the evaluation design, making it more realistic and naturalistic. This paper presents a field study that investigates the pattern of use of small devices. To be specific, our study mainly focuses on the input aspect of small device use. We ask: do small device users use the device while on the move? Do they look around while typing or just focus on the device screen with little attention to the surrounding environment? Do they correct their typing errors? Do they use abbreviations? What is the keyboard they prefer? The field study consists of a series of unobtrusive remote observations and interviews. In order not to disturb the users and thus alter their behaviours, the experimenter plays a passive and non-intrusive role during the observations. The interviews follow the observations and are used to confirm the observational results, and also to obtain details of small device users' long-term habit. In the interviews, we also ask questions about how small device users use the device to access the Internet. As the mobile Web has becoming increasingly popular, the investigation on this specific use case intends to reveal how people use the mobile Web. Is using-while-walking a valid scenario of using the mobile Web (see Methodology section).

The field study was conducted in December 2008. We have observed a total of 431 small device users, and have interviewed another 51 users. Results show that small device users normally type on their mobile phones or PDAs while they are walking. They normally type with one hand and press the keys with thumbs, and also correct their typing errors. When using the small devices while walking, people have rapid attention switches between the device screen and the surrounding environment. Regarding the use of the mobile Web, our study indicates that less than one third of small device users access the Web from their mobile phones. The main reasons of not using it are bad interface, high cost, and personal preference of laptop or desktop. For those who use the mobile Web, they normally prefer to use it while sitting or laying down whereas using-while-walking is not reported (see Results section).

Putting the results together, we find general patterns of small device users' behaviours. Small device users normally type on their mobile phones while they are walking alone and not talking. Comparing with typing while walking, small device users have significantly less attention switches when they are typing while standing or sitting still. In addition, small device users prefer a physical keyboard to a soft-keyboard. When using-while-walking, small device users normally use the basic functions of their devices, such as telephoning and text messaging. The use of mobile Web is rare. In the last few years, the landscape of small devices has changed by the emergence of touch-screen devices running on powerful operating systems, such as iPhone and Google phone. With these new devices, small device users' preference of physical keyboard over soft-keyboard may have changed because of the significant improvement on touch-screen technology. However, as the use of touch-screen systems are still far from prevalence and the vast majority of small device users are still using systems with physical keypads, the use patterns presented in this paper still represent use of small devices in general. In addition, as accessing the mobile Web becomes easier and less expensive, we would see more coverage of the mobile Web (see Discussion section).

BACKGROUND

Different methodologies have been used to investigate how people use their small devices. Widely used methods include field study, on-line survey, and user behaviour analysis with modern technology.

Field studies are characterized by researchers immersing themselves in the environment of their study, gathering data through observations and interviews (Kjeldskov & Graham, 2003). This method has been widely used to investigate the use of small devices (Petrie et al., 1998; Kristof-fersen & Ljungberg, 1999; Weilenmann, 2001). Kristoffersen and Ljungberg (1999) conducted two field studies with telecommunication service engineers and maritime consulting staff that were heavily involved in field work and used small devices for receiving orders and communicating with colleagues in the field. Their results illustrated the primary problem field workers faced when using small devices was that the interaction required too much visual attention and it required two hands for input. Pascoe et al. (2000) analyzed the fieldwork of a group of ecologists observing giraffe behaviour in Kenya. Weilenmann (2001) conducted a field study with 11 ski instructors during a one-week ski trip. Both studies generated similar results: they found that fieldworkers used a small device in very dynamic context (e.g., while standing, walking, crawling or skiing), with limited attention on the device. They also needed high-speed interaction where the device needed to be able to enter high volumes of data quickly and accurately. In addition, location awareness is also an important feature of the small devices used in outdoor environments. For instance, Sun et al.'s (2009) field studies which were undertaken at large sports events in UK and China showed that spatial context awareness is crucial for enabling the design of personally related mobile services for spectator's at large sports events. Similarly, Greaves et al. (2009) presented a formative field study in which over a period of three days, people were observed in using projector phones and pico projectors. Similar to other field studies, this study also showed the importance of location awareness and context. While field studies can generate rich amount of data in relatively short time, the major disadvantage of this method is the unknown bias and uncertainty of the representative ness of the data. It is possible that behaviours of the participants of the field study are specific to certain population and thus hard to generalize (Kjeldskov & Graham, 2003).

In addition to field studies, surveys are also conducted on how people use the mobile Web. Kim et al. (2002) studied 37 small device users in a period of two weeks, using survey-based method. Results of the study showed that use of mobile Web was highly concentrated in a few key contexts. The most frequently experienced context was when participants felt joyful, in a calm and quiet environment, not moving and used just one hand to manipulate the device. Similar results were also confirmed by Chang (2010) where a survey was conducted with 249 mobile users in Australia. Lee *et al.*'s survey (2005) extended Kim et al.'s survey by looking at more context factors such as time of the day, privacy of the content browsed, and crowdedness of the surrounding environment. Similar results also indicated that the mobile Web use was heavily clustered around a few key contexts, rather than dispersed widely over diverse contexts. In addition, Chae and Kim (2004) conducted an on-line questionnaire survey, the results of which suggested that small device users preferred to buy products with low risk when accessing shopping Web sites using small devices. Kaikkonen (2008) also conduced a global online survey with 390 people. This survey showed that people use the mobile Web in different contexts and for different activities including viewing pictures, vidoes, etc. Similarly, Schmiedl et al. (2009) conduced a face-to-face survey with 109 participants about the usage and usability of the

Web. This survey showed that their participants prefer to use touch screen devices, and they were also faster using the mobile tailored version. The survey-based method requires a participant to report their activities and mental status straight after using the device. However, using this method alone has the risk of being too subjective since the personal characteristics, working habits, and attitude to the study of a participant may mediate the survey and thus affect the results. For example, the participants may choose not to report some details that they think are irrelevant but in fact very important to researchers.

Finally, modern technologies are also used to investigate the behaviours of small device users. Cui and Roto (2008) investigated the use context of small devices using interviews combined with traffic log analysis. They found that the use context of small devices could be characterized with four factors: spatial factor, temporal factor, social factor, and access factor. For example, small device users preferred to access the Web when they were stationary, such as sitting at home or in a restaurant. They also tend to use the mobile Web during short breaks, such as waiting for a bus. A similar study was conducted by Heimonen (2009), which was a four-week diary study with experienced and active mobile Internet users. Participants were asked to use a Web form to keep a diary of their information needs. This study suggested that mobile information services should consider a wider context of use other than location based services, including social interactions and situated activities. A similar diary study was conducted by Amin et al. (2009). This study showed that people tend to stick closely to regularly used routes and to regularly visited places such as home and office. A more specific application oriented study was conducted by Chin & Salomaa (2009) where the context of the study was 2008 Beijing Olympics. The data in this study was collected by a logging application and survey. Even though the target of this study was identifying the most popular application in a suite of applications, this study shows the importance of context which was investigated in other studies such as by Oulasvirta et al. (2005). Oulasvirta et al. (2005) looked at how context affects small device users' attention. In their study, a participant was equipped with three small cameras: two mounted on the mobile phone and one attached to the participant's coat. These cameras were used to record the device screen, the surrounding environment, and the participant's face. In addition, an experimenter carried a fourth camera to record the whole scene. Results of the study showed that when walking in public areas, small device users had much rapid attention switches, and compared with that in a laboratory, the continuous span of attention to the small device was much shorter in public areas. The interview method used by Cui *et al.* has the drawback that it is limited to those who are accessible and will cooperate, and the responses obtained are produced in part by dimensions of individual differences irrelevant to the topic at hand (Webb, 2000). Oulasvirta et al.'s study (2005) was conducted in public areas without artificial setting. However the participants had to carry additional equipments with them, which would affect their performance.

Methodology

The overall methodology of our field study consists of two phases: phase one includes an observational study and phase two includes an interview study. The presence of the observer could change the behaviour of the subjects, thus affecting the validity of the results (Webb, 2000). In this study, we used remote observation and observed the subjects from a distance without acknowledging them. Five places in Manchester, the United Kingdom, were chosen for the observational study, including a train station, a shopping centre, a university bus stop, a business area and a market street. These places were chosen for following reasons: first

of all, compared to other public areas, the chosen locations, such as train station and market street, have higher volume of passengers. Secondly, we were likely to observe different classes of small device users at different locations. For example, at the university bus stop, we were expecting students and younger users; and at the business street, we expected more people using high-end business oriented smart-phones.

At each of these places, the observer first chose an observation spot, for example a seat near the window of a coffee house where the observer could look at people walking on the street. Figure 1 is a picture of the coffee shop we used as an observation spot on the market street. After settling down, the observer then spent about two hours taking notes on the small device users passing by. We deliberately chose different hours of a day to cope with the factor that an individual's behaviour may shift as the time changes (Webb, 2000). In addition, some environmental factors, such as lighting condition and congestion of the road also changed with time. For example, the market street would have less people than the business area in the morning, whereas it would be the other way around in the afternoon after working hours. The location sampling and time sampling techniques used in this study allow us

Figure 1. A coffee shop on the Market Street used as one observation spot

Table 1. Small device users' physical status and activities that have been observed

Activities	Explanations
Movement	Whether the small device user is walking, standing still, or sitting.
Company	Whether the small device user is alone or accompanied.
Talking	Whether the small device user is talking to others or being silent.
Hand Used	Whether the small device user manipulates the device with one hand or both hands.
Finger Used	Which finger a small device user uses to press the keys.
Device Usage	Whether the small device user is making phone calls, sending messages, or reading materials on the screen.
Attention Switch	The number of attention switches between the device screen and the surrounding environment.

to obtain a more representative sample of the small device users.

Table 1 lists the activities of small device users that we have observed. Besides the movement status and hand/finger usage, we also looked at attention switches of a small device user. An attention switch occurred when an obvious change of attention between the device screen and the surrounding environment was observed with a small device user. It is an indicator of the disturbance a small device user received from the environment. The observer counted the number of attention switches of every small device user in a period of 20 seconds. The assumption is that the more attention switches a user had, the less likely the user focuses on small device tasks. For example, a small device user walking on a busy road would have more attention switches than one walking on a quiet road, and would be disturbed more from using the small device. Note that due to the limitation of observation setup, it was difficult to precisely count a small device user's attention switches. In this study, attention switches were counted upon obvious head movements of small device users. Other subtle indicators, such as eye movement and change of walking speed, were not counted.

The observational study gives us a snapshot of small device users' behaviours. The interviews conducted in phase two seek to confirm the observational results and also obtain information on small device users' long-term habit. The interviews were conducted on the street where the interviewer stopped pedestrians randomly and proposed to have a conversation with them about how they used mobile phones or PDAs in their daily lives. Upon agreement, the interviewer asked questions and wrote down the answers on a notebook. On average, an interview lasted 5 to 10 minutes. Again, location sampling and time sampling techniques are used in the interviews to get a more representative sample. Table 2 lists questions used in the interview. Questions 1 to 13 were formed based on results of the observational study. These questions seek to confirm the observational study results and also revealed details of small device users' behaviours. Questions 14 to 19 are related to accessing the Internet from small devices. Although this study did not specifically focus on one functionality of small devices, we see use of Internet on a mobile device as a emerging usage pattern and thus devoted some effort to investigate.

RESULTS

This section presents the results of our field study. We start with the results of the observational study, and then move on to the results of the interviews.

Table 2. Questions asked in the interviews

Index	Questions
1	Do you normally use your mobile phone while you are walking?
2	When you are walking, what do you use your mobile phone for?
3	When using your mobile phone, do you normally type with one hand or both hands?
4	Do you use your thumb to press the keys or other fingers?
5	Do you normally correct your typing errors?
6	Do you normally use predictive text? (e.g. T9)
7	Do you normally use abbreviations?
8	Have you ever used a stylus and a touch-screen on a PDA before?
9	Do you normally type very long text messages?
10	Do you use your mobile phone to do other text editing tasks other than sending text messages?
11	When you are typing while walking, which keypad do you prefer, on-screen keypad or physical keypad, why?
12	Comparing with typing while standing still, do you think you have more attention switches between the mobile phone and the surrounding environments when you are typing while walking?
13	Comparing with typing while walking with friends, do you think you have more attention switches between the mobile phone and the surrounding environments when you are typing while walking alone?
14	Does your mobile phone have the function to access the Internet?
15	Do you use the Internet on your mobile phone?
16	How often do you use it?
17	What do you use it for, news, entertainment, emails, maps or something else?
18	Where do you normally use your mobile phone to access the Internet?
19	If you do not use the Internet on your mobile phone, why not?

Observational Study Results

We observed 431 small device users in total, 100 of whom were typing on their devices, and the other 331 were having phone conversations. Results presented in this section focus on the 100 small device users making text entry. Of the 100 small device users, 61 are male and the other 39 are female. Judged from their appearance, 63 small device users were in the age range of 15 to 35, and the other 37 were between 35 and 60.

Typing While Walking

Figure 2 shows that 83 of the 100 observed small device users were walking while typing on their mobile phones. The other 17 participants were standing or sitting still. In terms of company, 87 were alone while typing, and other people accompanied the other 13. In addition, 90 small device users were not silent while typing, and the other 10 were talking with others.

Based on our observation, the majority of small device users who made text input to the devices were making text entry. Figure 3 shows that 79 small device users were entering text using their devices. Due to the distance between the observer and small device users observed, it is hard to find out exactly what text entry task those small device users were doing. They were either sending text messages, or making entries into calendars, or writing emails. However, according to our interview results, 76% of small device users do text messaging as the only text entry task. We

Figure 2. Physical statuses of the observed small device users

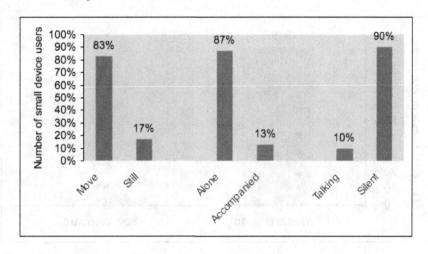

Figure 3. Activities of the observed small device users

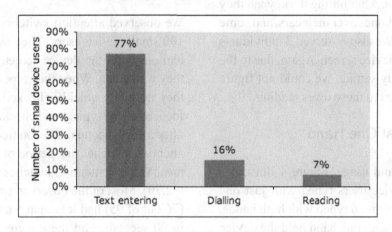

Figure 4. Hand usage of the observed small device users

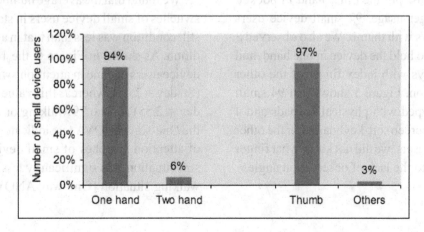

Figure 5. Keyboard usage of the observed small device users

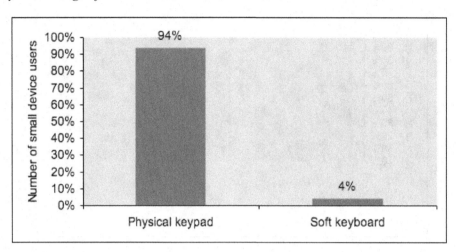

have also observed 16 small device users dialling telephone numbers. After hitting the keypad they directly held the phones over their ears, and some started speaking. We also observed 7 individuals reading from the device screen. Again, due to the observational study setting, we could not figure out what exactly were these users reading.

Typing with Just One Hand

With regard to hand usage, Figure 4 illustrates that 94 small device users typed with just one hand whereas the other 6 typed with both hands. Those who typed with one hand held the device in palm and press the keys with thumb. The other hand was normally used to carry bags or holding books. Some just put the other hand in pocket. Regarding finger usage, 97 small device users pressed the keys with thumb. We also observed 3 individuals who held the device in one hand, and pressed the keys with index finger of the other hand. In addition, Figure 5 shows that 94 small device users typed with physical keypads and 4 typed with on-screen soft-keyboard. For the other 2 small device users, we did not know what finger they used due to the lack of observation angle.

Rapid Attention Switches

We observed attention switches from 95 of the 100 small device users. They switched their attentions from the device screen to the path that they walked on. When they typed while walking, they normally quickly checked the path ahead, focused on the typing, and checked the path again after a few seconds. We counted the number of attention switches in a period of 20 seconds. The mean value of attention switches was 3.27 (St. dev = 2.70). Most of the observed small device users (76 out of 95) had less than 5 attention switches in 20 seconds, and there were 10 small device users who had more than 7 attention switches in 20 seconds.

We found that the average number of attention switches of small device users in standing/sitting still condition was less than that in a walking condition. As shown in Figure 6, the 16 static small device users had mean attention switches of 1.19 (St. dev = 2.05), whereas this value was 3.69 (St. dev = 2.55) for the 79 walking ones. Results of the One-way ANOVA test indicate that the mean of attention switches of small device users in a still situation was significantly less than that in a walking situation (One-way ANOVA, $\alpha = 0.05$,

Figure 6. Comparing the number of attention switches in four different conditions

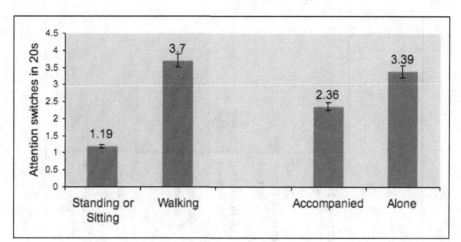

p = 0.04). We also found that the mean attention switches of the small device users who were accompanied or led by other people were less than that of those who were alone. Figure 6 shows that the mean value of attention switches of the 11 small device users who were accompanied was 2.36 (St. dev = 1.79); and the value for the other 84 small device users was 3.39 (St. dev = 2.80). However, the result was not statistically significant (One-way ANOVA, p> 0.05).

Interview Results

A total of 51 small device users were interviewed. Table 3 presents the demographic information of the interviewees. The rest of this section presents the interview results in details.

Table 3. Genders and age ranges of the interviews

Gender and Age Range	Affirmatives	Percentage
Male	29	56.85%
Female	22	43.14%
Age: 15-35	30	58.82%
Age: 35-50	11	21.57%
Age: 50-65	10	19.61%

Using Small Devices While Walking

75% of the small device users we interviewed indicated that they used their mobile phones or PDAs while they were walking. The main use, not surprisingly, was making phone calls. 47% of the interviewees also claimed that they had experiences in sending text messages while walking. In addition, 76% of the interviewees claimed that they did not do any text editing tasks other than sending text messages.

Typing with Just One Hand

76% of the interviewees claimed that they typed with just one hand, and 88% claimed that they pressed the keys with thumb. As illustrated in Figure 7, interview results also revealed other typing habits of small device users. 61% of the interviewees claimed that they normally used the predictive text function to assist their text entry. 59% indicated that they normally used abbreviations in text messages; and 86% said they would correct the typing errors in their messages if any.

Figure 7. Typing habits of small device users

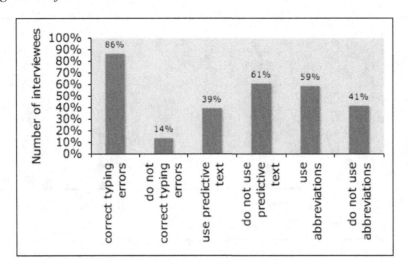

Using a Physical Keypad

Figure 8 shows that only 24% of the interviewees used a soft-keyboard for typing; whereas 76% of them used physical keyboards. This is possibly because there are more small devices with physical keyboards than those equipped with soft-keyboards in the market. And also small devices with soft-keyboard and touch screen are always high-end product and thus more expensive. However, for those who had experiences in using both soft-keyboard and physical keyboard, still 58% preferred physical keyboard because of the lack of tactile feedbacks on a soft-keyboard. One interviewee commented that as there was no tactile feedback on a touch-screen, he usually makes typing errors by pressing a key that is neighbouring the target key.

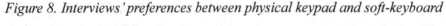

Figure 8. Interviews' preferences between physical keypad and soft-keyboard

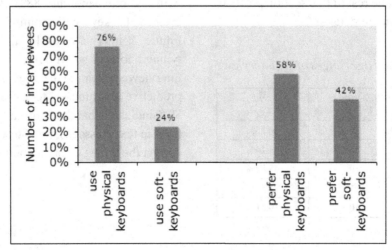

Figure 9. Interviewees' responses on their attention switches under 4 different conditions: walking vs. standing still, and accompanied vs. alone

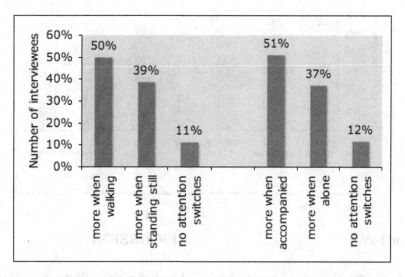

More Attention Switches While Walking or Accompanied

In the interview, we asked two questions regarding a small device user's attention switches under different conditions (see Question 12 and 13 in Table 2). As shown in Figure 9, 43% of the interviewees claimed they had more attention switches when using a small device while walking, and 33% thought they had more attention switches when using the device while standing still. The

other 24% responded that they always focused on the device and thus had no attention switch at all. Similarly, 43% claimed they had more attention switches when typing on a small device and walking with a friend; 31% reported they had more attention switches when tying and walking alone. Again, the other 26% of the interviewees claimed they had no attention switch when using a small device.

Figure 10. Interviewees' responses on how often and where they used the mobile Web

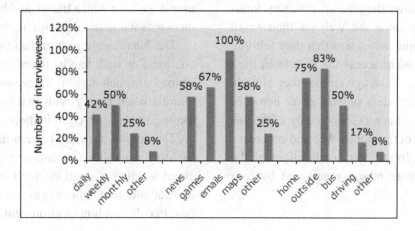

Figure 11. Interviewees' responses on why not using the mobile Web

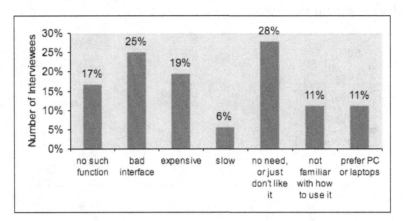

Using the Mobile Web

During the interview, we also asked questions about how small device users access the Web (i.e., using the mobile Web). 76% of the interviewees claimed that their small devices had the function to access the Web. However, only 29% of the interviewees reported using the function. Figure 10 shows that most of those who claimed using the mobile Web either used it daily or weekly. The main activities on the mobile Web were checking emails, reading news and playing games. With regard to location of using the mobile Web, responses shows that people either used it when they were at home or sitting outside (e.g., in a park or a coffee shop). We did not record any instance that a small device user using the mobile Web while walking.

Figure 11 shows the reasons given by interviewees for not using the Web via their small devices. The main reason was that they felt that there was no need to access the Web from their mobile phones or PDAs given that they already had desktop computers or laptops at home or working places. This was followed by criticisms on bad interface of the mobile Web and expensive data traffic cost. In addition, not familiar with the function is another reason mentioned by four interviewees.

DISCUSSION

A pattern of use reflects a typical scenario of how a device is used in real life. This section consolidates the results presented above and presents patterns of use of small devices. According to the results of the observational study, patterns of use of small devices can be grouped into three sets: mobility, hand usage, and attention switches. Each of these patterns has several attributes, describing use of a device from different aspects.

The mobility pattern consists of three attributes, each of which has two values: move or still, alone or accompanied, and silent or talking. When all three attributes are used, the mobility pattern has 8 distinct combinations (2×2×2). For example, a possible scenario based on one combination of the mobility pattern can be that a small device user is walking with a friend, and he replies a text messages while talking with his friend.

The hand usage pattern has three attributes: one hand or both hands, thumb or other finger, keypad or touch-screen. Since one can only use thumbs when typing with two hands, the hand usage pattern has 6 combinations.

The attention switch pattern has just one attribute: the number of attention switches. We care about whether a small device user would have more attention switches in a more visual cognitive resource demanding environment.

Table 4. Possible values of the mobility pattern and hand usage pattern

Pattern Index	Explanations
M1	Walking alone and not talking
M2	Walking while accompanied and talking
M3	Walking while accompanied and not talking
M4	Not walking and alone and not talking
M5	Not walking and accompanied and talking
M6	Not walking and accompanied and not talking
M7	Walking alone and talking
M8	Not walking and alone and talking
C1	One hand typing, using thumb, using physical keypad
C2	One hand typing, using index finger, using physical keypad
C3	Both hands typing, using thumb, using physical keypad
C4	One hand typing, using thumb, using soft keyboard
C5	Both hands typing, using thumb, using soft keyboard
C6	Writing on a touch-screen with a pen

Table 4 lists the possible values of mobility pattern and hand usage pattern. Figure 12 and Figure 13 reconstruct the observational study results based on these patterns. From Figure 12 we can see

that 72% of the observed small device users typed while they were walking, alone, and not talking. This observation is supported by the interview result that 75% of the interviewees claimed that they used their small devices while walking and 47% reported that they sent text messages while walking. On the other hand, Figure 13 shows that 89% of the observed small device users typed with a physical keypad, using just one hand to manipulate the device, and pressed the keys with thumb. This is also confirmed by the interview results that 76% of the interviewees typed with one hand, and 88% pressed the keys with thumb.

In terms of the attention switch pattern, observational study results show that when typing while walking, small device users had 3.27 attention switches in 20 seconds on average. They also had significantly less attention switches when standing still. Interview results indicate that more participants thought they had less attention switches when standing still, which confirms the observational study results. Interview results also suggest that more participants thought they had more attention switches when being accompanied than being alone. However, the observational study on this comparison yields an opposite but insignificant result. Oulasvirta et al.'s (2005) work suggests that a small device user's continuous

Figure 12. A sectorial breakdown of mobility pattern of small device users

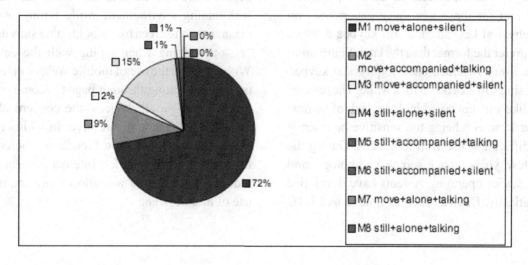

Figure 13. A sectorial breakdown of hand usage pattern of small device users

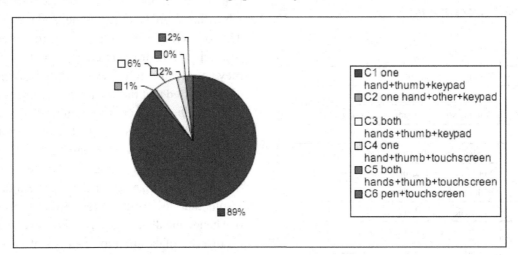

attention to device fragmented and broke down to spans of 4 to 8 seconds. Our results confirm with Oulasvirta et al.'s (2005) results. Since the count of attention switch in our study is based on estimate rather than precise recording, it is difficult to compare two result sets statistically. However, the author would expect the average number of attention switches go up if precise recording techniques was applied.

With regard to use of touch-screen, our results show that although only four touch-screen users were observed during the observation phase, 24% of small device users participating in the interviews claimed that they had used or were using a small device with touch-screen. In addition, we also found that for those who have experiences on both physical keypads and soft-keyboard, more users prefer the former than the later. Some small device users switched back to physical keypad after using soft-keyboard for a while. The reasons of dislike soft-keyboard include: lack of tactical feedback, screen being too sensitive or insensitive, difficult to use with long nails. During the past few years, touch-screen technology and small device operating systems have developed dramatically. Devices such as iPhone and HTC

smartphone which run on iOS and Android allow users to accessing vast amount of applications by simply tapping and sliding the screen. As new technologies emerge, behaviours and preferences of small device users may change. From the author's opinion, it is more convenient to hold a touch-screen device like iPhone with one hand, and type using the index finger of the other hand. With the improved touch-screen technology, and also larger screen size, a user can press a key on a screen easily. As touch-screen does not provide tactile feedbacks as a physical keyboard does, it would probably requires more visual attention to type with. Therefore, a touch-screen user would devote more attentions to the screen than to the surrounding environment while typing, which means a touch-screen user would either slow down or stop walking when typing with the device. With regard to the use of mobile Web, as entering url and navigating through pages become easier on emerging small devices, the concern about interface will gradually resolve. In addition, as the bandwidth of mobile broadband increases, the cost of accessing rich Internet content will reduce. These factors will allow more and more use of mobile Web.

CONCLUSION

This paper presents a field study that investigates the pattern of use of small devices. The field study consists of a series of unobtrusive remote observations and interviews. Following use patterns of small devices were found based on the field study results.

- Small device users type on their mobile phones or PDAs while they are walking alone and not talking (72% of small device users observed).
- Small device users type with just one hand, and press the keys with thumb (89% of small device users observed).
- Small device users use predictive text function (61%), use abbreviations (59%), and correct typing errors in their text messages (86%).

The patterns derived from this study apply to general small device users. Behaviours of users of advanced touch-screen systems might be different. However, the vast majority of small devices, are still far less sophisticated. Therefore, use of the majority of small devices followed, and will keep on following the patterns presented.

EXPERIMENTAL DATA

Further details of the study presented here, including the materials, data and its analysis can be found at the Web Ergonomics Lab's data repository, http://wel-eprints.cs.man.ac.uk/98/.

ACKNOWLEDGMENT

This paper is written as part of the RIAM project[1] funded by EPSRC (EP/E002218/1), whose support we are pleased to acknowledge. The authors would like to thank all the participants for their enthusiasm and help. The authors would also like to thank Alex Chen and Grace Mbipom for carrying out the field study as experimenters.

REFERENCES

Amin, A., Townsend, S., et al. (2009). Fancy a drink in Canary Wharf?: A user study on location-based mobile search. In *Proceedings of the Human-Computer Interaction (INTERACT 2009)*, Uppsala, Sweden (pp. 736-749).

Barnard, L., & Yi, J. S. (2005). An empirical comparison of use-in-motion evaluation scenarios for mobile computing devices. *International Journal of Human-Computer Studies*, *62*(4), 487–520. doi:10.1016/j.ijhcs.2004.12.002

Brewster, S. (2002). Overcoming the lack of screen space on mobile computers. *Personal and Ubiquitous Computing*, *6*(3), 188–205. doi:10.1007/s007790200019

Chae, M., & Kim, J. (2004). Do size and structure matter to mobile users? An empirical study of the effects of screen size, information structure, and task complexity on user activities with standard web phones. *Behaviour & Information Technology*, *23*(3), 165–181. doi:10.1080/01449290410001669923

Chang, P. (2010). Drivers and moderators of consumer behaviour in the multiple use of mobile phones. *International Journal of Mobile Communications*, *8*(1), 88–105. doi:10.1504/IJMC.2010.030522

Chin, A., & Salomaa, J. (2009). A user study of mobile Web services and applications from the 2008 beijing olympics. In *Proceedings of Hypertext 2009*, Turin, Italy.

Cui, Y., & Roto, V. (2008). How people use the Web on mobile divices. In *Proceedings of the 17th international conference on World Wide Web*, Beijing, China (pp. 905-914). New York: ACM.

Greaves, A., & Akerman, M. (2009). Exloring user reaction to personal projection when used in shared public places: a formative study. In *Proceedings of the Context-Aware Mobile Media and Mobile Social Networks Workshop in MobileHCI 2009 Conference*, Germany.

Heimonen, T. (2009). Information needs and practices of active mobile internet users. In *Proceedings of the 6th International Conference on Mobile Technology, Application & Systems*, Nice, France (pp. 1-8).

Kaikkonen, A. (2008). Full or tailored mobile Web - where and how do people browse on their mobile phones. In *Proceedings of the International Conference on Mobile Technology, Applications, and Systems*, Yilan, Taiwan (pp. 1-8).

Kane, S., Wobbrock, J., et al. (2008). TrueKeys: Identifying and Correcting Typing Errors for People with Motor Impairments. In *Proceedings of the 13th International Conference on Intelligent User Interfaces* (pp. 349-352). New York: ACM.

Kim, H., Kim, J., et al. (2002). An empirical study of the use contexts and usability problems in mobile Internet. In *Proceedings of the 35th Annual Hawaii international conference on system Sciences (HICSS'02)* (pp. 132). Washingotn, DC: IEEE Computer Society.

Kjeldskov, J., & Graham, C. (2003). A review of mobile HCI research methods. *Human-Computer Interaction with Mobile Devices and Services*, 317-335.

Kjeldskov, J., & Stage, J. (2004). New techniques for usability evaluation of mobile systems. *International Journal of Human-Computer Studies*, 60(5-6), 599–620. doi:10.1016/j.ijhcs.2003.11.001

Kristoffersen, S., & Ljungberg, F. (1999). "Making place" to make it work: emprical explorations of hci for mobile cscw. In *Proceedings of the international ACM SIGGROUP conference on supporting group work* (pp. 276-285). New York: ACM.

Lee, I., & Kim, J. (2005). Use contexts for the mobile internet: a longitudinal study monitoring actual use of mobile Internet Services. *International Journal of Human-Computer Interaction*, 18(3), 269–292. doi:10.1207/s15327590ijhc1803_2

Lin, M., & Goldman, R. (2007). How do people tap when walking? An empirical investigation of nomadic data entry. *International Journal of Human-Computer Studies*, 65(9), 759–769. doi:10.1016/j.ijhcs.2007.04.001

Lumsden, J., & Brewster, S. (2003). A paradigm shift: alternative interaction techniques for use with mobile wearable devices. In *Proceedings of the 13th annual IBM centres for advanced studies conference* (pp. 197-210). IBM Press.

Mizobuchi, S., Chignell, M., et al. (2005). Mobile text entry: relationship between walking speed and text input task difficulty. In *Proceedings of the 7th international conference on Human computer interaction with mobile devices & services* (pp. 122-128). New York: ACM.

Mustonen, T., Olkkonen, M., et al. (2004). Examning mobile phone text legibility while walking. In *Proceedings of Conference on Human Factors in Computing Systems* (pp. 1243-1246). New York: ACM.

Oulasvirta, A., Tamminen, S., et al. (2005). Interaction in 4-second bursts: the fragmented nature of attentional resources in mobile HCI. In *Proceedings of the SIGCHI conference on Human factors in computing systems* (pp. 919-928). New York: ACM.

Pascoe, J., & Ryan, N. (2000). Using while moving: HCI issues in fieldwork environments. *ACM Transactions on Computer-Human Interaction*, 7(3), 417–437. doi:10.1145/355324.355329

Petrie, H., Johnson, V., et al. (1998). Design lifecycles and wearable computers for users with disabilities. In *Proceedings of the First Workshop on Human-Computer Interaction with Mobile Devices*.

Pirhonen, A., Brewster, V., et al. (2002). Gestural and audio metaphors as a means of control in mobile devices. In *Proceedings of the SIGCHI conference on Human factors in computing systems: Changing our world, changing ourselves* (pp. 291-298). New York: ACM.

Schmiedl, G., Seidl, M., et al. (2009). Mobile Phone Web Browsing - a study on usage and usability of the mobile web. In *Proceedings of the 11th International Conference on Human-Computer Interaction with Mobile Devices and Services*, Bonn, Germany (pp. 1-2).

Sun, X., & May, A. (2009). The role of spatial contextual factors in mobile personliszation at large sports events. *Personal and Ubiquitous Computing, 13*(4), 293–302. doi:10.1007/s00779-008-0203-6

Webb, E. J. (2000). *Unobtrusive measures* (Rev. ed.). Thousand Oaks, CA: Sage Publications.

Weilenmann, A. (2001). Negotiating use: making sense of mobile technology. *Personal and Ubiquitous Computing, 5*(2), 137–145. doi:10.1007/PL00000015

ENDNOTE

[1] See http://riam.cs.manchester.ac.uk

This work was previously published in the International Journal of Mobile Human Computer Interaction, Volume 3, Issue 1, edited by Joanna Lumsden, pp.37-54, copyright 2011 by IGI Publishing (an imprint of IGI Global).

Chapter 4
Evaluating the Readability of Privacy Policies in Mobile Environments

R. I. Singh
University of Alberta, Canada

M. Sumeeth
University of Alberta, Canada

J. Miller
University of Alberta, Canada

ABSTRACT

Recent work has suggested that the current "breed" of privacy policy represents a significant challenge in terms of comprehension to the average Internet-user. Due to display limitations, it is easy to represent the conjecture that this comprehension level should drop when these policies are moved into a mobile environment. This paper explores the question of how much does comprehension decrease when privacy policies are viewed on mobile versus desktop environments and does this decrease make them useless in their current format? It reports on a formal subject-based experiment, which seeks to evaluate how readable are privacy policy statements found on the Internet but presented in mobile environments. This experiment uses fifty participants and privacy policies collected from ten of the most popular web sites on the Internet. It evaluates, using a Cloze test, the subject's ability to comprehend the content of these privacy policies.

INTRODUCTION

Computing is undergoing a trend-shift from an environment dominated by desktops and servers to mobile environments such as smart phones, PDA's, and various handheld devices. The success of mobile commerce has led to the large-scale development of applications for mobile users, providing easy access to the purchase of products using such devices. Most companies or organizations post one or more privacy policy documents on their websites to educate users and enhance

DOI: 10.4018/978-1-4666-2068-1.ch004

security. A privacy policy can be described as a comprehensive description of a website's practices on collecting, using and protecting user information. A privacy policy defines what information is collected, the purpose of information collection, and how this will be handled, stored, and used. Furthermore, it provides information on whether customers are allowed to access the collected information and to resolve privacy-related disputes with the website, etc (Story, 2007, para. 1). A poll conducted by CBS News (CBS) and The New York Times (Wobbrock, 2007) showed that 82% of respondents believe that the right to privacy is either under serious threat or is already lost. In addition, Internet users are concerned about companies collecting personal information and any risk that information may be shared with others inappropriately (Roberts, 2005, para. 1).

Unfortunately, the current privacy policies published on websites are usually long, complex and difficult for the end users to read and comprehend. Research has found that many online privacy policies lack clarity and most require a reading skill considerably higher than the Internet population's average literacy level (Jensen & Potts, 2004). For privacy policies to be useful, they must be readable by online users who visit and use the website. This means that the privacy notice should be written at an appropriate level, and should be easy to navigate for information. There is a need to improve the current policies to help Internet users to read and understand website privacy policies and increase privacy awareness. A privacy policy must contain the following information for users (Federal Trade Commission, 2007):

- How and where collected user information is used?
- Whether the information can be linked to an individual
- What happens to the information that has been collected?
- If the information is shared with other websites or companies

- Does the website install any software on the user system?

In this paper, we shall address the following fundamental question: Is readability affected in a mobile environment? More precisely, we shall focus on understanding the impact of display constraints of mobile devices on the readability of privacy policies. In this work, we shall study readability of policies in mobile devices and compare the results of mobile and desktop environments. The remainder of this paper is organized as follows: the next section provides an introduction to the issues and current directions associated with improving the readability of privacy policies. Next, the obvious readability challenge of mobile devices, namely screen size, is covered and others such as connection fees, lack of standards, and how user mobility affects readability. Then we explain some of the serious shortcomings of readability formulas and propose the Cloze test as a reliable replacement. The section concludes with a practical example of a Cloze test. Then, the readability of privacy policies on mobile devices is empirically explored. Next, we ask the question, do people prefer the mobile or desktop environments for readability? Not surprisingly, readers find the mobile environment significantly more difficult to comprehend privacy policies. Then suggestions are made in terms of what directions may be viable to resolve this problem; finally, the paper is concluded.

Readability Challenges for Privacy Policies

Privacy policies have a significant history of being considered difficult to implement. They seek to address the legal obligations of an organization to disclose its privacy practices and policies as they relate to the interactions of users using the organization's web-site, while attempting to describe this information in a form which is accessible to the average Internet user. The organization's legal obligations rather than the document's readability tend to be the principal driving force in terms of document construction often resulting in documents which are lengthy, unclear and contain sophisticated legal-oriented language.

Several initiates have been launched which seek to address this issue by attempting to define writing practices. For example, several US Government agencies have undertaken research on this topic (Kleimann Communication Group, 2006). Their key research finding is that users need a context for understanding the information contained in privacy policies; and that without this context, consumers were unable to make informed choices about the implications of supplying information to organizations. Similarly, the UK Office of the Information Commissioner states that users are unable to assimilate the information provided (Corporate Solutions Consulting, 2007). The OECD provides a Privacy Statement Generator (http://www2.oecd.org/pwv3/) in an effort to assist organizations to produce more comprehensible privacy policies. Despite these calls, privacy policies remain rather complex documents. According to McDonald and Cranor (2008) the median size for a privacy policy, on a "popular web site", is 2514 words resulting in a significant reading overhead causing many individuals to ignore these essential notices (Vila et al., 2003; Reay et al., 2009). Recently, the idea of layered notices has emerged (The Center for Information Policy Leadership, 2009); Lemos (2005) reports that a number of leading multi-national corporations have rushed to adopt this proposal. However, McDonald et al. (2009) report that layered notices are generally ineffective; and in fact result in subjects being less able to accurately describe privacy policies.

While all of this research points to the readability of privacy polices being problematic, it fails to "quantify" how problematic. Our results may seem predictable but we are unaware of any empirical examination in the literature. We believe that it is an important result to demonstrate the extent of difficulty in reading privacy policies on

mobile devices. Recent research by Sumeeth and Miller (2009) concludes that, on average, privacy policies, when presented in a desktop environment, are becoming more readable. However, it also concludes that they are still beyond the capability of a large section of Internet users; and that roughly 20% of the policies require an educational level approaching a post-graduate degree to support their comprehension. Hence, this paper sets out to evaluate the impact of reading these policies in a mobile environment which is likely to decrease the readability of the average policy. If this is in fact the case then many if not most of the policies may become inaccessible to the vast majority of Internet users. The result also provides a baseline against which new privacy policy initiatives can be compared.

Readability Challenges in Mobile Platforms

Most mobile devices have a common challenge to provide users with information through devices that have limited interfaces, small displays and limited input facilities. Approaches taken for desktop environments are less useful in mobile environments: as keyboards, mice, and monitors are replaced by small screens, touch pads and pens. Some of the challenges of mobile platforms (Wobbrock, 2006) that affect readability are:

- Limited screen size, display resolution, font size, color support, paging and scrolling affect readability. Furthermore, mobile devices have limited keys for interaction that often make it hard for readers to scroll and read long privacy statements.
- There is no standardized definition of input interfaces for mobile devices. In the desktop environment, the operation of keyboards and mice are consistent among various manufacturers. This is unlike the case in mobile devices, where input interface

varies from device to device, making it hard for mobile users to operate effectively.

- Such devices are often used when users are mobile, where conditions and environment are not as convenient as working in a desktop setting. User mobility also affects readability.

- Finally, there is a considerable cost associated with using mobile Internet. Users tend to be fast, looking for specific information and would want to skim through as quickly as possible; a different mindset while browsing the Internet on a mobile device.

Readability Assessment for Mobile Platforms

Two methods are commonly used to evaluate the readability of text documents: readability formulas (Chall, 1988; Davison, 1984; Klare, 1975) and the Cloze test (Taylor, 1953; Coleman & Blumenfeld, 1963). While much research has been undertaken in deriving readability formulas, current efforts in this direction still possess serious shortcomings; their primary drawbacks being:

- The formulas assume that longer sentences are harder to comprehend; this assumption is often false. Longer sentences could be syntactically less complex than shorter sentences. Consider the difficulty in parsing "That that is is that that is not is not." especially without appropriate punctuation.

- The formulas assume that shorter words imply easy comprehension and readability. Counter examples can easily be constructed; again, see above, this sentence has no word with more than 4 characters.

- Sentence and word complexities are weighed differently across these metrics. Moreover, the basic relation between word complexity and sentence complexity in the formulas is rather unclear and varies

between formulas (Zakaluk & Samuels, 1988; McConnell, 1982).

The formulas attempt to gauge only the semantic complexities of text passages. They, however, do not measure other factors that affect comprehension such as, writing style (which is often adapted for mobile devices), logical structure, display challenges (such as in a browser in a mobile device) and textual features (bullets, font size, color contrast, hyperlinks, etc.). Additionally, in privacy policies, Internet related issues such as readers' site-specific knowledge, acronyms, technical vocabulary, services, etc. cannot be assessed by these formulas, specifically:

- Objective evaluation of a text passage is not useful when the author's main consideration in writing a passage is oriented for specialized readers with domain specific knowledge and interest.

- Comprehension is both a textural and knowledge driven process, i.e., it requires understanding of the context of the passage (that can be achieved by simple words and sentences) and domain specific knowledge (technical vocabulary). Formulas can only attempt to evaluate text complexities.

- Readability formulas are not interactive and cannot gauge an audience's specific traits such as typography, layout, skills, interest, or usefulness.

- Readability formulas can offer quick feedback to authors to revise their material. It has been argued that they are only useful for grade-level texts. However, when used for a specialized material, such as privacy policies, they can make it less readable.

The issues surrounding readability formulas are overcome by the Cloze test. The main benefit of this test is that it uses the interaction between the text and a specific population from the target audience to evaluate readability. This interaction is both text and knowledge driven and allows authors to better gauge the readers' comprehension.

The Flesch-Kincaid readability formula is the authoritative test for the US Government Department of Defence. An application of this formula will illustrate the shortcomings of readability formulae. Start by calculating L = average sentence length (how many words ÷ how many sentences). Next, calculate N, the average number of syllables per word (number of syllables ÷ number of words). So then:

- Grade level = (L × 0.39) + (N × 11.8) - 15.59
- Reading Age = (L × 0.39) + (N × 11.8) - 10.59 years

In applying the Flesch-Kincaid formula to the following text, all these sentences score the same:

The cat sat on your mat.
The cat sat on the mat.
On the mat the cat sat.
Sat, on the mat: the cat
The cat on the mat sat.
On the man, sat the cat.
Sat: the cat on the mat.
Sat the cat on the mat?

The issue here is that the first sentence is probably the more readable and the formula does not detect this. The first sentence is more personal, names and describes the subject, and also has a definite order of subject, verb, and object. The nuances of the writer's style are not taken into account in reading formulae in general. (Johnson, 2004)

And lastly, one more simple example:

- He waved his hand.
- He waived his rights.

Both sentences score the same using readability formulas; the second one is more complicated. A child can be taught to wave his hand.

Introduction to the Design of Cloze Experiments

The Cloze procedure was introduced by Taylor (1953) to assess the relative comprehension and readability of written text documents. For those authors who have access to a cross-section of their target audience, the Cloze procedure is especially useful. Its functioning is straightforward. Word deletion according to a predetermined strategy is performed on a selected passage. The reader is then asked to reproduce the original passage. Thus, the Cloze procedure measures how accurately readers can select the words that fit the meaning of the missing words in the passage. The English language has certain characteristics that are exploited in the Cloze Test:

1. **Redundancy:** Core phrases make up an informative sentence, once the core phrase is identified, the redundant words in the sentence that support it are easily identified.
2. **Transitional Probability:** Certain words are commonly used in certain situations: the phrase "_ _ _ _ birthday, Bob" has a high probability of "happy" being the missing word, as opposed to "merry", due to its degree of common use.
3. **Reading Hypothesis:** Combining acquired information (reading) with accumulated information (stored knowledge) to draw a correct conclusion automatically. Stored knowledge is applied to whatever is being read and confirming words are accepted with ease, however, conflicting words or concepts will stop the reading process until they can be validated. This makes reading familiar subjects more easy to comprehend - and unfamiliar ones more difficult.

The Cloze test is also not an ad hoc formulation; instead, it is directly from theories in psychology, especially Gestalt psychology and the Law of Closure. Cloze tests have been the focus of considerable interest in recent years as easily constructed and scored measures of integrative proficiency. Although there has been debate as to whether all forms produced by the Cloze procedure are equally reliable and valid, as well as controversy over what is actually measured, the balance of evidence favors a positive view of the Cloze test as an effective testing instrument; see Fotos (2006) for a detailed discussion.

Cloze Test: A Practical Example

Consider the following passage outlining a copyright issue where every 5th word of the passage has been deleted

- "Organizations have used the _____ Millennium Copyright Act to _____ that Google remove references _____ allegedly copyrighted material on _____ sites."

The word the reader selects is an "educated guess" for each blank to make the sentence meaningful. The number of "educated guesses" provided by the reader is considered proportional to the reader's comprehension of the passage. There are two points worth noting:

1. In the example, the reader's capability to pick the word "digital" is based on technical familiarity with copyright acts. Thus solely understanding the context of the passage would not be enough to select the correct word.
2. The second, third and fourth omitted words i.e. "demand", "to", "other" build connections between phrases. Here, readers must understand the context and meaning of the passage to select the right word.

So the Cloze procedure is capable of measuring comprehension for a reader with a specialized vocabulary and readers with wide knowledge bases and vocabularies. Cloze tests are used in the appraisal of materials from many fields (Gemoets et al., 2004); and recent work by Fanguy et al. (2004) focused on assessing readability of privacy policies using a Cloze test. However, their work has several shortcomings that will be discussed later.

The Protocol Followed by the Cloze Test

The Cloze test can be viewed as having two tasks – the administration of the test itself and the subsequent scoring of the test results.

Administration of the Test

- **Word Deletion:** Based on every nth word, random word deletion based on a random interval or a predetermined deletion pattern (deletion of pronouns, articles, or technical words) words are removed from the text and a blank line left in its place.
- The reader is encouraged to fill in all the blanks.

Scoring the Test

- In most cases, the reader will have to provide the exact word. Synonyms are allowed in certain circumstances.
- The Cloze test score is the number of successes against blanks in the passage.
- Typographical errors and misspellings are not counted as errors

The Cloze test is easy to conduct and assess. Research consistently shows that the Cloze procedure is more suitable than readability formulas in determining the comprehensibility of technical material or with technical, specialized vocabulary (Klare, 1985).

DESIGN OF THE EXPERIMENT

Here, we outline the design of our experiment, specifically the administration of the test and the selection of privacy policies and study participants.

- **Selection of Privacy Policies**: The amount of traffic to a website is a good indicator of its popularity. Alexa (2008) provides an analysis of the top 500 web sites based on an analysis of unique visitors. The top ten listed web sites are listed in Table 1, and it's believed this set is a reasonable representation for the most popular websites on the Internet. The policies on these sites cover all of the activities of the site and while these sites may not be purely e-commerce nor mobile-oriented most of the sites do possess a significant commercial component. For example, Yahoo policies cover over 50 products including sub-sites such as Yahoo!Autos which is the site to buy, sell, maintain, research, and discuss cars. Google's policy covers Google Mobile which provides email and mapping services for mobile devices. It also includes mobile ads or Google for mobile advertisers and Android-specific components. Facebook's policy includes Facebook Beacon, a means of sharing actions you have taken on third-party sites, such as when users make a purchase or post a review and advertisements that appear on Facebook that are delivered directly to users by third-party advertisers. Hence, these sites do contain significant amounts of e-commerce activity which can be accessed via mobile devices. In addition, many of these sites are members of TRUSTe (www.truste.org) and are active members of the EU Safe Harbour Privacy Framework as set forth by the United States Department of Commerce. Hence, these sites can be

Table 1. Alexa's (2008) list of top ten most visited websites – all privacy policies were harvested on July 10, 2008

Web Site Name	Location of privacy policy
Yahoo	http://info.yahoo.com/privacy/us/yahoo/
Google	http://www.google.com/privacypolicy.html
Youtube	http://www.youtube.com/privacy
WindowsLive	http://privacy2.msn.com/en-ca/fullnotice.aspx
Microsoft	http://www.microsoft.com/info/privacy/fullnotice.mspx
Myspace	http://www1.myspace.com/index.cfm?fuseaction=misc.privacy
Wikipedia	http://wikimediafoundation.org/wiki/Privacy_policy
Facebook	http://www.facebook.com/policy.php
Orkut	http://www.orkut.com/html/en-US/privacy.orkut.html?rev=4
Ebay	http://pages.ebay.com/help/policies/privacy-policy.html

considered as reasonable proxies for good practice with respect to privacy policies.

- **Test Administration:** Each test participant was given two different sample policies (one presented for a desktop environment and one presented for a mobile environment) selected at random from the ten privacy policies. The two policies were guaranteed to be different. Assignments were random except that each policy was handed out the same number of times per device. The participants were from the general public between the ages of 25-60 as privacy policies are better understood by this age group and minors are out of consideration due to their level of comprehension. The selected group consisted of regular Internet users. Interviews were conducted on a one-on-one basis and the respondent's feedback given the utmost attention. The project was explained to the participant in a one-on-one white board presentation. A slide show about the project was also done on a lap top. The participants were further given an information sheet explaining the project, as well as a consent form. On the consent form, it was made clear to the participant that they were free to drop out at any time. A total of 60 minutes was allotted to each participant. Regardless of the environment, the participants were asked to record their "answers" on paper. The paper was blank except for a line number. Each line number corresponded to the deletions within the text. The "output" from the survey was structured in this way to avoid (a) results being impacted by potentially inferior input mechanisms available on the pseudo-embedded environment; and (b) to avoid built-in mechanisms (such as spell checkers, thesaurus facilities, etc) impacting the result. The Cloze procedure was done manually to check for misspellings, typographical errors, and synonyms. The Cloze score was calculated as the number of correct replications against the total number of blanks.

There is no one correct way to interpret a Cloze score. Bormuth (1966) selected a 75% threshold and another author Harris (1962) noted that a word replacement of 57% in the test presented a 90% reading comprehension level. In our manual scoring approach, which accepts synonyms and misspelled words, we demand a higher scoring threshold. Hence, we accept 60% as a compromise to those authors above. It is interesting to note that 60% was also used by Harr and Kossack (1990) when evaluating the readability of remuneration packages. We decided to use an every n^{th} word deletion strategy since this was most unlikely to introduce biases between "privacy policies". We set $n = 6$ as is common practice (Stevens et al., 1992; Bormuth, 1966). An example can be seen in Appendix A.

- **Subject Selection:** A total of 50 participants were chosen from the general population after a brief screening as to their availability and suitability for the experiment. The age range was from 19 to 60. There were 22 female and 28 male participants. The participants were regular users of the Internet and had no prior knowledge of these ten privacy policies.

It is argued that it is more important to derive a sample which is representative of the general Internet user population than the current Smartphone user-ship. Smart-phones are a relatively young technology, and like the Internet in its early years, are dominated by young, technically-aware individuals with a relatively high educational level. However, like the Internet, it is believed that this demographic is highly unstable and will evolve over time with new segments of the population joining smart-phone users as the technology matures. Hence, this study has elected to try to capture a cross-section of the population representative of this longer-term envisaged user community. For example, Geoghegan (2009) now reports that "78% of adults over the age of .... 60 are now mobile." Crum (2010) report that in February 2010 14% of iPhone users were over 55 and 53% were over 35; Stroud (2010) reports that in January 2010 that 17% of iPhone users were over 55 and 53% were over 35. Whereas Marketing Vox (2008) reports that in the first Quarter of 2008 "55 percent of iPhone owners are under the age of 35 (compared with 34 percent of all mobile subscribers)". These figures clearly imply that the demographics of iPhone users may be changing and hence we have adopted the position that long-term demographics of Internet users, regardless of platform, will coincide. Simply put, it is unfortunate that demographics on the exact composition are difficult to locate. For the U.S., Rainie

(2010) based upon a survey of 2258 U.S. adults presents the following demographic breakdown of the Internet (Table 2).

For Canada, Statistics Canada (2010) provides this table for people who use the Internet at home for personal use (i.e., non business use). Note that respondents are over the age of 16 and are listed as not concerned about their privacy (Table 3).

- **Environment:** Participants were given the Cloze test in a mobile emulator on a desktop computer. An Apple iPhone simulator developed by Aptana (2008) was used for this study. Display specifications of an iPhone are as follows: 0.5-inch (diagonal) widescreen Multi-Touch display 480-by-320-pixel resolution at 163 ppi. Websites displayed on the Aptana simulator appear the same as on the iPhone. Clearly, this decision represents a validity threat to this study. The study selected this option to enable it to recruit a wide selection of individuals. Initial investigations into recruiting active iPhone users strongly suggested that the majority of these individuals would be young, technically-aware individuals with a relatively high educational level. This group again would represent a threat to the validity of the study – being a poor representation of the general Internet user population. Secondly, this environment was selected in an effort to keep as many experiment variables constant as possible to allow us to isolate the effect of interest (impact of output device on the readability of privacy policies). While, we accept that other factors associated with mobile device use (such as reading in high distraction environments or reading while "on the move") are likely to have an impact on readability, if we allowed wildly different environments between the desktop and mobile components of the experiment, then we would have no prospect of isolat-

ing the effect and associating it with a single cause. It is argued that this approach is relatively common in experimental design; and even in reading experiments involving mobile devices – see Öquist et al. (2004) as an example of an experiment which broadly takes a similar approach in designing their experimental environment. While, this approach leads to a "safe" design, we would accept that it probably over-estimates the efficiency of reading privacy policies in mobile environments.

Hence, in the face of competing threats, it was decided that using the simulator on a more "representative" cross-section of the population minimized these threats to validity. A snapshot of the Cloze test experiment represented on the simulator is provided in Figures 4 and 5 in the Appendix.

Table 2. These figures clearly indicate that all age groups, perhaps with the exception of adults over 65, are well represented (Rainie, 2010)

Demographic Breakdown -- Age	Internet Users (%)
18 – 29	93
30 – 49	81
50 – 64	70
65+	38

Table 3. Demographics for the year 2009. Note that before 2005, Statistics Canada target population was 18 years old as opposed to 16 years

Demographic Breakdown -- Age	Year: 2009 __% users
All Users	24.9
34 years and under	30.2
35 to 54 years	21.2
55 to 64 years	19.5
65 years and older	27.1

There may still be some questions in the readers mind as to design of the study. A possible question might be why it would have been necessary to recruit active iPhone users to use actual iPhones in this study. Instead of recruiting non iPhone users and giving them an actual iPhone to acquaint themselves as to its operation, the experimenters decided to use an emulator. There are number of reasons for this design. It was decided that people would be too unfamiliar with mobile devices and so it would be better to go with emulators. Consider the findings of a mobile device study done in Trelease (2008). The study found that users change their cell phone every 18 months and the findings are suggestive that this period is decreasing. The issue that can be raised is when asking a participant if they are familiar with the iPhone, would it be clear what version they would be familiar with? While the iPhone manufacturer may have a definition for iPhone versions how and why should it apply to iPhone users? What about firmware upgrades, do they constitute a new version?

Investigating this situation quickly becomes a sequence of decisions that more often than not are made without a solid basis to guide decision making. They are also based on series of compromises and a series of guesses (Singh, 2010). To add another dimension to this decision making, how do reading behaviors differ between desktop and reading on a mobile device? Tewksbury and Althaus (1999) found that readings behaviors between newspaper reader and net paper readers vary considerably. Newspaper reading is done linearly, and is done as a pleasurable undertaking. Net paper reading (on a desktop) is done furtively, between answering emails, etc. and net paper readers tend to read more than scan. How do these behaviors transfer over to reading privacy policies on a mobile device?

At this point it is beneficial to remind the reader that while readability is a well researched scientific area, many of their results remain precise so long as they deal with abstract concepts. Trying to make ideas concrete dissipates precision and the

investigator must make a large number of choices in circumstances that are largely unrepeatable by a second investigator; an understandable situation considering the objective is to arrive at an explicit solution (Singh, 2010).

Cloze Score Results for Mobile Environment

Results of the Cloze test are presented in Table 4. It provides the website, the calculated Cloze score, and the number of correct restatements out of the length of the website (given in Table 5) divided by six. (every sixth word was replaced).

Table 4. Cloze test results of privacy policies in a mobile device environment

Websites	CLOZE SCORE	Words	Websites	CLOZE SCORE	Words
Orkut	0.4	44	Wikipedia	0.3125	50
	0.5090	56		0.1187	19
	0.1818	20		0.2875	46
	0.1181	13		0.2375	38
	0.3363	37		0.3875	62
Yahoo	0.3916	56	Youtube	0.2105	32
	0.4195	60		0.3882	59
	0.3356	48		0.1184	18
	0.1398	20		0.1776	27
	0.1818	26		0.1250	19
Windows Live	0.3358	45	Google	0.1237	23
	0.1119	15		0.1613	30
	0.1417	19		0.2419	45
	0.1641	22		0.1505	28
	0.1194	16		0.1022	19
Myspace	0.1428	23	Ebay	0.1005	22
	0.1304	21		0.1050	23
	0.2111	34		0.1781	39
	0.0745	12		0.1461	32
	0.3167	51		0.0548	12
Microsoft	0.0894	33	Facebook	0.1552	54
	0.0569	21		0.0977	34
	0.1192	44		0.0603	21
	0.0325	12		0.0833	29
	0.1653	61		0.1178	41

Table 5. Average Cloze test scores of privacy policies

Website	Length (words)	Avg. Cloze Score	Website	Length (words)	Avg. Cloze Score
Microsoft	1847	0.0926	Ebay	1099	0.1169
Myspace	806	0.1751	Wikipedia	803	0.2687
Yahoo	715	0.2937	Google	930	0.1559
Youtube	761	0.2039	Orkut	552	0.3090
Facebook	1740	0.1029	Windows Live	673	0.1746
			Average		**.1893**
			Std. Dev.		**.1138**

Table 5 provides the domain name, administered policy length, and average Cloze test scores. An average is taken over 5 participants answering Cloze test of a website's policy.

The following conclusions can be drawn from these tables:

- Microsoft's policy is most difficult to read followed by Facebook's policy. Orkut's policy was easy to comprehend. The reason for the low readability of Microsoft's policy may be its length.

- The standard deviation of the Cloze scores is found to be 0.1138: the minimum score was 0.032 and the maximum was 0.509.

- From Table 2, we can see that the scores of all policies fall below 0.6. Therefore, based on the conclusion in Harr and Kosack (1990) we can conclude that the web policies are difficult to comprehend in a mobile environment. The Cloze score distribution is given in Figure 1.

- In informal debriefings with the subject, they postulated that the primary cause for low readability scores in mobile environment is that the mobile display can only accommodate two or three sentences per page. This forces readers to use the scroll bar more often to move back and forth between sentences. With this, readers lose paragraph-context and interest in further reading, thereby resulting in lower scores. This proposition corresponds to the observed behaviour of the subjects during the study.

Figure 1. Frequency of occurrence of Cloze scores

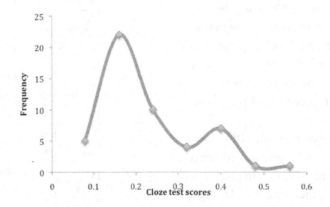

To explore the conjecture that the length of a privacy policy directly impacts its readability and hence the observed Cloze scores, a correlation analysis was conducted and the results are tabulated in Table 6. This suggests that the Cloze scores and policy length are dependent.

Figure 2 illustrates the distribution of number of correct word restatements in a Cloze test. Notice that most of the readers were able to correctly answer between 20 and 80 words in a Cloze test. The average number of correct restatements is found to be 32.62 with a standard deviation of

Table 6. Correlation analysis on the policy length and Cloze score

Pearson correlation	-.5135
Coefficient of determination	.2636
Significance of correlation	p < .0001

Figure 2. Frequency of correct word fill-ins

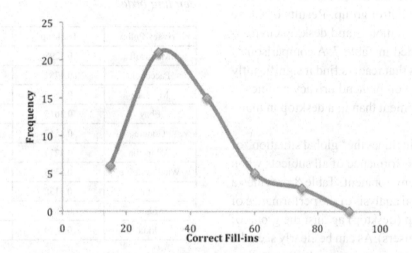

Table 7. Comparison of Cloze scores for desktop and mobile, and t-test results

Website	Avg. Cloze Score		Result of t-test	
	Desktop	Mobile	t	p-value(one-tailed)
Microsoft	0.1298	0.0926	1.23	0.1250
Myspace	0.2577	0.1751	0.91	0.1950
Yahoo	0.5895	0.2937	3.75	0.0028
Youtube	0.5138	0.2039	3.55	0.0038
Facebook	0.1781	0.1029	1.69	0.0650
Ebay	0.3015	0.1169	3.71	0.0030
Wikipedia	0.4274	0.2687	1.72	0.0618
Google	0.3517	0.1559	2.28	0.0256
Orkut	0.6579	0.3090	3.45	0.0043
Windows Live	0.5115	0.1746	2.99	0.0085
Mean	**.3918**	**.1893**		
Std. Dev.	**.2202**	**.1138**		

14.32. The average number of correct answers (32.62) corresponds to a policy length of 163 words.

A COMPARISON STUDY: DESKTOP VERSUS MOBILE ENVIRONMENTS

The results reported above were also collected for privacy policies viewed on a typical desktop environment. The desktop environment is considered to represent the status quo of the readability of privacy policies and hence can be considered as effectively a control group. Results of Cloze tests carried out in mobile and desktop environments are provided in Table 7. A comparison of the scores shows that readers find it significantly more difficult to comprehend privacy policies in a mobile environment than in a desktop in many situations.

This finding includes the "global situation" of comparing the performance of all subjects when using a mobile environment. Table 8 provides a detailed statistical analysis of the performance of the control group (desktop) against the group of interest (mobile users). As can be clearly seen the control group performs statistically better than the group of interest.

In Table 9, the Cloze scores for desktop environment are shown in ascending order; a comparison with mobile Cloze scores reveals that the results are "loosely" dependent on each other. The nature of dependence is captured by a correlation analysis. The results are provided in Table 10.

Table 11 illustrates average Cloze scores for different subject age-groups; and Table 12 shows readability based on gender in mobile and desktop environments. Notice that in the desktop environment, the average Cloze score decreases with age. The performance of subjects in the age-groups 19-23, 24-30 and 31-40 is nearly equal with the performance of the older groups decreasing significantly. On the other hand, in a mobile environment, a general trend does not exist; subjects in

Table 8. Summary of t-test results

	Desktop	Mobile
Samples	50	50
Average	.3918	.1893
$Ave_{Desktop} - Ave_{Mobile}$	0.2025	
T	5.78	
degree of freedom	98	
p-value	< 0.0001	

Table 9. Comparison of Cloze scores of desktop and mobile environments – in desktop score ascending order

Privacy Policy	Desktop	Mobile
Microsoft	0.1298	0.0926
Facebook	0.1781	0.1029
Myspace	0.2577	0.1751
Ebay	0.3015	0.1169
Google	0.3517	0.1559
Wikipedia	0.4274	0.2687
Windows Live	0.5115	0.1746
Youtube	0.5138	0.2039
Yahoo	0.5895	0.2937
Orkut	0.6579	0.3090

Table 10. Correlation analysis of Cloze scores in mobile and desktop environments

Pearson correlation	0.4377
Coefficient of determination	.1916
Significance of correlation	p < .0015

Table 11. Average Cloze scores for different age-groups in different environments

Age-Group	Desktop	Mobile
19 - 23	0.45	0.14
24 - 30	0.44	0.18
31 - 40	0.42	0.25
41 - 55	0.30	0.23
55 - 69	0.18	0.12

Table 12. Average Cloze scores for male and female in mobile and desktop environments

Gender	Desktop	Mobile
Female	0.37	0.20
Male	0.43	0.22

age-group 31-40 performed the best. From Table 12 we can see that the average Cloze scores of male participants are slightly higher than the female participants. However, the difference is not significant.

DISCUSSION

Our results imply that natural language based privacy policy statements are unlikely to ever be an effective communication mechanism for mobile devices. While they present a full statement on an organization's privacy policies, it is at the expense of making the policies inaccessible to the average Internet user. Hence, it is believed that a radical departure from the format is required if this information, or a subset of this information, is to be accessible to the average user. Fortunately, perhaps, the Internet has already seen research in this direction. The Platform for Privacy Preferences Project (P3P) (www.w3.org/TR/P3P) can be viewed as an initial attempt to resolve this conundrum. P3P enables Websites to express their privacy practices in a standard format that can be retrieved automatically and interpreted by user agents. These agents seek to allow users to be informed of site practices and to automate decision-making based on these practices when appropriate. Thus, users need not read the privacy policies at every site they visit. Unfortunately, P3P seems to have failed for a number of reasons. A lack of adoption (Reay et al., 2007) on web sites, a lack of acceptance by users (Reay et al., 2009), a lack of characteristics which have been found to drive technology adoption (Beatty et al., 2007)

and limited support within mainstream browsers (Cranor et al., 2008) have implied that P3P is unlikely to be the answer. More recently, Kelley et al. (2009) has sought an alternative formulation for P3P – the P3P expandable grid. This new formulation has significant promise as it resolves some of the aforementioned problems by moving to a simplified graphical representation of a subset of the information present in current privacy policies. However, the grid is still a "sizable" structure and is still unlikely to conform to the demanding limitations found in mobile environments.

The mobile community is also actively seeking solutions in how to present web-sites. In general, research approaches mainly seek to split the information up into hierarchal structures (Muhanna, 2007). While these approaches may have merit in other contexts, it is difficult to see that these approaches are directly applicable to our situation. Oquist et al. (2004) report on their experiments on reading on a mobile device. In their results, on what they estimate to be Grade 6 reading material, they quote on average reading rate amongst adults of 216.9 words per minute. Our trials indicate that privacy policies have a much higher level of difficulty, which is likely to influence the reading rate; however, if we accept their average figure – what does this mean for privacy policies? This question is explored in Table 13.

Given that these average reading times are likely to be underestimations, clearly an important question is – how likely is the average user to spend this volume of time reading a privacy policy is clearly? Given the average Internet user's willingness to make large amounts of their personal information public on forms such as Facebook, Twitter, and LinkedIn, a strong case can be made that the likelihood is low. More formally, Reay et al. (2009) have investigated this value proposition and concur with this position. Assuming that these observations are valid, this implies that we need to completely rethink the presentation of privacy policies to radically increase the value proposition for the average user.

Table 13. Estimation of the time required by the average individual to read the selected privacy policies

Website	Length	Avg. Time to Read (minutes)	Website	Length	Avg. Time to Read (Minutes)
Microsoft	1847	8.52	Ebay	1099	5.07
Myspace	806	3.72	Wikipedia	803	3.70
Yahoo	715	3.30	Google	930	4.29
Youtube	761	3.51	Orkut	552	2.54
Facebook	1740	8.00	Windows Live	673	3.10

Hence, it is believed that we need to move to even more minimalistic graphic designs representing only the truly essential information to have a chance of finding a realistic mechanism for mobile environments. We believe that a privacy statement based upon something like the mechanisms explored by the U.K. Food Standards Agency represents a potential viable option. The U.K. Food Standards Agency has undertaken two sets of trials to investigate user reaction to a number of "Signpost labeling" options (Synovate, 2005a; Synovate, 2006b). These trials suggest that the "multiple traffic lights" formulation was viewed as simple, and easy to use and understand. Figure 3 gives an example of this formulation.

Clearly, much research needs to be done to transform this concept into a working privacy policy model of mobile environments. However, we believe that it represents a good starting point; we are currently embarking upon further work in an attempt to realize such a model.

Consider a possible scenario as an example. Joe Shopper goes to Movies.all to buy a movie. Assume that Movies.all has a privacy policy for each page. His browser supports P3P and retrieves the privacy policy from the opening page. Broadly speaking Joe has four main concerns: what information is being collected, who gets it, how it is used, and its shelf life. More specifically, the concerns are anonymity of information, what is collected (e.g., IP address, email address), the ultimate receiver of the information, how long it will be kept for, and how the user can access this

information, e.g., read-only, opt-out, opt-in. Joe's browser decides it safe to proceed after checking with Joe's privacy preferences. Joe buys a movie and the site asks for his telephone number. Joe's privacy preferences notify him of this and he decides to proceed. Joe has made purchases from this site before and relaxes stringent privacy rules for the credit card information needed by this site. After Joe completes the transaction, the site asks him to take a third party survey and does not indicate how the information will be used. Joe's preferences decline this request (Cranor et al., 2002).

With a "multiple traffic lights formulation", one possible scenario is that the four lights could

Figure 3. Multiple Traffic Lights Formulation. Available at: http://www.food.gov.uk/foodlabelling/signposting/siognpostlabelresearch/

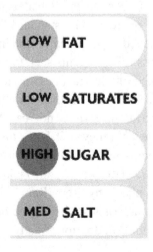

signal what is being collected, who gets it, how it's used, and the shelf life of the information. So Joe arrives at the site, makes his purchase and receives a yellow light when asked for his phone number. When he completes his purchase, since he has dealt with the company before, he gets a green light. The only red light occurs when he is asked to take a third party survey.

CONCLUSION

We conducted a Cloze test experiment to assess readability of privacy policies in a mobile environment. A mobile emulator called Aptana was used for the study. This experiment was part of a bigger study conducted to assess readability of privacy policies in a desktop environment. Several results are worth noting from the Cloze test experiment on mobile devices:

- Cloze test results of all subjects were below 0.6. Therefore, based on the conclusion drawn in Harr and Kosack (1990), no readers were able to comprehend the privacy policy. The maximum score was 0.509 and the average was found to be 0.1823
- The low average readability estimates indicate that mobile displays are not convenient for reading and comprehending "unfamiliar" technical materials as privacy policies. Furthermore, there is a positive correlation between mobile and desktop Cloze scores.
- Scores from readability formulas correlated with subject-based performances (as measured by the Cloze test) in both environments. However, the association between the formulas and the Cloze test scores seems stronger in the desktop environment than in the mobile environment. This perhaps suggests that desktop environment is "neutral" in terms of its impact on the read-

ability of a passage or text. Whereas, the mobile environment has a significant negative impact upon the readability of, at least, the presented privacy policies.

- An inverse dependency between length of privacy statement and Cloze test scores shows that readers' comprehension drops with length; a similar conclusion was drawn for the desktop environment.
- Participants in the Cloze test experiment indicated several concerns:
  - Participants indicated that the privacy documents are too technical and require too much focus and attention in reading, when generally the importance is in downloading or using services provided by the website.
  - The privacy statements are too long to read and comprehend both for desktop and mobile environments. Most users complained about the time needed for a thorough read.
  - In a mobile environment, participants noted the display screen and scrolling to be a major concern. Furthermore, some users pointed out that the Internet rates are exorbitant for browsing through mobile devices. Hence, it was less likely that they would spend time and money in browsing privacy policies.

Our results are extremely pessimistic in terms of the applicability of traditional privacy policies to a mobile environment. While possibilities exist to improve these policies, it is believed that the improvements cannot be substantive enough to justify representing this information using natural language. Hence, exploring initial possibilities for presenting this information in a much more minimal, graphical-oriented format which provides a more realistic basis for privacy policies in mobile environments.

REFERENCES

Alexa. (n.d.). *The Web Information Company*. Retrieved August 8, 2008, from http://www.alexa.com/site/ds/top_sites?ts_mode=lang&lang=en

Aptana. (n.d.). *Apple iPhone Emulator*. Retrieved July 10, 2008, from http://www.aptana.com/iphone/

Beatty, P., Reay, I., Dick, S., & Miller, J. (2007). P3P Adoption on E-Commerce Websites: A Survey & Analysis. *IEEE Internet Computing, 11*(2), 65–71. doi:10.1109/MIC.2007.45

Bormuth, J. R. (1966). Readability: A new approach. *Reading Research Quarterly, 1*, 79–132. doi:10.2307/747021

Chall, J. S. (1988). The beginning years . In Zakaluk, B. L., & Samuels, S. J. (Eds.), *Readability: Its past, present, and future*. Newark, DE: International Reading Association.

Coleman, E. B., & Blumenfeld, P. J. (1963). Cloze scores of nominalization and their grammatical transformations using active verbs. *Psychological Reports, 13*, 651–654.

Columbia Broadcasting System (CBS). (n.d.). Retrieved March 3, 2008, from http://www.cbsnews.com

Corporate Solutions Consulting (UK) Limited. (2007). *Research Report Fair Processing Notifications: Current Effectiveness and Opportunities for Improvement*. Retrieved from http://www.ico.gov.uk/upload/documents/library/corporate/research_and_reports/ic_final_report_version_1.1_final.pdf

Cranor, L., Egelman, S., Sheng, S., McDonald, A., & Chowdhury, A. (2008). P3P Deployment on Websites. *Electronic Commerce Research and Applications, 7*(3), 274–293. doi:10.1016/j.elerap.2008.04.003

Cranor, L., Langheinrich, M., Marchiori, M., Presler-Marshall, M., & Reagle, J. (2002). *The Platform for Privacy Preferences 1.0 (P3P1.0) Specification*. Retrieved from http://www.w3.org/TR/P3P/#intro_example

Crum, C. (2010). *Consumer Demographics and Their Wireless Devices. B2B Publications*. Retrieved July 21, 2010 from http://www.webpronews.com/topnews/2010/02/25/consumer-demographics-and-their-wireless-devices

Davison, A. (1984). *Readability formulas and comprehension. Comprehension instruction: Perspectives and suggestions*. New York: Longman.

Evaluation of Online Privacy Notices. (2004). In *Proceedings of ACM Conference on Human Factors in Computing Systems (CHI 2004)*, Vienna, Austria (pp. 471-478).

Fanguy, B., Kleen, B., & Soule, L. (2004). Privacy policies: Cloze test reveals readability concerns. *Issues in Information Systems, 5*(1), 117–123.

Federal Trade Commission. (2007). *Fair Information Practice Principles*. Retrieved from http://www.ftc.gov/reports/privacy3/fairinfo.shtm

Fotos, S. S. (2006). The Cloze Test as an Integrative Measure of EFL Proficiency: A Substitute for Essays on College Entrance Examinations? *Language Learning, 41*(3), 313–336. doi:10.1111/j.1467-1770.1991.tb00609.x

Gemoets, D., Rosemblat, G., Tse, T., & Logan, R. (2004). Assessing readability of consumer health information: An exploratory study. In *Proceedings of the 11th World Congress on Medical Informatics*, San Francisco, CA (pp. 869-874). Amsterdam, The Netherlands: IOS Press.

Geoghegan, E. (2009). *Experian Simmons Fall 2009 Consumer Study/National Hispanic Study*.

Harr, J., & Kosack, S. (1990). Employee benefit packages: How understandable are they? *Journal of Business Communication*, 27(2), 185–200. doi:10.1177/002194369002700205

Harris, Z. S. (1962). *String Analysis of Sentence Structure*. The Hague, The Netherlands: Mouton.

Jensen, C., & Potts, C. (2004). *Privacy Policies as Decision-Making Tools: an evaluation of online privacy notices*.

Johnson, K. (2004). *Readability*. Retrieved July 17, 2010, from http://www.timetabler.com/reading.html

Kelley, P. G., Bresee, J., Cranor, L. F., & Reeder, R. W. (2009). A "Nutrition Label" for Privacy. In *Proceedings of the Symposium on Usable Privacy and Security*.

Klare, G. R. (1963). *The Measurement of Readability*. Ames, Iowa: Iowa State University Press.

Klare, G. R. (1975). Assessing readability. *Reading Research Quarterly*, 10, 62–102. doi:10.2307/747086

Kleimann Communication Group, Inc. (2006). *Evolution of a Prototpye Financial Privacy Notice*. Retrieved from http://www.ftc.gov/privacy/privacyinitiatives/ftcfinalreport060228.pdf

Lermos, R. (2005). *MSN sites get easy-to-read privacy label, CNET News*. Retrieved from http://news.cnet.com/2100-1038_3-5611894.html

Marketing Vox. (2008). *The Latest Data about Demographics of iPhone Users*. Retrieved from http://www.futurelab.net/blogs/marketingstrategyinnovation/2010/01/latest_data_about_demographics.html

McConnell, C. R. (1982). Readability formulas as applied to college economics textbooks. *Journal of Reading*, 14–17.

McDonald, A. M., & Cranor, L. F. (2008). The Cost of Reading Privacy Policies. *Journal of Law and Policy*.

McDonald, A. M., Reeder, R. W., Kelley, P. G., & Cranor, L. F. (2009). A comparative study of online privacy policies and formats. In *Proceedings of the Privacy Enhancing Technologies Symposium*.

Muhanna, A. (2007). *Exploration of human-computer interaction challenges in designing software for mobile devices*. Unpublished Master's thesis, University of Nevada, Reno, USA.

Öquist, G., Hein, A. L., Ygge, J., & Goldstein, M. (2004). Eye Movement Study of Reading on a Mobile Device Using the Page and RSVP Text Presentation Formats. In *Proceedings of the Mobile Human-Computer Interaction (MobileHCI 2004)* (pp. 108- 119).

Rainie, L. (2010). *Internet, broadband, and cell phone statistics*. PewResearchCenter.

Reay, I., Beatty, P., Dick, S., & Miller, J. (2007). A Survey and Analysis of the P3P Protocol's Agents, Adoption, Maintenance, and Future. *IEEE Transactions on Dependable and Secure Computing*, 5(2), 151–164. doi:10.1109/TDSC.2007.1004

Reay, I., Dick, S., & Miller, J. (2009). An Analysis of Privacy Signals on the World Wide Web: Past, Present and Future. *Information Sciences*, 179(8), 1102–1115. doi:10.1016/j.ins.2008.12.012

Roberts, J. (2005, October 2). Poll: Privacy Rights under Attack. *CBS News*. Retrieved from http://www.cbsnews.com/

Singh, R. I., & Miller, J. (2010). Empirical Knowledge Discovery by Triangulation in Computer Science. *Advances in Computers*, 80, 163–190. doi:10.1016/S0065-2458(10)80004-X

Statistics Canada. (2010). *Internet use by individuals, by Internet privacy concern and age*. Retrieved from http://www40.statcan.gc.ca/l01/cst01/comm31a-eng.htm

Stevens, K. T., & Stevens, K. C. (1992). Measuring the Readability of Business Writing: The Cloze Procedure vs. Readability Formulas. *Journal of Business Communication, 29*, 367–382. doi:10.1177/002194369202900404

Story, L. (2007, November 1). FTC to review online ads and privacy. *The New York Times.* Retrieved from http://www.nytimes.com/

Stroud, D. (2010). The Latest Data about Demographics of iPhone Users. *FUTURELAB.* Retrieved July 21, 2010 from http://www.futurelab.net/blogs/marketing-strategy-innovation/2010/01/latest_data_about_demographics.html

Sumeeth, M., & Miller, J. (2009). Are on-line privacy policies readable? *International Journal of Information Security and Privacy.*

Synovate. (2005a). *Quantitative Evaluation of Alternative Food Signposting Concepts, Report of the U.K. Food Standards Agency, No. 265087.* Retrieved from http://www.food.gov.uk/multimedia/pdfs/signpostquanresearch.pdf

Synovate. (2005b). *Qualitative Signpost Labelling Refinement Research, Report of the U.K. Food Standards Agency, No. 951968.* Retrieved from http://www.food.gov.uk/multimedia/pdfs/signpostqualresearch.pdf

Taylor, W. L. (1953). Cloze procedure: A new tool for measuring readability. *The Journalism Quarterly, 30*, 415–433.

The Center for Information Policy Leadership. (2009). *Multi-layered Notices Explained.* Retrieved from http://www.hunton.com/files/tbl_s47Details/FileUpload265/1303/CIPL-APEC_Notices_White_Paper.pdf

The New York Times. (n.d.). Retrieved March 3, 2008, from http://www.nytimes.com/

Trelease, R. B. (2008). Diffusion of innovations: smartphones and wireless anatomy learning resources. *Anatomical Sciences Education, 1*, 233–239. doi:10.1002/ase.58

Vila, T. R., & Greenstadt, D. Molnar. (2003). *Why we can't be bothered to read privacy policies models of privacy economics as a lemon market.* New York: ACM International

Wobbrock, J. O. (2006). The future of mobile device research in HCI. In *Proceedings of the Workshop on What is the Next Generation of Human-Computer Interaction?* Montréal, Canada (pp. 131-134).

Zakaluk, B. L., & Samuels, S. J. (1988). *Readability: Its past, present and future.* Newark, DE: International Reading Association.

APPENDIX

PRIVACY POLICY DISPLAYED ON DESKTOP

Google Privacy Policy

Last Modified: October 14, 2005

At Google we recognize that privacy {1}_____ important. This Policy applies to all {2}_____ the products, services and websites offered by {3}_____ Inc. or its subsidiaries or affiliated {4}_____ (collectively, Google's "services"). In addition, where {5}_____ detailed information is needed to explain {6}_____ privacy practices, we post separate privacy {7}_____ to describe how particular services process {8}_____ href="http://www.google.com/privacy_faq.html#personalin fo"{9}_____ information, which are accessible from the {10}_____ bar to the left of this {11}_____.

Google adheres to the US {12}_____ harbor privacy principles of Notice, Choice, {13}_____ Transfer, Security, Data Integrity, Access and {14}_____, and is registered with the U.S. {15}_____ of Commerce's safe harbor program.

{16}_____ you have any questions about this {17}_____, please feel free to <u>contact us</u> through our {18}_____ or write to us at Privacy {19}_____, c/o Google Inc., 1600 Amphitheatre Parkway, {20}_____ View, California, 94043 USA.

We offer a number of services {21}_____ do not require you to register {22}_____ an account or provide any personal {23}_____ to us, such as Google Search. {24}_____ order to provide our full range {25}_____ services, we may collect the following {26}_____ of information:

When you sign {27}_____ for a <u>Google Account</u> or other {28}_____ service or promotion that requires registration, {29}_____ ask you for personal information (such {30}_____ your name, email address and an {31}_____ password). For certain services, such as {32}_____ advertising programs, we also request credit {33}_____ or other payment account information which {34}_____ maintain in encrypted form on secure {35}_____. We may combine the information you {36}_____ under your account with information from {37}_____ Google services or third parties in {38}_____ to provide you with a better {39}_____ and to improve the quality of {40}_____ services. For certain services, we may {41}_____ you the opportunity to opt out {42}_____ combining such information.

Google Cookies

When you visit {43}_____, we send one or more cookies {44}_____ a small file containing a string {45}_____ characters - to your computer that {46}_____ identifies your browser. We use cookies {47}_____ improve the quality of our service {48}_____ storing user preferences and tracking user {49}_____, such as how people search. Most {50}_____ are initially set up to accept {51}_____, but you can reset your browser {52}_____ refuse all cookies or to indicate {53}_____ a cookie is being sent. However, {54}_____ Google features and services may not {55}_____ properly if your cookies are disabled. {56}_____

Log Information

{57}_____ you use Google services, our servers {58}_____ record information that your browser sends {59}_____ you visit a website. These server {60}_____ may include information such as your {61}_____ request, Internet Protocol address, browser type, {62}_____ language, the date and time of {63}_____ request and one or more cookies {64}_____ may uniquely identify your browser {65}_____

User Communications

When {66}_____ send email or other communication to {67}_____, we may retain those communications in {68}_____ to process your inquiries, respond to {69}_____ requests and improve our services.

Affiliated Sites

We {70}_____ some of our services in connection {71}_____ other web sites. Personal information that {72}_____ provide to those sites may be {73}_____ to Google in order to deliver {74}_____ service. We process such information in {75}_____ with this Policy. The affiliated sites {76}_____ have different privacy practices and we {77}_____ you to read their privacy policies. {78}_____

Links

Google {79}_____ present links in a format that {80}_____ us to keep track of whether {81}_____ links have been followed. We use {82}_____ information to improve the quality of {83}_____ search technology, customized content and advertising. {84}_____ more information about links and redirected {85}_____, please see our FAQs.

Other

This Privacy {86}_____ applies to web sites and services {87}_____ are owned and operated by Google. {88}_____ do not exercise control over the {89}_____ displayed as search results or links {90}_____ within our various services. These other {91}_____ may place their own cookies or {92}_____ files on your computer,

collect data {93}_____ solicit personal information from you. {94}_____
only processes personal information for the {95}_____ described in the applicable Privacy Policy {96}_____ privacy notice for specific services. In {97}_____
to the above, such purposes include:{98}_____ Providing our products and services
{99}_____ users, including the display of customized {100}_____ and
advertising; Auditing, research and {101}_____ in order to maintain, protect and
{102}_____ our services; Ensuring the technical {103}_____ of our network; and Developing {104}_____ services. You can find {105}_____ information about how we process personal {106}_____ by referring to the privacy notices
{107}_____ particular services.

Google processes personal {108}_____ on our servers in the United
{109}_____ of America and in other countries. {110}_____ some cases,
we process personal information {111}_____ a server outside your own country.
{112}_____ may process personal information to provide {113}_____
own services. In some cases, we {114}_____ process personal information on behalf of
{115}_____ according to the instructions of a {116}_____ party, such as
our advertising partners.{117}_____

Choices for Personal Information

When {118}_____ sign up for a particular service {119}_____ requires registration, we ask you to {120}_____ personal information. If we use this
{121}_____ in a manner different than the {122}_____ for which it was collected, then {123}_____ will ask for your consent prior {124}_____ such use.

If we propose {125}_____ use personal information for any purposes
{126}_____ than those described in this Policy {127}_____ in the specific
service notices, we {128}_____ offer you an effective way to {129}_____ out of
the use of personal {130}_____ for those other purposes. We will {131}_____
collect or use sensitive information for {132}_____ other than those described in this
{133}_____ and/or in the specific service notices, {134}_____ we have
obtained your prior consent. {135}_____

You can decline to submit personal {136}_____ to any of our services, in
{137}_____ case Google may not be able {138}_____ provide those services to you.

{139}_____ sharing:

Google only shares personal {140}_____ with other companies or individuals outside
{141}_____ Google in the following limited circumstances: {142}_____

We have your consent. We require {143}_____ consent for the sharing of any
{144}_____ personal information.

Privacy Policy Displayed on iPhone

Figure 4. Horizontal view

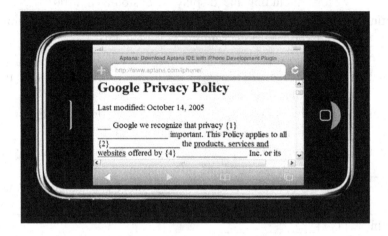

Figure 5. Vertical view

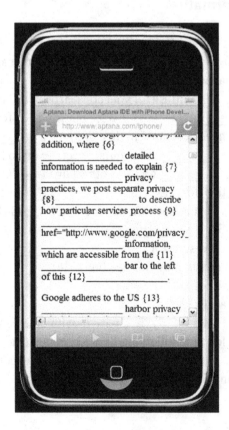

This work was previously published in the International Journal of Mobile Human Computer Interaction, Volume 3, Issue 1, edited by Joanna Lumsden, pp.55-78, copyright 2011 by IGI Publishing (an imprint of IGI Global).

Chapter 5
Remote and Autonomous Studies of Mobile and Ubiquitous Applications in Real Contexts

Kasper Løvborg Jensen
Aalborg University, Denmark

ABSTRACT

As mobile and ubiquitous applications become increasingly complex and tightly interwoven into the fabric of everyday life it becomes more important to study them in real contexts. This paper presents a conceptual framework for remote and autonomous studies in the field and two practical tools to facilitate such studies. RECON is a remote controlled data capture tool that runs autonomously on personal mobile devices. It utilizes the sensing and processing power of the devices to capture contextual information together with general usage and application specific interaction data. GREATDANE is a tool for exploration and automated analysis of such rich datasets. The presented approach addresses some key issues of existing methods for studying applications in situ, namely cost, scalability and obtrusiveness to the user experience. Examples and experiences are given from remote and autonomous studies of two mobile and ubiquitous applications where the method and tools have been used.

INTRODUCTION

As mobile and ubiquitous applications become increasingly complex and tightly interwoven into the fabric of everyday life it becomes more important to study them in real contexts. While facing

the constraints of running on mobile devices with small form factors, small screens and limited input, the applications are also subject to the challenges arising from mobility and a range of dynamic environments and situations of use (Gorlenko & Merrick, 2003). They are often sensitive or even

DOI: 10.4018/978-1-4666-2068-1.ch005

dependent on the context in which they are used. Thus, context will in many cases be an important factor to consider when studying the interaction with such applications. For some applications the user experience changes over time due to habituation, learning or personalization. Because of the situational nature of how these applications are used it has become increasingly important to take them out the controlled settings in which they are developed and study them in real contexts.

The Importance of Context

Although the importance of context is commonly agreed upon, the exact meaning of the word as a concept is less clear. Throughout the literature many views have been expressed on what "context" is and how it can be used for computing purposes, such as (Dey, 2001; Dourish, 2004; Schmidt, Beigl, & Gellersen, 1999).

Because the main focus of this paper is the study of applications and their use in real contexts the following definition will be used: Context is the sum of relevant factors that characterize the situation of a user and an application, where relevancy implies that these factors have significant impact on the user's experience when interacting with that application in that situation.

This definition of context reflects the perspective of the evaluators of applications who are interested in the user experience. This is somewhat different from the more used perspective of those applications being context-aware, as discussed by Dey (2001).

Examples of factors that can potentially impact the user experience of applications are: location, physical environment aspects such as lighting conditions and background noise, mobility and activity of the user, social setting, availability of computing resources and network conditions.

"Capturing context" in this paper refers to capturing information about the abovementioned factors; while acknowledging that such information may never be complete or even entirely

correct. The focus is on contextual factors that can be sensed, processed and captured digitally. Contextual information can be considered at various levels of abstraction ranging from raw sensor data like accelerometer readings to high level concepts like user activity. Often it is the higher level representations that are of interest to the evaluators.

Taxonomies have been proposed representing such contextual data e.g. (Schmidt et al, 1999); however, Dourish (2004) argues that treating context like a representational problem might not be realistic since it is not a static concept that can be neatly captured, modeled and represented. Due to the dynamic nature of context a factor may be relevant in one instant of time and irrelevant the next, just as the significance to the user experience may change depending on the situation. It will be up to the individual evaluators to specify what is of relevance for their specific studies.

Reality Traces

Due to the conceptual fuzziness of context, it makes sense to talk about capturing *traces* of context. The captured information is not a complete record of how it was, but rather a trace giving evidence to how it might have been. Capturing user interaction together with traces of the context in real life settings produces *reality traces*: information about what happened and how it happened. Such data can form the basis for a broad range of studies aiming at investigating usability and user experience of mobile and ubiquitous applications in situ. Obtaining and analyzing such rich datasets is the goal of remote and autonomous field studies.

Essentially reality traces can be seen as log files augmented with contextual information about the situation in which the interaction occurred. In other words, capturing reality traces is like taking the fingerprints that reality leaves on the mobile device, and analyzing them is the detective work of piecing together what it really happened.

STUDYING MOBILE AND UBIQUITOUS APPLICATIONS IN REAL CONTEXTS

A central discussion within the field of mobile human-computer interaction concerns whether to evaluate applications in the laboratory or in the field. Intuitively, they should be studied in situ under realistic conditions, yet this is not always done (Kjeldskov & Graham, 2003). Arguments against field studies are that that data collection is difficult, costly and that such experiments lack control. Some like Kjeldskov et al. (2004) claims that it might not be worth the hassle which is supported by Kaikkonen, Kekäläinen, Cankar, Kallio, and Kankainen (2005) and while others claim that it is (Nielsen, Overgaard, Pedersen, Stage, & Stenild, 2006).

Usability laboratories allow for controlled experiments and high quality data capture, but they are generally lacking the realism and ecological validity attributed to field studies. There are obvious constraints to the scale of such studies in terms the number of participants and duration.

Field observation studies can be used to gain understanding of usage in real contexts, but the data collection is difficult and necessitates evaluators to be present during the study. Self report methods such as diaries and experience sampling as discussed in Consolvo and Walker (2003) can be used to gain insights without presence of researchers, but the biggest problems of these approaches are that they put a great burden on the users; who are constantly reminded that they are participating in a study.

The distinction between laboratory and field settings has become more fuzzy as new approaches have tried to combine the best of both worlds either by taking the laboratory out in the field or by moving the field into the laboratory.

Moving the Laboratory Out In the Field

A lot of effort has been put into moving the capturing apparatus of the laboratory into the field by constructing mobile usability laboratories. These setups use small cameras attachable to the mobile devices, handheld camcorders etc. to capture rich data about both usage and context in the field (Kjeldskov et al., 2004). These approaches however tend to be expensive with regard to time, manpower and logistics, to be obtrusive to the user's experience as they rely on the user to actively report data or observers to be physically present, and to not scale well with the number of users, duration of study, and geographic area in which the study is conducted. Self reporting may methods scale well from a data capture perspective since most of the work is done by the users, but they are very obtrusive to the user experience and might create a constant awareness of the fact that they are participating in a study.

Crossan, Murray-Smith, Brewster, and Musizza (2009) discuss how instrumentation with sensors such as accelerometers can be used infer about the context of use. Hoggan and Brewster (2010) is an example where a mix of self report and hardware instrumentation is used to gather data in a longer duration field study of a mobile game.

Moving the Field Into the Laboratory

Other studies have explored simulation of contextual factors in laboratory settings (Kjeldskov & Skov, 2007). This allows for experimental control and easy collection of high quality data, but unfortunately maintains all the disadvantages of laboratory evaluation mentioned above. The strength of the method lies in isolating and studying specific aspects of interaction in context under controlled circumstances, where experiments can be meaningfully replicated. In the end the gain of this approach depends on how realistic a context can actually be created in a laboratory setting.

Problems of Existing Methods

The following summarizes key problems with the effectiveness and efficiency for these methods for doing so:

- They are obtrusive to the user experience as they rely on the user to actively report data or observers to be physically present.
- They are costly in terms of time, manpower and other resources.
- They do not scale well with the number of users, duration of study, and geographic area in which they can be conducted.
- It is difficult to study long term usage.
- It is difficult to study interaction in many real world settings.

New methods are thus needed to evaluate mobile and ubiquitous applications in real contexts, especially for investigating long term usage and interaction in context.

REMOTE AND AUTONOMOUS FIELD STUDIES

The following will present a conceptual framework for remote and autonomous field studies (Figure 1) and some practical tools needed to facilitate such studies. The presented framework is mainly concerned with studies where the evaluators, both researchers and practitioners, want to study specific applications running on mobile devices. While the experimental design and choice of overall methodology may differ, the capture and analysis data is common to such studies and forms the basis for answering questions regarding the users' interaction and experience with the given application.

Figure 1 illustrates the conceptual framework. Capturing software is installed on a mobile device together with the given application. The users in the experiment will interact with the application in natural settings for a period while the evaluator is spatially and temporally remote. Reality traces are automatically captured on the device and reported to a central server where they can be retrieved and analyzed by the evaluator during the study. Some control over the experiment is possible through remote configuration of the

Figure 1. Remote and autonomous field experiments

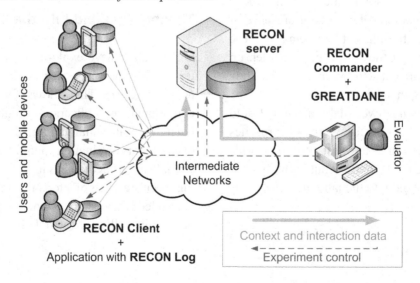

capturing software. Figure 1 shows RECON and GREATDANE in the framework. These tools will be used through the paper to discuss how studies can be facilitated, however the concept of remote and autonomous field studies is not bound specifically to these tools.

As reality traces are datasets describing the users' interaction with an application and the context in which it occurred, the goal of remote and autonomous field studies is to obtain such datasets while the users are interacting with the given application in situ. Synchronizing these interactions and contextual events into one coherent dataset allows the evaluator to analyze and reconstruct the situations and form a richer basis for evaluation than standard log files.

In a large survey of automated usability evaluation methods, Ivory and Hearst (2001) broke the process of conducting usability studies down to three phases which can each be automated: data capture, data analysis and critique of the application. The framework presented here aims at automating significant parts of the first two phases to give a more effective and efficient way for researchers to obtain and process data from field studies.

Personal Mobile Devices as Data Capture Platforms

A key to realizing remote and autonomous evaluations is using the users' personal mobile devices as a data capturing platform. Mobile personal devices are becoming increasingly powerful with regard to processing power, memory, permanent storage, network and sensor capabilities. These features can be utilized to capture and process a wide range of information about the usage and context of both the device and its applications as shown in Figure 2.

Another important characteristic of such devices is their personal nature and how they almost always follow the user in his/her everyday life and activities. This means that these devices

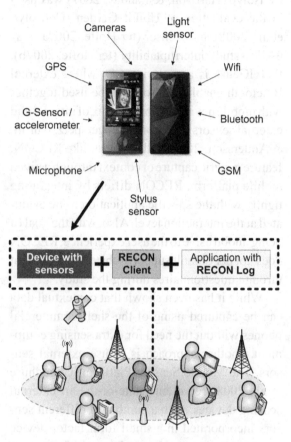

Figure 2. User with RECON-instrumented mobile device and application in the field

traverse through the same contexts as their respective users do. Obviously this assumption will not always be true, but when studying the use of applications executed on the device itself, it makes sense to make the link between device and user.

Capturing Context

Several tools have emerged in recent years for capturing context and user experience data from mobile personal devices. The following systems are representative of the state-of-the-art and have all been successfully used in field studies: ContextPhone (Raento, Oulasvirta, Petit, & Toivonen, 2005) was used in the reality mining study (Eagle & Pentland, 2006) and for investigating mobile awareness cues (Oulasvirta, Petit, Raento, &

Tiita, 2007), MyExperience (Froehlich, Chen, Consolvo, Harrison, & Landay, 2007) was used in the evaluation of UbiFit Garden (Consolvo et al., 2008) and Xensor (ter Hofte, 2007a) was used to study interruptability (ter Hofte, 2007b). BeTelGeuse is a capturing tool where external Bluetooth enabled sensors can be used together with mobile devices enabling a mix of internal and external sensors (Kukkonen, Lagerspetz, Nurmi, & Andersson, 2009). All of these, like RECON, feature generic capture of context through a given mobile platform. RECON differs by integrating tightly with the specific application to be evaluated at the interaction level. Also, with the goal of being unobtrusive to the user experience, RECON does not per default support subjective user input through questionnaires during the study.

While it has been shown that contextual data can be captured using of-the-shelf commercial phones without the need for extra sensing equipment, another approach is to use external sensors. The Mobile Sensing Platform (Choudhure et al., 2008) represents state-of-the-art external sensing devices, with a number of different sensors incorporated in a small form factor device with storage, processing and networking power. When worn by users, it can be used for sensing their physical activity. The biggest downside is cost and the fact that users have to wear an extra device, which besides being a hurdle to manage can be forgotten.

Conducting Remote and Autonomous Field Studies

Through the use of tools like RECON and GREAT-DANE researchers and practitioners can setup, deploy and manage large scale remote studies of mobile and ubiquitous applications in real contexts. By automating the data collection and reporting it is possible for even a single researcher to do relatively large experiments with many users over large periods of time. If data is continuously uploaded from the devices during the period the

researcher may start to work on early data to spot promising tendencies or to prepare interviews and questionnaires for the test participants.

Deploying the Studies

Generally there are three main strategies for setting up and deploying such experiments:

- Handing out devices to the users with everything installed and set up.
- Putting the application and capture tool on the users' own devices.
- Letting the users download and install the instrumented application and capture software themselves

The two first strategies necessitate the user and the evaluator to be physically present at the same place which puts some constraints on the geographical scalability of the studies. Using the users own device has obvious benefits, as they are already familiar with its use and it keeps the cost of the study down in terms of needed devices. The last mentioned strategy opens up for massive scale studies with large numbers of users spread out across the globe. The distribution could be done through websites, social networks and app stores.

RECON

RECON (REmote CONtrolled RECONnaissance in REal CONtexts) is a tool for automated capture of application specific interactions, general usage of the device and a wide range of contextual factors. This makes it perfectly suited for remote and autonomous studies.

RECON is a continuous development effort and although it has been used in several studies, it must still be considered a prototype. The most current information is available on the tool's website (RECON, n. d.). It is not within the scope of this paper to thoroughly detail all the technical aspects of the tool itself. Rather the aim is to present the tool at a technical level necessary to understand its use and benefits specifically for remote and autonomous studies. This means explaining how applications can be instrumented for capturing reality traces and how this data can be made accessible to the evaluator. Also, how configurability and remote control of the tool can be used in longitudinal studies.

RECON has been designed and developed based on six main criteria which state how the tool should operate during studies:

- **Safe**: As it will reside on mobile devices possibly containing sensitive data and capture personal contextual information it is important that the safety of the data and the privacy of the user are secured.
- **Invisible**: The tool should be executing in the background to ensure that it does not interfere with the user experience of the application or with normal use of the device.
- **Efficient**: It should seek to minimize its footprint on the mobile device with regard to CPU and memory usage. At the same time battery time should be maximized. Efficiency also applies to limit use of local storage and network communication.
- **Robust**: As unforeseen events can happen during this type of experiments the tool should be fault tolerant and be able to recover from system errors and crashes.
- **Remote Controllable**: To give the evaluator some degree of experiment control it should allow for change the logging and reporting policies remotely during the experiment.
- **Autonomous**: In general the tool should act intelligently both under normal and unforeseen circumstances. This entails opportunistic reporting of data when network conditions are good, power management, and active self-recovery if needed, etc.

RECON Architecture

RECON consists of four main components: RECON Log, RECON Client, RECON Server and RECON Commander. Figure 1 illustrates how they are distributed in the conceptual framework. RECON Client is the core component and essentially the only one necessary to run studies. However, to fully gain the benefits when evaluating applications the RECON Log component is important. The RECON Server and RECON Commander components are not strictly necessary but become important for large scale studies over longer duration. The following elaborates on these with the main focus on the first two components which reside on the mobile device.

RECON Log

The purpose of RECON Log is to empower the application developer with an easy to use logging interface and to allow for some experiment control to be built into the application. RECON Log is a dynamic-link library (DLL) which is linked and compiled together with the application. Interaction is captured through source code instrumentation and a simple log API is used to report events based on a predefined interaction model. Basically, it offers two methods: *reportEvent(stateID, eventID, data)* and *requestAction(actionID, parameters)*.

Examples of actions could be screen capture, scan for Bluetooth devices or a status of current network connectivity. It could also be requesting a change to the current logging policy, e.g. increasing the sampling rate for the accelerometer or turning on the GPS.

Using source code instrumentation gives much more power to the application developers than general event capture. General capture can be done through hooking into the UI event streams of the operating system from which higher level behaviors can be inferred. Source code instrumentation is more desirable because the designers and programmers may know if certain contexts will be of particular interest at certain points during the execution of the application or at special events such as system errors. From a logging perspective this also gives access to internal state information of the application, which general logging would not.

Another important feature of is that the events are time stamped in the RECON Log module using the same clock as the RECON Client, thus the events will be synchronized with the context when reported to the RECON Client.

RECON Client

RECON Client is the most complex component of the tool. Figure 3 shows the overall architecture with emphasis on the internal structure. As can be seen RECON Client consists of a number of modules. The modules handle the steps of transforming raw sensor data and interaction input into reality traces saved in a database. The following is a bottom-up walkthrough of the modules as presented in Figure 3.

The Context Capture Manager controls the low level sampling of sensors based on logging policies dictating e.g. sampling rates of sensors. Mobile platforms are highly heterogeneous from a hardware point-of-view. To accommodate for this, RECON consists of a generic base configuration that can be extended with both general and device-specific CCMs (Context-Capture Modules). This enables effective and efficient context capture. Each module is basically a wrapper around hardware and software APIs ranging from sensors to the operating system level. RECON supports the capture of a wide range of contextual factors, e.g., GPS position, acceleration, device orientation, battery level, light intensity, nearby Wifi spots and Bluetooth devices, incoming and outgoing communication etc.

The Interaction Capture Manager handles all incoming events and requests from applications through their respective ICM (Interaction Capture Module). Any number of applications can be studied at the same time, as long as each one is instrumented with a RECON Log component and set up with a dedicated ICM in the RECON configuration.

Both context and interaction events are streamed to the Context Processor, where they are filtered, abstracted, aggregated and/or transformed into contextual information of same or higher abstraction levels. The Context State Manager is updated from the Context Processor which communicates changes to the RECON Manager and active GUI. When using RECON as a platform for context-aware applications, the Context State Manager notifies the application of context events.

All data is saved locally in a database until manually collected or successfully uploaded to the Server. The captured reality traces are based on customizable context models and interaction models. This enables a compact and efficient coding of the captured data in the database. These models are then used for later decoding and analysis of the data by using the GREATDANE tool.

RECON Manager is the core control module ensuring that the whole system starts up and runs optimally at all times. It is also the module in charge of initiating data upload or checking for new configurations.

With the aim of invisibility during studies RECON should be run in stealth mode where no GUI is shown on the phone. However it is possible to switch into a number of other modes such as observer, command and debug mode as part of setting up and configuring studies.

RECON Server

The server is mainly a repository for uploading data from devices in experiments. Depending on the upload policy, the evaluator can retrieve and inspect data during the study instead of waiting until the end to collect it. The server also hosts the current configuration file for the given study which can be changed remotely by the evaluator

Figure 3. RECON architecture

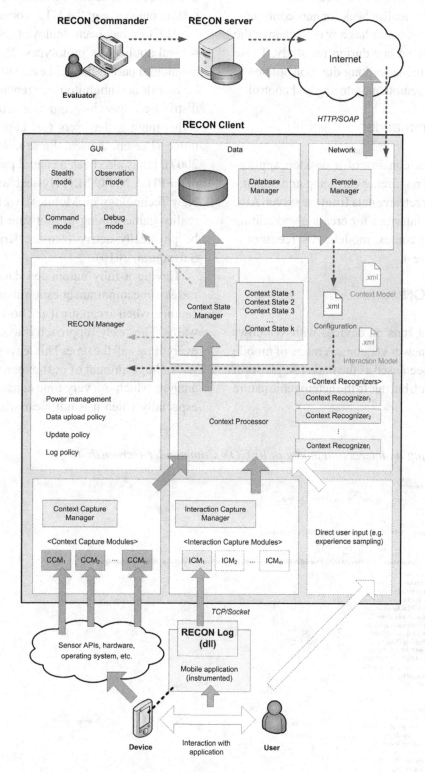

during an experiment. Devices running RECON Client will periodically check for new configurations, thus the evaluator has a way to change the logging scheme ad hoc during the study. These features are suitable for long duration studies as this allows for remote monitoring and control.

RECON Commander

The component consists of a desktop application where the researcher can setup and manage experiments or retrieve data from the server. Also it serves as an interface for creating and editing interaction and context models. A screenshot is shown in Figure 4.

Using RECON

RECON Client runs on devices with Windows Mobile 5.0 or newer versions. A range of mobile devices have been used as the main test platform for development, but due to the heterogenic nature of phones some capturing modules only work on certain phones, e.g. the HTC Touch series.

RECON has been deployed and tested with several application prototypes. Most notably as an integral part of the field evaluation of μCARS – a mobile and ubiquitous car rental service using distributed speech-recognition combined with stylus input to let users rent cars while on the move (Larsen, Jensen, Larsen, & Rasmussen, 2007). It has also been a central part in the study of the PH.A.N.T.O.M. (PHysical Activity through New Technology on Mobiles), which is a mixed reality game aiming to motivate the players to be physically active (Jensen, Krishnasamy, & Selvadurai, 2010).

Having a fully automated data capture tool raises some important question about what data to capture, when to capture it and how to use it afterwards. The "easy" approach is to simply capture everything - all the time. This leaves the evaluator with a huge amount of post-experimental data to analyze which is a very time consuming activity, especially when it is not clear what to look for.

Figure 4. Editing an interaction model in RECON Commander (screenshot)

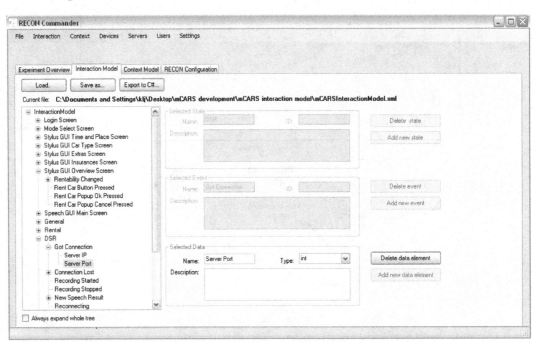

Automating the analysis to some degree through tools like GREATDANE is necessary step to take full advantage of these rich datasets, and also to ensure scalability of the studies.

GREATDANE

GREATDANE (Generic REAlity Trace Data ANalysis Engine) is, as implied by the name, a generic tool for analyzing the data contained in a captured set of reality traces. It is an evolving tool still in the early phases, but the most up-to-date information can be found on the dedicated website (GREATDANE, n. d.). GREATDANE

is designed to be a generic tool independent of the way in which the reality traces are captured. Thus any capturing tool could be utilized to obtain the dataset, as long as the data is coded with an interaction model and a context model.

The tool is an interface for browsing and interacting with the captured reality traces. The goal is to transform these traces into meaningful constructs and metrics from which the user experience can be evaluated or research questions answered. The aim is to save the evaluator time and effort on analysis by at least partially automating it.

Figure 5. Conceptual overview of GREATDANE

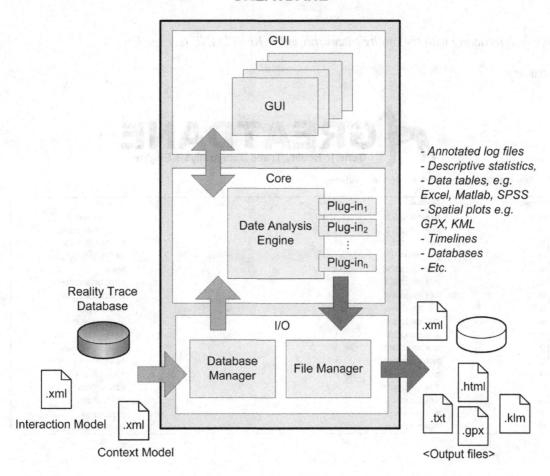

GREATDANE Architecture

Figure 5 shows a conceptual overview of the architecture of GREATDANE. After the reality trace database has been loaded together with the corresponding interaction models and context models, the analysis engine can begin processing the dataset using the models for decoding the data from the compact representation used in RECON.

GREATDANE can be seen as a filter for the evaluator to select data of interest and then to apply analysis procedures to those selections. The analysis algorithms can be considered as "plug-ins" which can be applied to selected subsets of the data. Some of these plug-ins produce output for direct data visualization and exploration within the GREATDANE application while others export the results to specific file formats for visualization or further analysis with external software tools. The automated analysis algorithms are either aimed at the interaction data, the context data or a combination of both. Figure 6 shows a screenshot from the GUI.

For the analysis of interaction, the analysis engine uses the predefined interaction model to abstract low-level events and actions into higher level concepts such as sessions, states and activities. From the processed data a wide range of metrics and statistics can be computed at each level, such as duration, time-on-task, task success, etc. This would normally require some manual work by the evaluator, but can be automated because the interaction is already coded through the interaction model. For large scale studies this is a essential to the scalability.

As with the interaction, the captured contextual data can be processed into higher level concepts by using context recognition techniques. A plethora

Figure 6. Interacting with the reality traces through GREATDANE (screenshot)

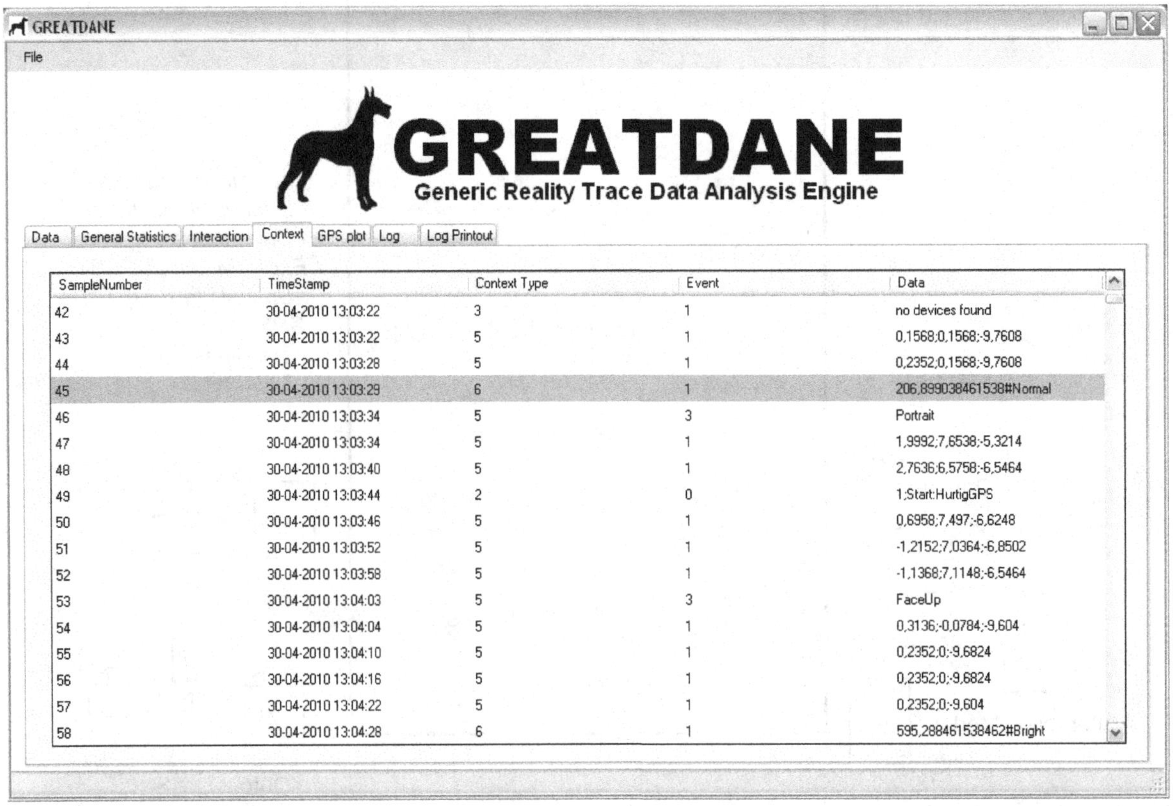

of methods exists for this, but the details of these are outside the scope of this paper. Conceptually the important part is that they can be encapsulated and used in GREATDANE to allow evaluators to automatically transform lower level data into more useful constructs.

Much of the value of reality traces is gained when the interaction and context is viewed together. The context of use can be recreated at any time of the study based the capture data and the contextual model. As the context events are synchronized with the interaction it is possible for the evaluator to see how the context might have factored in at a time of interaction when evaluating the usability and user experience of applications.

In general the partially automated approach for analyzing data in GREATDANE is inspired by ESDA (Exploratory Sequential Data Analysis) as proposed by Fisher and Sanderson (1996), LSA (Lag Sequential Analysis) as used in (Consolvo, Arnstein, & Franza, 2002) and methods for automated extraction of usability information from user interface events(Hilbert & Redmiles, 2000).

Using GREATDANE

The first version of GREATDANE was used for data processing and analysis of usage patterns in a longitudinal pilot evaluation a mobile diabetes application that ran for three months (Jensen & Larsen, 2007), although this particular study did not capture context but focused on the interaction model. Currently it is being used as part of the aforementioned studies of µCARS and PH.A.N.T.O.M.. In both studies the correlation between interaction with the applications and the context in which it occur is very important.

STUDIES

RECON and GREATDANE have been used in remote and autonomous studies of real applica-

tions in the wild. In the following the µCARS study is used to exemplify how the application can be instrumented and set up with RECON and the PH.A.N.T.O.M. demonstrates how RECON can be used as a platform for context-aware applications. This is neither a full account of the actual studies nor the results from these. It is to demonstrate through example how various aspects of mobile and ubiquitous applications can be investigated through remote and autonomous field studies.

The µCARS Study

µCARS is representative of the broad range of context-sensitive applications can be used anytime and anywhere. The application combines speech and stylus input into a multimodal interface for renting cars on mobile devices while mobile (Larsen et al., 2007). In short, the aim of the study was to explore how this mobility and real contexts of use would affect the performance of the system and the usability and user experience of the application.

Instrumentation of the Application

Figure 7 shows the system architecture for the application as reported in (Larsen et al., 2007) and how it has been modified with RECON for the field study. The integration of RECON Log allows for logging of internal system states and events which is then sent to the RECON Client via a local TCP socket. For this study it was important to investigate the performance of the speech recognition system and map this to contextual factors such as location, movement of the user during interaction, background noise and network conditions. Also, it was a goal to investigate how the user would react to specific situation where e.g. the speech modality would not be available or not be performing well.

Figure 7. The µCARS system architecture instrumented with RECON Log

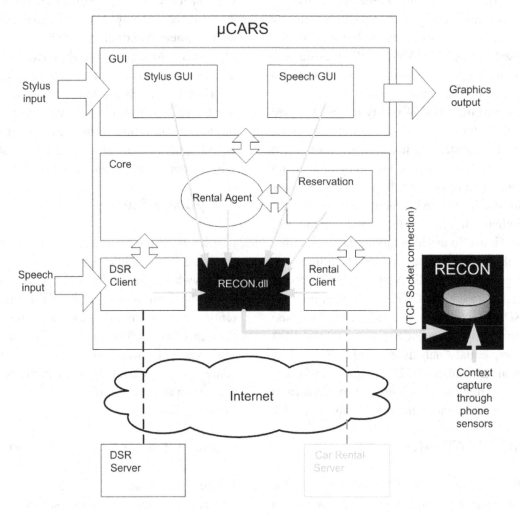

Modeling the Interaction

Figure 8 shows the application interface and how the user can switch between two modes of interaction represented through either the Speech GUI or Stylus GUI. The interaction model for this study was a mapping of these screens to states and user actions such as speech input or widget interaction would be events. Other events of interest are internal events in the speech recognition system, switching between modes, car rental events, etc. Figure 4 shows how the interaction model was created with the RECON Commander.

Setup and Configuration

A pilot study was conducted with three simultaneous users over the course of three days. The setup is shown in Figure 9. Each user was given an instrumented phone and briefed on the study. During the period a number of tasks were sent a randomized times through SMS to the users. These messages was automatically sent with HERMES, a small tool for dispatching messages automatically from a mobile phone based on preset (e.g. randomly generated) time schedules (HERMES). Thus the main execution part of the experiment was fully automated and data was collected from the devices after the study.

Figure 8. The interface of µCARS showing the various screens and modes of interaction

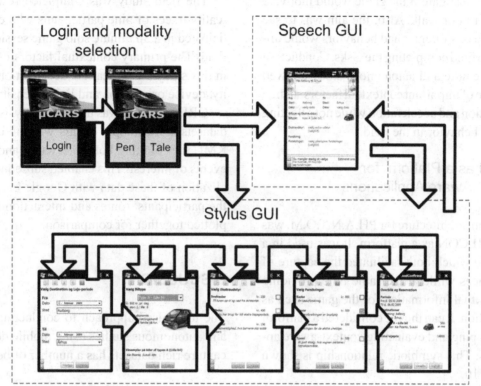

The PH.A.N.TO.M. Study

This PH.A.N.T.O.M. application is a persuasive mixed reality game aimed at motivating people with a sedated life style to exercise. The game's objective is for the players to move around in a geographical area using GPS while solving different tasks with the mobile device. An experiment was

Figure 9. Experiment setup for the µCARS pilot study

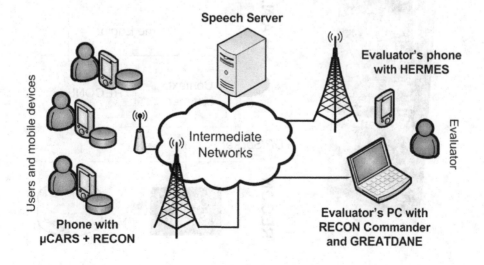

set up to investigate if the game would motivate people to run or walk. Also, the aim was to see what kinds of strategies and behaviors would appear in the field completing the tasks. Conducting this as a remote and autonomous study allowed for capture of important contextual factors such as GPS position and acceleration while not affecting the users' behavior in the field.

RECON as a Platform for Context -Aware Applications

This system architecture for PH.A.N.T.O.M. was built on RECON as a platform. It was used in a dual mode which both facilitated the capture of reality traces while at the same time providing the contextual information for the game such as GPS position. It can thus be considered a platform for developing and evaluating context-aware applications. This symbiotic relationship is shown in Figure 10.

The field study was completed resulting in reality traces of nine participants who each participated in a 30 minute solo game session in the field. The primary contextual factors of interest in this study were location, acceleration, mobility, device orientation and light intensity.

GREATDANE was used for analysis of these datasets. An example of this was the export to KML files plotting both interaction and context events of interest. This enabled subsequent visualization of these data with Google Earth where the participants' routes and interaction could be plotted together for comparison.

DISCUSSION

The tool-based approach to conducting remote and autonomous studies using mobile devices to capture rich datasets has a number of benefits:

Figure 10. PH.A.N.T.O.M. system architecture showing the dual use of RECON for both logging and notifications of context events (e.g. GPS position updates)

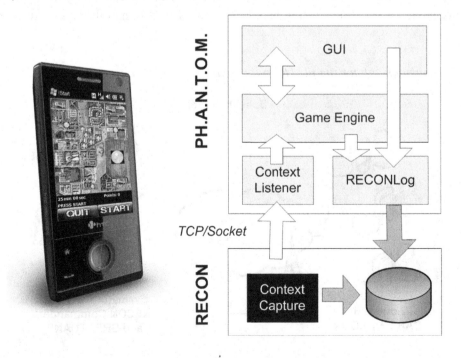

- Enables studies in real contexts.
- Enables the study of long term usage and interaction.
- No need for external sensors.
- Unobtrusive to the user experience.
- Cost-efficient with regard to time and manpower.
- Scales well with many users and long duration studies as well as studies that are geographically spread out
- Can be monitored and managed remotely.
- Customizable and extensible to fit heterogeneous and evolving mobile platforms.

However there are also a number of limitations and issues associated with conducting such studies.

Issues

The experimental control of such studies is limited even though RECON itself can be reconfigured remotely, and the reality traces collected are subject to noise, uncertainties and even errors. Also, for remote and autonomous field experiments to work a functional application prototype must be available. It should ideally be robust enough for deployment in the field without the need for maintenance.

Any technical issues arising during the study must be handled by the user at the risk of compromising the study. This problem can be addressed in several ways. For the μCARS study it has proven to be effective to let the users do a soft reset of the phone if the encounter any problems and let RECON start up automatically.

This type of studies raises privacy, security and other ethical issues. Strategies must be made for ensuring anonymity and security of personal and sensitive data during and after the study. This is a crucial problem to solve for these kinds of methods to gain widespread use. Data encryption, automated anonymization algorithms and other measures can be taken but this is a complex problem to work around.

Using only the phone for context capture requires a certain level of sophistication of the device. Smartphones with sensors and high speed data connections are not common devices in many parts of the world. This limits the use of the method. An interesting question raised is how less advanced mobile devices could be used to capture reality traces and facilitate remote and autonomous studies.

FURTHER WORK

Several technical issues and challenges still need to be investigated: How to tradeoff processing, storage and network usage? Which parts of the context processing should be done on the device and which should be saved for server-side? (Figo, Diniz, Ferreira, & Cardoso, 2010) investigated many methods for extracting information from low level sensor data. Essentially they looked at which techniques could be used on the mobile device and which are better run no more powerful PCs.

Another big challenge is optimizing performance with regard to battery life, which is also a tradeoff with the resolution of captured data? Dynamic logging policies that change based on the situation seems like a promising part of the overall solution.

Due to the size of the dataset from remote and autonomous experiments using capturing tools like RECON, the scalability can only be ensured with the use of analysis tools like GREATDANE, which can aid evaluators in the analysis process through some degree of automation. This requires consistency in representation of both context and interaction data within the study. It is important to address the representational problem of captured data and how it can be synthesized with many users, many contexts, heterogeneous platforms, sensors, etc. If a rich, flexible and uniform representation format of such data was developed and agreed upon, it would enable sharing of datasets and pave the way for generic analysis and data visualization tools, such as GREATDANE.

Remote and autonomous field studies are in many ways complementary to the wide range of qualitative methods available (interviews, questionnaires, etc.). Designing experiments to use of mix of these methods can help to address some of the uncertainties in the data. E.g. data captured by RECON can be grounded using these methods during or after the study. This was done in the µCARS and PH.A.N.T.O.M. studies.

Although the unobtrusiveness and invisibility of the studies conducted has been a main point in this paper, the merits and needs for user input is important in many experiments. Thus it would however be interesting to combine such subjective input with reality traces e.g. by use the knowledge of the context state to do context-aware experience sampling.

It is envisioned that reality traces can be used for many purposes, besides supporting the evaluation of the usability and user experience of mobile and ubiquitous applications. For example the creation and training of user models and context models for personalization.

Another important thing to consider is the meta-usability issue of whether tools like RECON and GREATDANE are easy and intuitive enough to use for evaluators to even consider doing field studies based on them. Future work will include studies with evaluators using the tools to setup and deploy studies.

CONCLUSION

A tool-supported framework has been presented for remote and autonomous field studies. It was shown how this approach can be used by researchers and practitioners to conduct large scale and unobtrusive studies of mobile and ubiquitous applications in real contexts. The use of automated capture and analysis tools such as RECON and GREATDANE enables efficient studies that allow scalability with regard to number of participants, duration and area of study which otherwise cannot easily be obtained.

While the proposed type of study address key issues of existing methods it also comes with some limitations and unresolved issues. These must be resolved for the method to gain widespread use. Remote and autonomous field studies should be considered as an addition to the existing toolbox for in situ studies and not a replacement of current practices. The data capture paradigm is complementary to subjective and more qualitatively oriented methods such as interviews and questionnaires; thus they should be used together to gain an even richer picture.

ACKNOWLEDGMENT

The author wishes to thank the researchers and students who have contributed in the projects in which the method and tools have been used: Lars Bo Larsen, Søren Larsen and Christian Fischer Pedersen and Morten Højfeldt Rasmussen for their work on DiasNet Mobile and µCARS, Rameshnath Krishnasamy and Vashanth Selvadurai for their work on the PH.A.N.T.O.M. game and Tais Holland Mogensen and Christian Ølholm for their work on an early prototype of RECON. Also, thanks to the reviewers whose valuable input helped improve the paper.

REFERENCES

Choudhure, T., Consolvo, S., Harrison, B., Hightower, J., LaMarca, A., & LeGrand, L. (2008). The mobile sensing platform: An embedded activity recognition system. *IEEE Pervasive Computing*, 7(2), 32–40. doi:10.1109/MPRV.2008.39

Consolvo, S., Arnstein, L., & Franza, B. R. (2002). User study techniques in the design and evaluation of a ubicomp environment. In *Proceedings of the 4th International Conference on Ubiquitous Computing* (pp. 73-90).

Consolvo, S., McDonald, D. W., Toscos, T., Chen, M. Y., Froehlich, J., Harrison, B., et al. (2008). Activity sensing in the wild: a field trial of ubifit garden. In *Proceedings of the 26th Annual SIGCHI Conference on Human Factors in Computing System* (pp. 1797-1806).

Consolvo, S., & Walker, M. (2003). Using the experience sampling method to evaluate ubicomp applications. *IEEE Pervasive Computing, 2*(2), 24–31. doi:10.1109/MPRV.2003.1203750

Crossan, A., Murray-Smith, R., Brewster, S., & Musizza, B. (2009). Instrumented usability analysis for mobile devices. *International Journal of Mobile Human Computer Interaction, 1*(1), 1–19. doi:10.4018/jmhci.2009010101

Dey, A. K. (2001). Understanding and using context. *Personal and Ubiquitous Computing, 5*(1), 4–7. doi:10.1007/s007790170019

Dourish, P. (2004). What we talk about when we talk about context. *Personal and Ubiquitous Computing, 8*(1), 19–30. doi:10.1007/s00779-003-0253-8

Eagle, N., & Pentland, A. S. (2006). Reality mining: Sensing complex social systems. *Personal and Ubiquitous Computing, 10*(4), 255–268. doi:10.1007/s00779-005-0046-3

Figo, D., Diniz, P., Ferreira, D., & Cardoso, J. (2010). Preprocessing techniques for context recognition from accelerometer data. *Personal and Ubiquitous Computing, 14*(7), 645–662. doi:10.1007/s00779-010-0293-9

Fisher, C., & Sanderson, P. (1996). Exploratory sequential data analysis: Exploring continuous observational data. *Interaction, 3*(2), 25–34. doi:10.1145/227181.227185

Froehlich, J., Chen, M. Y., Consolvo, S., Harrison, B., & Landay, J. A. (2007). MyExperience: A system for in situ tracing and capturing of user feedback on mobile phones. In *Proceedings of the 5th International Conference on Mobile Systems, Applications, and Services* (pp. 57-70).

Gorlenko, L., & Merrick, R. (2003). No wires attached: Usability challenges in the connected mobile world. *IBM Systems Journal, 42*(4), 639–651. doi:10.1147/sj.424.0639

GREATDANE. (n. d.). Generic reality trace data analysis engine. Retrieved from http://www.loevborg.com/tools/greatdane

HERMES. (n. d.). *HERMES tool*. Retrieved from http://www.loevborg.com/tools/hermes

Hilbert, D. M., & Redmiles, D. F. (2000). Extracting usability information from user interface events. *ACM Computing Surveys, 32*(4), 384–421. .doi:10.1145/371578.371593

Hoggan, E., & Brewster, S. A. (2010). Crosstrainer: Testing the use of multimodal interfaces in situ. In Proceedings of the 28th International Conference on Human Factors in Computing Systems (pp. 333-342).

Ivory, M. Y., & Hearst, M. A. (2001). The state of the art in automating usability evaluation of user interfaces. *ACM Computing Surveys, 33*(4), 470–516. .doi:10.1145/503112.503114

Jensen, K. L., Krishnasamy, R., & Selvadurai, V. (2010). Studying PH. A. N. T. O. M. in the wild: a pervasive persuasive game for daily physical activity. In *Proceedings of the 22nd Conference of the Computer-Human Interaction Special Interest Group of Australia on Computer-Human Interaction* (pp 17-20).

Jensen, K. L., & Larsen, L. B. (2007). Evaluating the usefulness of mobile services based on captured usage data from longitudinal field trials. In Proceedings of the 4th International Conference on Mobile Technology, Application, and Systems and the 1st International Symposium on Computer Human Interaction in Mobile Technology (pp. 675-682).

Kaikkonen, A., Kekäläinen, A., Cankar, M., Kallio, T., & Kankainen, A. (2005). Usability testing of mobile applications: A comparison between laboratory and field testing. *Journal of Usability Studies, 1*(1), 4–17.

Kjeldskov, J., & Graham, C. (2003). A review of mobilehci research methods. In L. Chittaro (Ed.), *Proceedings of the 5th International Symposium on Human Computer Interaction with Mobile Devices and Services* (LNCS 2795, pp. 8-11).

Kjeldskov, J., & Skov, M. B. (2007). Studying usability in sitro: Simulating real world phenomena in controlled environments. *International Journal of Human-Computer Studies, 22*, 7–37.

Kjeldskov, J., Skov, M. B., Als, B. S., & Høegh, R. T. (2004). Is it worth the hassle? Exploring the added value of evaluating the usability of context-aware mobile systems in the field. In S. Brewster & M. Dunlop (Eds.), *Proceedings of the 6th International Symposium on Mobile Human-Computer Interaction* (LNCS 3160, pp. 529-535).

Kukkonen, J., Lagerspetz, E., Nurmi, P., & Andersson, M. (2009). BeTelGeuse: A platform for gathering and processing situational data. *IEEE Pervasive Computing / IEEE Computer Society [and] IEEE Communications Society, 8*(2), 49–56. doi:10.1109/MPRV.2009.23

Larsen, L. B., Jensen, K. L., Larsen, S., & Rasmussen, M. H. (2007). A paradigm for mobile speech-centric services. *Proceedings of the Abstract Interspeech, 1*(4), 2344–2347.

Nielsen, C. M., Overgaard, M., Pedersen, M. B., Stage, J., & Stenild, S. (2006). Its worth the hassle! The added value of evaluating the usability of mobile systems in the field. In *Proceedings of the 4th Nordic Conference on Human-Computer Interaction: Changing Roles* (pp. 272-280).

Oulasvirta, A., Petit, R., Raento, M., & Tiita, S. (2007). Interpreting and acting on mobile awareness cues. *Human-Computer Interaction, 21*, 97–135.

Raento, M., Oulasvirta, A., Petit, R., & Toivonen, H. (2005). ContextPhone - a prototyping platform for context-aware mobile applications. *IEEE Pervasive Computing, 4*(2), 51–59. doi:10.1109/MPRV.2005.29

RECON. (n. d.). *Remote controlled reconnaissance in real contexts*. Retrieved from http://www.loevborg.com/tools/recon

Schmidt, A., Beigl, M., & Gellersen, H. W. (1999). There is more to context than location. *Computers & Graphics Journal, 23*(6), 893–902. doi:10.1016/S0097-8493(99)00120-X

ter Hofte, G. H. (2007a). *What's that hot thing in my pocket? SocioXensor, a smartphone data collector.* Paper presented at the Third International Conference on e-Social Science.

ter Hofte, G. H. (2007b). Xensible interruptions from your mobile phone. In *Proceedings of the 9th International Conference on Human Computer Interaction with Mobile Devices and Services* (pp. 178-181).

This work was previously published in the International Journal of Mobile Human Computer Interaction, Volume 3, Issue 2, edited by Joanna Lumsden, pp.1-19, copyright 2011 by IGI Publishing (an imprint of IGI Global).

Chapter 6
Nudging the Trolley in the Supermarket:
How to Deliver the Right Information to Shoppers

Peter M. Todd
Indiana University, Bloomington, USA

Yvonne Rogers
The Open University, UK

Stephen J. Payne
University of Bath, UK

ABSTRACT

The amount of information available to help decide what foods to buy and eat is increasing rapidly with the advent of concerns about, and data on, health impacts, environmental effects, and economic consequences. This glut of information can be overwhelming when presented within the context of a high time-pressure, low involvement activity such as supermarket shopping. How can we nudge people's food shopping behavior in desired directions through targeted delivery of appropriate information? This paper investigates whether augmented reality can deliver relevant 'instant information' that can be interpreted and acted upon in situ, enabling people to make informed choices. The challenge is to balance the need to simplify and streamline the information presented with the need to provide enough information that shoppers can adjust their behavior toward meeting their goals. This paper discusses some of the challenges involved in designing such information displays and indicate some possible ways to meet those challenges.

DOI: 10.4018/978-1-4666-2068-1.ch006

INTRODUCTION

Increasingly we are told about the risks, costs, and benefits of particular food choices. A flood of information is becoming available from a variety of sources, online, on food labels, in information leaflets and books, aimed at informing the consumer so that better decisions can be made while shopping. But all this information risks overwhelming and overloading the shopper trying to navigate the complex store environment in a hurry, leading to the opposite outcome: poor decisions made without the proper input. How can all this information be consolidated, pruned, and presented to supermarket shoppers in an easy to understand and meaningful form that will actually help them make better choices in terms of the values they care about?

Technology pundits and researchers are beginning to promote 'augmented reality' that uses Smartphones and other ubiquitous technologies as the latest solution to this problem. Kuang (2009), for example, marvels at the possibility: "What if all the food in your grocery store was marked with a QR code—you could compare the carbon footprints of two batches of produce... without having to spend any time or effort looking it up..." He continues by claiming it is "The best chance we have to speed crucial information about our world to the people living in it". This vision, however, begs the research questions: Will people be able to read and act upon such 'instant information'? Will just throwing more information at people have the desired galvanizing effect of encouraging and empowering people to act upon various social causes (e.g., reducing carbon emissions) or improve their well-being (e.g., changing their diet)? Or do we need to tailor that information glut into simple nudges that make behavior change easy to achieve? And if so, what kind of nudges will work?

Having instant information at one's fingertips is certainly a promising technological approach, but for it to succeed in changing people's behavior we need to understand how new forms of augmented reality are interpreted and used, especially when *in situ*. While the capabilities of the emerging technologies are impressive in how they can project contextualised information, there is a paucity of research into whether people can process and exploit that extra information profitably. It is easy to imagine soda drinkers enjoying the surprise of being presented with a new branded game or a funny website on their mobile phone, but it is less clear whether people will make greener and healthier choices while keeping to their weekly budget when presented with extra information of one form or another in the middle of their busy shopping trip. Thus, research is needed, first, to determine whether instant information will enable people to make better-informed choices when shopping, and second, to ascertain whether and how such information is able to change people's behavior in the longer term.

Technology for ubiquitous information delivery must balance giving people enough new information to improve their decisions against overwhelming them with new things to consider. Ambient information displays, as already used in homes and offices to provide feedback about energy consumption and nudge users toward greater conservation, may strike the right balance in food purchase and consumption as well—for instance, lighting up a shopping trolley (cart) handle in a color representing the fat content of the products a consumer has chosen so far. However, as we discuss in this paper, moving beyond momentary nudges toward long-term behavior change requires providing detailed-enough feedback to enable learning what to do in the future, for instance on the next shopping trip. We argue that we must improve our (currently limited) understanding of whether and how people attend to and learn from visualizations of multi-dimensional information while engaged in an ongoing activity such as food shopping. This can be done using cognitive science models of decision-making and learning together with design principles for information visualization and interaction design.

BACKGROUND

Research on Decision Making Strategies

Technology designed to deliver an ever-increasing amount of information to consumers is intended to help them make better decisions or otherwise influence their behavior. But without knowing how people actually process the information they are presented with in service of decisions and actions, we cannot say how to help decision makers make better decisions, nor what and how much information would best accomplish this goal. While it is obvious that we must take human psychology into account in figuring out what and how to communicate to consumers, we first have to settle on an appropriate view of that psychology, which means adopting one of a number of competing views of human rationality.

The traditional view of unbounded rationality says that decisions should be made by gathering and processing all available information, without concern for the human mind's computational speed or power. According to this view, found surprisingly commonly in the fields of economics, psychology, and consumer behavior, information technologies should either shower people with all the information that could possibly be relevant for making a particular decision and let the consumer work out the optimal inference for themselves, or the technology should gather as much information as possible and then make the decision for the consumer by weighing and adding it all into a final recommended choice. This view of unbounded rationality at work can be seen in various decision aid sites on the Web, such as selectsmart.com, which helps users make choices about everything from what kind of beer to

purchase to what kind of pet to buy, by gathering extensive data on dozens of questions about the user's preferences and the strength of each preference, and then processing all that information into a final ordered list of possibilities for the user to buy. This information-intensive approach to choice does not match how most people make decisions most of the time, particularly in settings with high time pressure and relatively low consequence such as food shopping; as a consequence, traditional views of rationality provide a poor basis not only for building psychological models of choice, but also for creating decision tools meant to be used or understood by real people (Katsikopoulos & Fasolo, 2006).

In contrast, the perspective of *bounded rationality* studies how people (and other animals) can make reasonable decisions given the constraints that they naturally and commonly face, such as limited time, information, and computational abilities. Instead of needing to process all the available information and consider all the options, people can often make surprisingly good decisions using simple "fast and frugal" heuristics, which are rules of thumb or short-cut choice strategies that ignore most of the available information. The trick is to ignore the unnecessary pieces of information, and just search for the few pieces of information that will be most useful, or the few most appropriate options, and process them appropriately. Herbert Simon championed this view of cognition, arguing that because of the mind's limitations, humans "must use approximate methods to handle most tasks" (Simon, 1990, p. 6). These methods include recognition heuristics that largely eliminate the need for information and just make choices on the basis of what is recognized (Goldstein & Gigerenzer, 2002), search heuristics that look for options only until one is found that is "good enough" (Todd & Miller, 1999), and choice heuristics that seek as little information as possible to determine which option should be selected (Payne, Bettman, & Johnson, 1993; Gigerenzer, Todd, & the ABC Research Group, 1999).

Simon's notion of bounded rationality, originally developed in the 1950s, had great influence on psychologists and economists who followed, in two distinct ways. Both camps agreed that the mind is limited in what it can accomplish and what information it will use. But one set of researchers argued at the same time that the decisions people make are often flawed as a consequence: We would, and should, all be unboundedly rational, if only we could. Under this view, the simple heuristics that we so often use can often lead us astray, making us reach biased decisions, commit fallacies of reason, and suffer from cognitive illusions. The very successful "heuristics-and-biases" research program of Slovic, Tversky, and Kahneman (1982) has followed this interpretation of bounded rationality and led to much work on how to "debias" people so they could overcome their erroneous heuristic decision making.

In stark contrast, a second set of researchers has found that people can and often do make good decisions with simple rules or heuristics that use little information and process it in quick ways (Payne, Bettmann, & Johnson, 1993; Gigerenzer et al., 1999; Gigerenzer & Selten, 2001). This second view of bounded rationality argues that our cognitive limits do not stand in the way of adaptive decision making; in fact, these bounds can even be beneficial in various ways (Hertwig & Todd, 2003), because the mind is adapted so that its bounds often match the structures of information available in the environment. This leads to a new conception of bounded rationality, termed *ecological rationality*, which emphasizes the importance of the environmental information structures and how they fit to mental decision structures (Todd, Gigerenzer, & the ABC Research Group, in press; Todd & Gigerenzer, 2007). The implication of this perspective for information display systems is that if people typically use fast and frugal heuristics to process only a few pieces of information when making decisions, then striving to deliver them greater and greater amounts of information may not achieve the desired end of aiding good deci-

sions—or at least not as cheaply and effectively as could otherwise be possible. Thus this view of bounded and ecological rationality indicates that we should figure out what information people will actually use and focus on delivering just those items—a less-is-more, simplicity-based approach that is appearing in applications in business and marketing (Fasolo, McClelland, & Todd, 2007), medical communication (Todd et al., in press), and elsewhere. This is akin to structuring the information environment in subtly different ways that can easily and even unconsciously influence people's choices and behaviors in desired directions, achieving "informational nudges" (Thaler & Sunstein, 2008).

In line with bounded rationality, people often rely on a single reason to make decisions, choosing an option because just one thing about it leads them to select it over other options. This approach can be quick and simple, avoiding the need to make trade-offs between multiple, possibly conflicting, attributes for different options (e.g., if option A is cheaper, but option B will last longer, making the decision solely on the basis of price and ignoring longevity, or vice-versa, means that no trade-off need be made between the two). But can such a heuristic approach ever be reasonable? To find out, we must first define specific models of heuristics that use a single reason, and then compare them with traditional approaches using several pieces of information, both applied to decisions in different types of environments to see where each can work best.

A standard rational decision making approach is to weigh all the available information about each option by how important it is for the choice to be made, and then add all those weighted factors together to arrive at a total value for each option, and finally to choose the option with the greatest summed value. In contrast, a well-studied "fast and frugal" heuristic that relies on the principle of one-reason decision making is the take-the-best heuristic (Gigerenzer et al., 1999). Take-the-best and other one-reason decision heuristics are frugal in that they do not look for any more information than they need to make an inference, and they are fast because they do not involve any complex computation—not even the multiplication and addition required by weighted additive mechanisms. In comparisons with weighted additive models, take-the-best has been found to work well in particular types of environments, and not in others (Todd et al., in press). Specifically, take-the-best is not ecologically rational compared to weighted additive mechanisms in environments where the available pieces of information are roughly equally useful. But many environments, including that of consumer choice, are characterized instead by a distribution of information usefulness that is highly skewed or "J-shaped" (i.e., falling off rapidly, so that the most useful piece of information is considerably more important than the second-most useful piece, which is considerably more important than the third, etc.). In such environments, take-the-best can outperform the traditionally rational weighted additive model, particularly when generalizing to somewhat new situations (Gigerenzer et al., 1999).

Thus, empirical observations of actual quick and simple decision making, along with theoretical arguments that human minds have evolved to act quickly and make 'good enough' decisions by using fast and frugal heuristics, support the contention that humans typically ignore most of the available information and make choices using only a few important cues. In the supermarket, this can be seen as shoppers make snap judgments based on a paucity of information, such as buying brands they recognize, are low-priced, or have attractive packaging, seldom reading other package information (Todd, 2007). But at the same time, recent consumer surveys reveal that shoppers are demanding more information about the products they buy and are becoming increasingly aware of the global consequences of the decisions they make (EDS IDG, 2007). This raises the question of whether it is possible to encourage shoppers to pay attention to new (and possibly more) types

of information, such as nutritional, ethical, and environmental aspects, when making their food purchases and subsequently deciding how to use what they have bought to make healthy meals that have a low carbon footprint.

Research on Information Visualization

Despite the literature reviewed above, there is a scarcity of research on how people use multidimensional information under time pressure and the extent to which it affects rapid decision making (Feunekes et al., 2008). Visualization research has tended to adopt an unbounded rationality perspective, assuming that people have the time and cognitive capacity to pull out and use whatever information the displays provide. Within the field of Information Visualization there have been a number of tools that have been developed specifically to represent multidimensional data that allow for comparisons (Card, Mackinlay, & Shneiderman, 1999). Other simple canonical forms such as tables and trend graphs have been developed for web-based decision-making activities, including online shopping, making investments, choosing insurance policies or buying a house. An innovative approach has been to develop interactive visualizations that show some aspects of the performance of objects for a range of different parameter values. An early example was the Influence Explorer (Tweedie, Spence, Williams, & Bhogal, 1994) that allowed a user to compare how products (e.g., different light bulbs) perform on core values (e.g., brightness and working life) when varying multiple parameters (e.g., diameter, length, material and number of coils). More recently, bargrams have been developed for e-commerce applications. For example, EZChooser helps consumers choose one item from many (e.g., cars) through selecting attributes that are visualized as parallel horizontal interactive histograms along a number of dimensions (Wittenburg, Lanning, Heinrichs, & Stanton, 2001).

But even though these kinds of visualizations are mostly targeted at non-expert users, they are essentially visual query languages that require considerable cognitive effort to interpret.

A different approach to visualizing information has been to create ambient displays that represent invisible dynamic processes, such as changes in weather, stock, currency, and the amount of human presence or activity in a building. The idea is that people need only to glance at them when walking past to glean the current state. Types of ambient displays that have been used include lights that glow in intensity like Hello.Wall (Prante et al., 2003), water that ripples like ambientROOM (Wisneski et al., 1998), water fountains that vary in height like Datafountain (van Mensvoort, 2005), and bottles or other objects that jiggle like ambientROOM (Ishii et al., 1998). Part of their appeal lies in how they make the invisible visible through being aesthetic, public, fun, informative, and compelling. Their *glanceability* is also considered key; just as clocks on a wall are momentarily looked at, these displays, too, are intended to be glanced at occasionally and peripherally without distracting people from their ongoing activities.

More recently, ambient displays have been designed to do more than inform; by depicting certain kinds of information that can influence people's behavior. An example is a sculpture that sits near a person's computer monitor and slumps over if that person continues to sit without taking a break (Jafarinaimi, Forlizzi, Hurst, & Zimmerman, 2005). After the person takes a break, the sculpture sits upright and is assumed to be healthy. The way this kind of ambient display is assumed to influence is by raising people's awareness of a particular behavior that they normally overlook or try not to think about.

More extensively, Rogers et al. (2010) investigated whether a community's behavior could be changed by situating various forms of ambient displays in their workplace. The aim was to influence the prevalence of a socially desired behavior, namely, taking the stairs when moving between floors. A combination of abstract lures and aggregate representations was designed to entice and reveal stair/elevator usage for different time periods. The abstract lure display comprised white LED lights that were embedded in a carpet near the stairwell and elevator. Whenever someone approached them they started to twinkle, forming an aesthetically pleasing flowing pattern that suggested organic spread toward the entrance of the stairwell. The aggregate representation was designed as a large installation that was hung in the atrium of the building. Two moving 'clouds' of spheres were used to depict the aggregate number of people taking the stairs and the elevator over time. The information provided by both displays was intended to inform in subtle and playful ways, and in doing so nudge people to change their behavior. A two-month in situ study showed that the two displays elicited much intrigue and discussion from both the inhabitants of and visitors to the building. Moreover, although few people admitted to changing their behavior in response to seeing the displays, logged data of people's actual movements showed a statistically significant increase in the proportion of stair usage after installation. The findings suggest that the ambient displays had the effect of increasing awareness about stairs and elevators that, in turn, may have unconsciously nudged some people to take the stairs at *choice moments*, which they may subsequently not have remembered. Choice moments are the times when people might have previously taken the elevator, because they were feeling lazy or tired, and where they switched to taking the stairs without necessarily being aware of doing so.

In contrast to these implicit ambient nudges, a number of persuasive technologies have been developed that explicitly encourage people to take more exercise or reduce energy consumption. Examples of representations intended to motivate people to exercise more include Fish'n'Steps (Lin et al., 2006), Chick Clique (Toscos, Faber, An, & Gandhi, 2006), and UbiFit (Consolvo et al., 2008), where various types of graphic representations (e.g., butterflies, flowers, bar charts) are used to represent amount of exercise-type performed (e.g., cardio, strength training, and walking). Findings from a three-month field trial of UbiFit showed that these display systems can be motivating, encouraging participants to maintain fitness levels that were significantly higher than for a control group without the representations (Consolvo, McDonald, & Landay, 2009).

Another approach is to design representations that show the average usage of some resource and the extent to which people deviate from it, with the aim of helping people to alter their resource use to be closer to or better than that average. For example, an energy monitor/display that is commercially available in many countries, called Wattson, shows both the watts or cost of how much electricity someone is using in their home at a given time and how this compares with their own average over time. The former is conveyed using LEDs on the topside, while the latter is shown via a colored light from the underside which glows blue when the householder is using less energy than normal, purple when use is average, and red when use is higher than usual. The idea is that on seeing red people will start to turn off lights and appliances to change the display back to purple or even blue.

Most dramatically, Shultz et al. (2007) have shown how emoticons can have a powerful effect on changing energy consumption behavior. In the first part of their study, a number of householders were told exactly how much energy they had used and the average consumption of energy by others in their neighborhood. The above-average energy users then significantly decreased their energy use while the below-average energy users significantly increased theirs (presumably because they felt they had more room to increase their consumption). But then the researchers tested the effect of additionally giving householders who consumed more energy than average an unhappy smiley icon—suggesting it was socially

disapproved—and those who consumed less than the norm a happy smiley icon—suggesting their energy consumption was socially approved. The impact of providing these two representations was dramatic: The big energy users showed an even *larger* decrease in their energy use than in the first condition, while the below-average users did not change their energy consumption upward (presumably because the addition of the happy emoticon suggested the socially positive aspect of their behaviour).

These studies suggest that there are benefits of using different kinds of representations and ambient displays to encourage people to change or adhere to certain kinds of desired behaviors, such as taking the stairs, taking more physical exercise or reducing energy consumption. Here, we are interested in how best to design representations for different decision-making activities. In the following sections, we first consider how best to present information that can be glanced at and perceived rapidly to guide supermarket shopping decisions *in situ*. In particular, how can we help people make informed decisions when confronted with many possible options, such as 50 different kinds of yoghurt to select from? Second, we consider how to design aggregate displays of cumulative information that is not otherwise available in the store, such as a summary of the total fat content of products placed in a shopping trolley relative to some desired norm. Third, we consider what kind of information visualizations can support collaborative planning activities, such as a family deciding together how to make a healthy meal with a low carbon footprint.

Displaying Informational Nudges

We propose that rather than providing ever more information to enable consumers to compare products in minute detail when making a choice, a better strategy is to design technological interventions that provide just *enough* information and in the *right* form to facilitate good choices. One

Figure 1. GoodGuide display using numbers and colored icons to represent overall rating and other dimensions for a category of product (e.g., sliced whitebread) and a single item in that category (e.g., Arnold Country Whole Grain). The ratings are scored out of 10 and displayed besides an emerald green icon when most 'beneficial', brown-orange when 'controversial' and red when 'concerning'

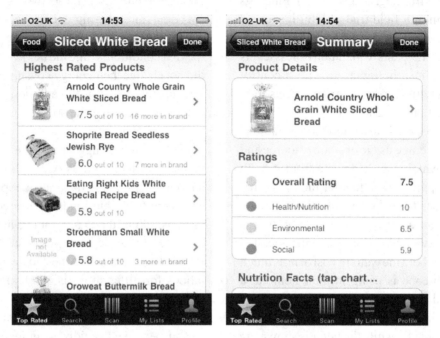

solution is to exploit new forms of augmented reality technology that enable 'information-frugal' decision making, in the context of an intensive activity replete with distractions (i.e., shopping in a supermarket or deciding at the kitchen table what to have for dinner).

An important consideration when representing multiple dimensions that can be glanced at and perceived rapidly is to enable comparisons to be made and cumulative information inferred *in situ*. For example, simple contrasting icons (e.g., thermometer icons, percentage bars, balls that change in color) can be presented which increase or decrease on some dimension in relation to the values being represented. An instance of such a display approach is the GoodGuide iphone app that rates products using a scale from 1-10 together with colored icons to represent where the products fall on the dimensions of health, environment and society. A score of 10 represents the best perfor-

mance while a score of 0 the worst, based on a number of indicators they have developed. The icons can be viewed at a glance and can be listed in order of highest rated products or expanded to provide more detail for individual products (Figure 1).

Another approach is to fuse relative measures on different dimensions (e.g., greenness, price, fat level) into singular displays where shape carries the salient information, such as a rectangle that gets wider to convey price and taller to convey a nutritional dimension that is general (healthiness) or specific (e.g., salt content) (Figure 2). A third dimension, such as 'greenness', could be added by filling in the rectangle with a shade from red to green to show the quantity of carbon emissions for that product. Similar to the idea behind Chernoff faces, the visualizations will be placed side by side to enable quick comparisons.

Figure 2. An example of a fused visualization for multi-dimensional data where nutrition is represented by height and its carbon footprint by width. In this example product A is 'good' (the rectangle is taller than its width) whereas product B is 'bad' (it is wider than its height)

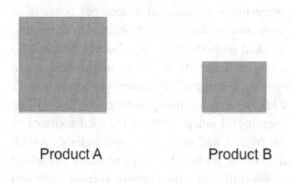

Product A   Product B

Another important question is whether emotive visualizations can be designed to persuade people to select food items they might not otherwise choose. For example, will smilies—indicating approval—placed on products that pass a 'greenesss' standard encourage people to buy them more than products that do not bear the smiling label, even if the former are more expensive? Is color-coding, such as the traffic light labeling promoted by the Food Standards Agency in the UK, emotive?

Design Considerations for an In-Store Trolley Interface

What then is a good way to provide appropriate information quickly and simply to shoppers in order to aid their decision-making during the hectic, distracting setting of a trip to the supermarket? Here we assume the shoppers have selected a particular dimension that they care about and want to be more responsive to in terms of their buying behavior—for instance, choosing products that are lower fat, or more sustainably grown. To inform shoppers about how they are doing in achieving this particular goal during their shopping expedition, cumulative values of the dimensions of interest across all products chosen so far could be summed up and displayed in an ambient manner as the current ongoing overall score "projected" onto the handle of the shopping trolley as a color. For example, a green handle could signify that the shopper has obtained a 'carbon footprint' or 'fat content' score below their target (or below some population average), while a red handle would indicate that the trolley's contents are above the desired level, with intermediate levels indicated by intermediate colors (Figure 3). This simple core idea allows a range of possibilities, with likely advantages and disadvantages; in particular there

Figure 3. Two hypothetical shopping trolleys with red and green glowing handles, indicating aggregate 'healthiness' of products selected relative to the average for a weekly shop for a family of four

is an important issue in terms of which summary statistics of a trolley's contents that might most helpfully be displayed.

- **Cumulative Versus Running Average Displays:** A cumulative display might, for example, use the color of the trolley's handle to indicate the total value of the contents on the dimension of interest. For example, a green handle could signify that a shopper has obtained a 'carbon footprint' or 'fat content' score below some target (or below some population average), provided the goal has been given in terms of total number of grams or whatever else is being measured. In contrast, a red handle would indicate that the trolley's contents are above the desired level. Depending on display technology, it may be possible to represent intermediate levels by intermediate colors.

Such a cumulative display is in accord with our low-information nudge principle, but has several limitations. First, assuming N colors can be displayed (and perceived by users), the trolley will pass through at most N-1 state changes during the shop as it accumulates the growing total value of what has been selected. In the extreme case, a successful shop may be indicated by no display changes, for instance when the empty trolley contents start out 'green' (below the maximum level for e.g. fat) and stays 'green' because the maximum level is never reached. The display uses a substitutive dimension to display an underlying additive scale, meaning changes will not be continuous and may be difficult to attribute to particular items. Second, the very nature of a cumulative property means that the display will be correlated with the simple amount of stuff in the trolley. This becomes a salient issue when one considers that different shoppers may have various goals for a single shop—for instance, they may be shopping for one person for a few days, or for

several people for a week or more. If cumulative displays are to be used, issues of normalization must be addressed.

The second type of simple trolley display would show a running average of the items, as they are placed in the trolley. Once the idea of by-item properties is considered it becomes possible to normalise the dimension-of-interest by correcting for item properties (e.g. number of portions). So a simple display of this nature might show, say, a running average of 'fat content per person-meal'. Clearly, such a running-average display will be meaningful independent of the total contents of the trolley, and may vary continuously, in both directions, during the shop, as items are added to the trolley and the running average rises and falls. This may make it more compelling and more informative for users without increasing its intrinsic complexity.

A possible disadvantage of a simple running-average display is that it seems to explicitly encourage the kind of rebound or boomerang behavior noted by Schulz et al. (2007) and others: Whenever the display indicates an unsatisfactory running average, it can be 'improved' by the addition of a 'good' item. However, this may be desirable. What is less desirable is if the boomerang flies in the opposite direction, as for instance when a shopper has achieved a 'green' display by purchasing many healthy items and decides to have a reward of a tub of ice cream, which still does not drive the display into the red zone (Figure 4).

- **Fusing Cumulative Total and Running Average Displays:** With certain display designs it is possible to simultaneously reflect running average and cumulative total, perhaps sidestepping the disadvantages of each. Imagine a display made up of many simple components—for example, a row of lights along the handle and edges of the shopping trolley—each of which can change in some property, such as color (see Figure 4). When an item is added to the trolley, a new light could be illuminated (or several lights, e.g., one per serving) with a color that reflects that item's properties. Now the total quantity of red or 'redness' indicates (to some approximation) the cumulative total of the shop (at least in terms of its overall quality), whereas the proportion of red to green indicates the running average. The simple number of illuminated lights—for instance as indicated by the length of the illuminated strip along the handle—would indicate the overall size of the shop-so-far. Alternatively, it would be possible to have a separate indication of the cumulative total of the property of interest—a larger single light that changes color, say—so that the total on that dimension could be displayed more clearly. This cumulative total display could also be set up to come on only when the shopper explicitly requests it, as for instance activating the button shown in Figure 4; the idea here is to give the shopper the chance to have a private peak at how well they are doing rather than displaying this continuously for other shoppers to see.

The fused display has advantages in terms of the issues raised above. It changes with every item, but not in a way that encourages rebound effects. It is meaningful at all stages of the shopping trip, and independently of the intended size of the shop. With colleagues we are currently exploring such fused displays in laboratory and *in situ* experiments.

Such an ambient and publicly visible display must first be studied to see if it fits with how people want to shop, or engenders unexpected side-effects. Will people be more or less likely to change their behavior when information about the contents of their shopping trolley is publicly visible for all to see rather than being privately displayed? Would shoppers try to fill their trolley

Figure 4. A simulation of a shopping trolley with LEDs that light up depending on the size of each product placed in it and the dimension chosen by the shopper (e.g. fat content per person-meal); green is below the average on that dimension, yellow-orange is average and red is above the average. The white pieces in the shopping trolley represent products and are moved into the trolley (simulating a purchase decision) from an adjacent palette where the size is known but the fat content unknown until scanned. The average for all items in the trolley can be compared against some norm, such as the average for a weekly shop for a family of four, can be displayed by pressing a button on the handle.

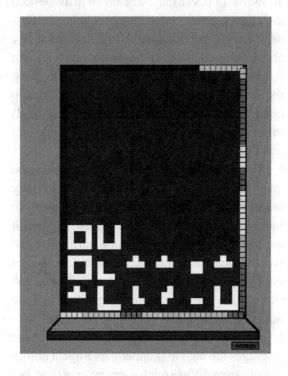

with healthy and green foods and on finding they were under the average then treat themselves to luxury goods high in fat and food miles? Would having their shopping trolley glow green at the check-out, indicating the contents were well below the average, make them feel good in front of other shoppers (Shultz et al., 2007)? Would the prospect of others seeing just how much butter and cheese they are buying make shoppers think about buying less, or just thinking about shopping elsewhere?

Other Tools to Promote Learning About Shopping

Assuming such an ambient information display *Cumulative Tool* achieves the desired features of providing some feedback without overloading the decision maker, without undesired effects of scaring shoppers off or making them "boomerang" and offset their good behavior with poorer choices, the question remains whether this kind of simple display provides *enough* feedback to allow the shopper to adjust behavior in the desired direction, e.g. reduced sodium or enhanced greenness. Seeing that one's entire trolley is red-lining above the goal level may motivate behavior, but it does not directly indicate what to do to bring the level back down. Thus, we must develop and test methods for ensuring that the (minimal) information delivered is actually actionable and conducive to behavior change.

There are at least three approaches that can be taken to solving this problem, which is essentially one of allocating global feedback appropriately to individual choices of products (akin to the "credit assignment" problem in machine learning). First, we could leave it all up to the users, and assume (or hope) that when they end their shop with a "green" trolley, they will buy more things like those the next time around, and when they get a "red" trolley, they will buy different things next time. This leverages the human shopper's intelligent ability to learn from diffuse reinforcement over time, but it will probably be slow, requiring many shopping outings before reliable change occurs. Second, to speed up this process, we could provide more specific feedback about each product that goes into the trolley, for instance momentarily flashing the ambient display with a color corresponding to the box of sugar-frosted chocolate bombs or bag of figs being chosen. This will allow shoppers to

make more targeted decisions about each product, provided they remember that individual feedback.

Third, to remove the need for such memory, a further interface can be developed to let shoppers query how they should adjust their purchases to come closer to their goal. This could take two main forms. A *Comparative Tool* could run as a 'private' mobile application on a smartphone or PDA and be displayed on the device or somewhere in the environment, such as the shopper's hand or the product package itself. After identifying the product via a photo or code scanner, the tool will show the product values on the dimensions of interest, and indicate whether this product helps or hinders the achievement of the current shopping goal. This interface could also be used in a comparative manner, scanning two or more products while they are still on the shelf and then showing at a glance which product is best based on the selected dimensions.

As a second 'off-line' method of providing more explicit feedback, a *Collaborative Tool* running on a home computer or surface display would allow shoppers to find out further information about the products they have bought once they get them home, along with input from their families. Multiple users could reflect and discuss together the decisions behind their food purchases with a view to attaining their goals at their next weekly shop, exploiting collaborative planning and social pressures that take place in a family setting. An interactive planner application would enable family members to find out more about particular dimensions (e.g., nutritional values) on a product, meal, or weekly-shop basis, and provide recipe-specific visualizations enabling items to be swapped. For example, a suggestion by dad to cook coq-au-vin for dinner will show it is low on 'greenness' (because of a large carbon footprint). This is a dimension the son has selected as an informational layer. Alternative items can be swapped with the chicken, such as tofu, which may then be shown by the application to have a higher greenness value (i.e., smaller carbon footprint). Finally, specific shopping lists could be generated that would achieve the goals set by the shopper and others involved.

FURTHER DIRECTIONS

To test whether any of these approaches succeeds in nudging shoppers' behavior in specific directions within a reasonable time-span, we have planned both lab-based experiments and field studies. One line of investigation will assess how the different information displays for the tools described above affect user decision-making strategy, focusing on when and how the interactive display of information enables fast and frugal decisions. This will then be tested further in supermarket studies, using techniques such as mobile eye tracking, observation and talk aloud methods to determine what people look at and how they use the comparative and cumulative tools. Longitudinal studies are also planned to determine whether the tools proposed have long-term impact on behavior, and how quickly such change occurs. Various kinds of households (e.g., family, young people, retired single) will be compared in terms of whether and how their shopping patterns and meal planning behavior change when using the tools—different groups of people may be more or less influenced by different types of nudges, and we cannot assume a one-size-fits-all approach.

Whether these various kinds of information delivery can help move people in the direction of better decisions—in the food shopping domain, or in other applications—remains to be seen. Our initial research suggests that simple visualizations can be designed to be information-frugal and emotive – encouraging people to change their behavior at the point of decision-making. But the trick will be balancing frugality and simplicity with *enough* feedback detail to allow people to change their choices at a pace that is sufficiently rapid and noticeable to be rewarding and motivating for long-term behavior change.

ACKNOWLEDGMENT

Thanks to Ricky Morris for creating Figure 1 and Stefan Kreitmayer for developing the simulation shown in Figure 4. We would also like to thank our colleagues Jon Bird from the Open University and Johannes Shöning and Antonio Krüger from Globus Innovative Retail Lab in Germany. The research is partially funded by the EPSRC grant "CHANGE: Engendering Change in People's Everyday Habits Using Ubiquitous Computing Technologies" (number XC/09/043/YR).

REFERENCES

Card, S. K., Mackinlay, J. D., & Shneiderman, B. (1999). *Readings in information visualization: Using vision to think*. San Diego, CA: Academic Press.

Consolvo, S., Klasnja, P., McDonald, D. W., Avrahami, D., Froehlich, J., LeGrand, L., et al. (2008). Flowers or a robot army? Encouraging awareness & activity with personal, mobile displays. In *Proceedings of the 10th International Conference on Ubiquitous Computing* (pp. 54-63).

Consolvo, S., McDonald, D. W., & Landay, J. A. (2009). Theory-driven design strategies for technologies that support behavior change in everyday life. In *Proceedings of the 27th International Conference on Human Factors in Computing Systems* (pp. 405-414).

EDS IDG Shopping Report. (2007). *Shopping choices: Attraction or distraction?* Retrieved from http://www.eds.com/industries/cir/downloads/EDSIDGReport_aw_final.pdf

Fasolo, B., McClelland, G. H., & Todd, P. M. (2007). Escaping the tyranny of choice: When fewer attributes make choice easier. *Marketing Theory, 7*(1), 13–26. doi:10.1177/1470593107073842

Feunekes, G., Gortemaker, I., Willems, A., Lion, R., & van den Kommer, M. (2008). Front-of-pack nutrition labelling: Testing effectiveness of different nutrition labelling formats front-of-pack in 4 European countries. *Appetite, 50*, 57–70. doi:10.1016/j.appet.2007.05.009

Gigerenzer, G., & Selten, R. (Eds.). (2001). *Bounded rationality: The adaptive toolbox*. Cambridge, MA: MIT Press.

Gigerenzer, G., & Todd, P. M.ABC Research Group. (1999). *Simple heuristics that make us smart*. New York, NY: Oxford University Press.

Goldstein, D. G., & Gigerenzer, G. (2002). Models of ecological rationality: The recognition heuristic. *Psychological Review, 109*, 75–90. doi:10.1037/0033-295X.109.1.75

Hertwig, R., & Todd, P. M. (2003). More is not always better: The benefits of cognitive limits . In Hardman, D., & Macchi, L. (Eds.), *Thinking: Psychological perspectives on reasoning, judgment and decision making* (pp. 213–231). Chichester, UK: John Wiley & Sons.

Ishii, H., Wisneski, C., Brave, S., Dahley, A., Gorbet, M., Ullmer, B., & Yarin, P. (1998). ambientROOM: Integrating ambient media with architectural space. In *Proceedings of the International Conference on Human Factors in Computing Systems* (pp. 173-174).

Jafarinaimi, N., Forlizzi, J., Hurst, A., & Zimmerman, J. (2005). Breakaway: An ambient display designed to change human behavior. In *Proceedings of the 7th International Conference on Human Factors in Computing Systems* (pp. 1945-1948).

Kahneman, D., Slovic, P., & Tversky, A. (Eds.). (1982). *Judgment under uncertainty: Heuristics and biases*. Cambridge, UK: Cambridge University Press.

Katsikopoulos, K. V., & Fasolo, B. (2006). New tools for decision analysts. *IEEE Transactions on Systems, Man, and Cybernetics . Part A, 36*, 960–967.

Kuang, C. (2009). *Better choices through technology*. Retrieved from http://www.good.is/post/better-choices-through-technology/

Lin, J. J., Mamykina, L., Lindtner, S., Delajoux, G., & Strub, H. (2006). Fish 'n' Steps: Encouraging physical activity with an interactive computer game. In *Proceedings of the International Conference on Ubiquitous Computing* (pp. 261-278).

Payne, J. W., Bettman, J. R., & Johnson, E. J. (1993). *The adaptive decision maker*. Cambridge, UK: Cambridge University Press.

Prante, T., Rocker, C., Streitz, N. A., Stenzel, R., Magerkurth, C., van Alphen, D., & Plewe, D. A. (2003). Hello.Wall – beyond ambient displays. In *Proceedings of the International Conference on Ubiquitous Computing* (pp. 277-278).

Rogers, Y., Hazlewood, W., Marshall, P., Dalton, N. S., & Hertrich, S. (2010). Ambient influence: Can twinkly lights lure and abstract representations trigger behavioral change? In *Proceedings of the 12th International Conference on Ubiquitous Computing*, Copenhagen, Denmark (pp. 291-300).

Rogers, Y., Lim, Y., Hazlewood, W., & Marshall, P. (2009). Equal opportunities: Do shareable interfaces promote more group participation than single users displays? *Human-Computer Interaction, 24*(2), 79–116. doi:10.1080/07370020902739379

Shultz, W., Nolan, J., Cialdini, R., Goldstein, N., & Griskevicius, V. (2007). The constructive, destructive and reconstructive power of social norms. *Psychological Science, 18*, 429–434. doi:10.1111/j.1467-9280.2007.01917.x

Simon, H. A. (1990). Invariants of human behavior. *Annual Review of Psychology, 41*, 1–19. doi:10.1146/annurev.ps.41.020190.000245

Thaler, R. H., & Sunstein, C. R. (2008). *Nudge: Improving decisions about health, wealth, and happiness*. New York, NY: Penguin.

Todd, P. M. (2007). How much information do we need? *European Journal of Operational Research, 177*, 1317–1332. doi:10.1016/j.ejor.2005.04.005

Todd, P. M., & Gigerenzer, G. (2007). Environments that make us smart: Ecological rationality. *Current Directions in Psychological Science, 16*(3), 167–171. doi:10.1111/j.1467-8721.2007.00497.x

Todd, P. M., Gigerenzer, G., & the ABC Research Group (in press). *Ecological rationality: Intelligence in the world*. New York, NY: Oxford University Press.

Todd, P. M., & Miller, G. F. (1999). From pride and prejudice to persuasion: Satisficing in mate search . In Gigerenzer, G., & Todd, P. M.ABC Research Group (Eds.), *Simple heuristics that make us smart* (pp. 287–308). New York, NY: Oxford University Press.

Toscos, T., Faber, A. M., An, S., & Gandhi, M. (2006). Chick clique: Persuasive technology to motivate teenage girls to exercise. In *Proceedings of Extended Abstracts on Human Factors in Computing Systems* (pp. 1873-1878).

Tweedie, L., Spence, B., Williams, D., & Bhogal, R. (1994). The attribute explorer. In *Proceedings of the Conference Companion on Human Factors in Computing Systems* (pp. 435-436).

van Mensvoort, K. (2005). *Datafountain*. Retrieved from http://infosthetics.com/archives/2005/08/datafountain.html

von Neumann, J., & Morgenstern, O. (1944). *Theory of games and economic behavior*. Princeton, NJ: Princeton University Press.

Wisneski, C., Ishii, H., Dahley, A., Gorbet, M., Brave, S., Ullmer, B., & Yarin, P. (1998). Ambient displays: Turning architectural space into an interface between people and digital information. In *Proceedings of the First International Workshop on Cooperative Building, Integrating Information, Organization, and Architecture* (pp. 22-32).

Wittenburg, K., Lanning, T., Heinrichs, M., & Stanton, M. (2001). Parallel bargrams for consumer-based information exploration and choice. In *Proceedings of the 14th Annual ACM Symposium on User Interface Software and Technology* (pp. 51-60).

This work was previously published in the International Journal of Mobile Human Computer Interaction, Volume 3, Issue 2, edited by Joanna Lumsden, pp.20-34, copyright 2011 by IGI Publishing (an imprint of IGI Global).

Chapter 7
Speech for Content Creation

Joseph Polifroni
Nokia Research Center, USA

Imre Kiss
Nokia Research Center, USA

Stephanie Seneff
MIT CSAIL, USA

ABSTRACT

This paper proposes a paradigm for using speech to interact with computers, one that complements and extends traditional spoken dialogue systems: speech for content creation. The literature in automatic speech recognition (ASR), natural language processing (NLP), sentiment detection, and opinion mining is surveyed to argue that the time has come to use mobile devices to create content on-the-fly. Recent work in user modelling and recommender systems is examined to support the claim that using speech in this way can result in a useful interface to uniquely personalizable data. A data collection effort recently undertaken to help build a prototype system for spoken restaurant reviews is discussed. This vision critically depends on mobile technology, for enabling the creation of the content and for providing ancillary data to make its processing more relevant to individual users. This type of system can be of use where only limited speech processing is possible.

INTRODUCTION

A couple are visiting Toronto and have just finished a meal at a small Chinese restaurant. The wife makes a habit of scouting out Chinese food in any city she visits and this restaurant was particularly good. As she walks out of the restaurant, she pulls out her mobile phone, clicks a button on the side, and speaks her thoughts about the meal she's just eaten. She then puts her phone away, having recorded her impressions of the restaurant. Her location and the time of day have been recorded as part of the interaction. Our hypothetical user then hails a cab and goes off to the theater. Figure 1 shows what a user might say in this context.

The scenario we describe above is the first-stage interaction with an overall system that uses speech for content creation, social media, and

DOI: 10.4018/978-1-4666-2068-1.ch007

Figure 1. A representation of how a user might create content via speech

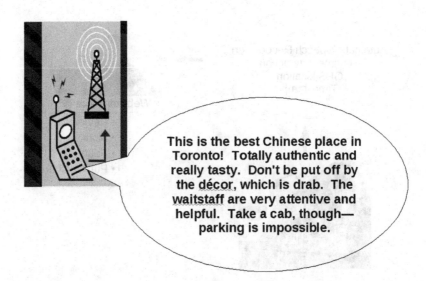

recommender systems. In subsequent sections, we will enlarge upon this scenario, with further glimpses into the user interaction and the underlying technology required for each step. We argue that these technologies are sufficiently advanced to enable the convenience of recording thoughts and impressions on the go, indexing the results, and extracting enough information to make the interaction useful for others.

One of the most important aspects of this scenario, and the ones that follow, is that the user is in charge of the interaction the entire time. Users do not have to worry about getting involved in an interaction when they're busy, in a noisy environment, or otherwise unable to devote time to the interface. Users can describe an experience while it is fresh in their memory through an interface that is always available to them. When they have the time and the inclination to make further use of the information, they can examine, review, and, ultimately, share it. The spoken input takes the form of a "note to self," where the user does not have to plan carefully what to say (Figure 2).

In this initial scenario, the user's interaction with the system stops after the review is spoken. Either immediately, or when connectivity is reestablished, speech is uploaded to a cloud-based system. With a combination of automatic speech recognition (ASR) and natural language processing (NLP) technologies, the system goes to work on indexing and deriving meaning from the dictated review. In the best case scenario, information about individual features, such as food quality or service, are extracted and assigned a scalar value based on user input. These values are used to populate a form, combined with other online sources of information (derived from GPS coordinates associated with the speech at the time of data collection), and made available to the user to review, modify, and share. Various other fallback levels of analysis are always available, so that the information is never completely lost or ineffectual. For example, the system may be able to only assign a single overall polarity to the entire review, or just extract some keywords for indexing. In the worst case, a simple audio file is saved and associated with a time-stamp and GPS location.

Figure 2. A schematic representation of data capture and processing in the restaurant review scenario

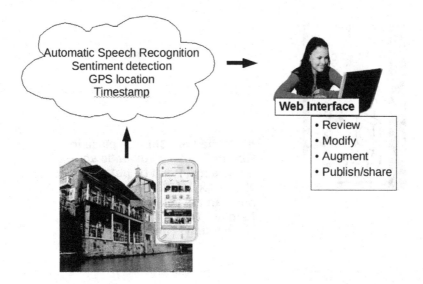

The user remains unaware of this processing, which need not be real-time. Further input will come later, at the discretion of the user. Figure 2 shows how this process might unfold.

Speech for content creation has several characteristics that make it attractive from a technological perspective:

- It does not have to be real-time. As our scenarios illustrate, the user simply speaks to a mobile device to enter content. Any further interaction takes place at the convenience of the user.
- It does not involve a detailed word-by-word analysis of the input. As we will show, further processing of the text can be done using just keywords/phrases in the user's input.
- It can be designed with multiple fallback mechanisms, such that any step of the process can be perceived as useful and beneficial to the user.

The key components that make this vision possible are large-scale ASR; NLP for keyword/named entity extraction, as well as opinion mining and sentiment detection; and content creation and information presentation informed by user modelling and research in recommender systems.

In the following sections, we describe, in order:

- The current state of the art in the technologies to be used in acquiring content via speech
- Recent research in recommender systems and user modelling that show how content collected via a mobile device fits into the emerging paradigms in these fields.
- A data collection effort currently underway at Nokia Research to enable us to build a prototype system for spoken restaurant reviews.
- Extensions of the concept of speech for content creation for developing markets.

ACQUIRING CONTENT VIA SPEECH

It is the following morning. The visitor to Toronto described above has had her morning coffee and is thinking back on the previous evening. She goes to a website and sees a map of Toronto with an icon at the location of the Chinese restaurant she went to the day before. The system has inferred the name of the restaurant from positioning data and has added further information from online restaurant review sites, including address, phone number, and summaries of what other people have said about the place. The user clicks on the icon and sees a display of what the system has done with the input (Figure 3). She looks it over, makes a small change, and decides to add a recommendation (via speech or typing) for a dish she particularly liked. After spending a few minutes on the review, she clicks Share and makes it available to her friends.

Extracting Meaning From Speech

The technologies described in this section demonstrate that extracting meaning from spontaneous speech is possible, and does not necessarily involve a complete analysis of the input utterance(s). As we describe below, there is value in simply extracting keywords and phrases from speech data. More sophisticated analysis for sentiment detection, although currently only applied to text data, also involves processing only parts of the text.

Keyword/phrase spotting has long been used to perform at least partial understanding of spontaneous speech in spoken dialogue systems when a complete understanding of spoken input is impossible (Seneff, 1992; Ward, 1989; Kawahara et al., 1997). In the context of an interactive dialogue system unfolding in real time, however, partial understanding must be accompanied by some mechanism to remember context, incorporate new information correctly, and draw inferences among varied and possibly competing input data. Although the systems that use it are typically automatically trainable, using this technology

Figure 3. A representation of the user interface for reviewing, accepting, and sharing content

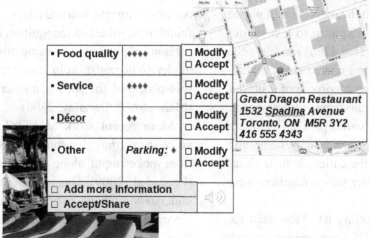

to sustain an interaction requires some heuristic component.

We feel that partial understanding is especially valuable for "one-shot" type applications, where there is no need for a more detailed analysis that will drive an ongoing dialogue. If an immediate and detailed interaction is not required, or even desired, the system has the luxury of performing computationally expensive processing while, at the same time, not having to completely understand everything the user spoke or involve them in a tedious confirmation dialogue. Many of the underlying technologies are currently available in frameworks that allow for easy exploration and experimentation (Bontcheva et al., 2004).

The extraction of named entities from speech has been used with large vocabulary ASR, most notably Broadcast News, associated with the DARPA HUB-4 task (Appelt & Martin, 1999; Miller et al., 1999; Horlock & King, 2003), as well as with similar corpora in Chinese (Zhai et al., 2004) or French (Favre et al., 2005). Although the speech in these corpora is not, for the most part, spontaneous, the extraction of proper names, locations, and organizations represents a significant advancement in the processing of this type of data.

Béchet et al. (2004) extracted named entities from spontaneous speech within the HMIHY corpus, concentrating on the extraction of phone numbers from utterances spoken to a customer care application. Huang et al. (2001) and Jansche and Abney (2002) perform named entity extraction on two separate speech corpora of a similar nature, i.e., voicemail transcripts. In both cases, they were looking for "caller phrases," phrases within the voicemail, typically near the beginning of the message, where the caller identifies him/herself. In addition, caller phone numbers were extracted.

Keyword/phrase-spotting has been used for partial understanding of spontaneous speech in systems where the interaction component is

restricted to a single turn. The How May I Help You (HMIHY) system at AT&T (Gorin et al., 1997) was one of the first to show the viability of extracting salient phrases from unconstrained speech input to perform call routing. Text categorization technology, applied to the same data, showed its applicability to the problem (Schapire & Singer, 2000).

The output of large vocabulary ASR has also been shown to be accurate enough to be used to segment and index audio recordings. In developing the SpeechBot system, van Thong et al. (2000) found that information retrieval performance using text derived via ASR was actually higher than would be expected given the error rate of the ASR engine.

Suzuki et al. (1998) found they were able to perform keyword extraction on radio news using ASR and use that information to identify domains of individual segments.

Lecture browsing is another domain in which the output of ASR engines has been shown to be sufficiently accurate to provide real value (Glass et al., 2007, Munteanu et al., 2009). To provide entree into hours of otherwise unsegmented spontaneous speech, topics must be discovered and delimited automatically, using keywords and phrases. One of the models used to partition these data, the minimum-cut segmentation model, was originally developed on text data and has been subsequently found to be robust to recognition errors. In comparisons of performance using this model, only a moderate degradation in classification accuracy was observed for speech transcription vs. text (Malioutov & Barzilay, 2006).

More recent work at AT&T, focussing on voice search, has also sought to extract locations from spoken input, along with query search terms (Feng et al., 2009). This level of natural language understanding helps in both making local search more precise and also in inferring a user's intent. In work on extracting named entities from utterances derived from a spoken corpus, the identification of named entities was found to significantly improve a later stage of parsing in which a more complete analysis is done (Polifroni & Seneff, 2010).

Dowman et al. (2005) describe a framework for annotating, segmenting and searching audio from television and radio news sources. An interesting component of this system is the use of key phrases extracted from ASR output to find related text documents on the Web, providing a more complete view of a particular news story from both text and audio sources. By unobtrusively offering additions of this sort to searches, the system enhances the information seeking activity of the user. Augmentations such as these need not be limited to keyword searches; GPS coordinates can provide place names that can also be used in such an information mash-up.

Whittaker et al. (2002) coin the phrase *What You See Is Almost What You Hear (WYSIAWYH)* for a new principle they propose for speech UI design. They provide empirical evidence to show that, though errorful, ASR transcripts contain enough information to make them useful for users as they navigate through voicemail messages. One key in instrumenting this principle is a graphical user interface that helps make the ASR output more comprehensible.

All of these studies show that a completely accurate transcription of speech, and a complete analysis of that transcript, is not necessary to build systems that benefit users. Given that we are not engaging the user in a lengthy interaction, we anticipate that the level of detail available from ASR transcripts will be sufficiently robust to make the service we propose a convenient and useful way to annotate one's life.

Opinion Mining/Sentiment Detection

Opinion mining and sentiment detection represent an extension of the paradigm of performing useful NLP in the absence of a complete parse of the input. These technologies work from tokenized collections of individual words. Many of the techniques employed for opinion mining, sentiment detection, and feature extraction make use of words and phrases found within the text rather than on a complete analysis of each word as it functions within a sentence. Their power comes from the amount of data used to draw inferences, data from consumer-generated media on the Web.

Unlike text written by professional journalists, consumer-generated media cannot be relied on to be grammatical, free of typos, or coherent within the context of an utterance. We anticipate that spoken reviews will be equally informal, but, as long as users speak, in general, the same words and phrases they use in informal written reviews, the technology should be portable, as was shown with text categorization data (Schapire & Singer, 2000).

The initial work in sentiment detection focussed on extracting the polarity of a given text, applied to written reviews for a variety of consumer products, as well as entities such as movies and restaurants (Pang et al., 2002; Nigam & Hurst, 2004). As the technology matured, it became possible to determine a more fine-grained rating, indicating a scale of sentiment (Dave et al., 2003; Goldberg & Zhu, 2006).

For completely unstructured text, such as that found in user-generated content on the Web, it is also useful to automatically extract information about individual features mentioned in reviews. This process of automatic knowledge acquisition can be complemented with sentiment detection to enable a more nuanced understanding of users' opinions (Branavan et al., 2009; Carenini et al., 2005; Hu & Liu, 2004). In a further refinement, automatically extracted features are combined with gradient rating of attributes to enable even deeper insight into consumer-generated media (Snyder & Barzilay, 2007; Titov & McDonald, 2008; Liu & Seneff, 2009; Gupta et al., 2010). Di Fabbrizio et al. (2010) have combined the output of their automated restaurant rating system (Gupta et al., 2010) with relevant utterances automatically extracted from reviews to produce a compact summarization of restaurants for display on mobile devices. These efforts show that it is possible to approach the insights of guides such as Zagat's or Consumer Reports through automated means, using information derived from a broad spectrum of opinions.

Some techniques make use of individual utterances and, therefore, utterance boundaries, for either computing the overall polarity of a given text (Pang & Lee, 2005) or to subset a larger text into just segments of interest for a market department, for example (Hurst & Nigam, 2004). The concept of a sentence is evident in dictated speech, as well, and we anticipate it will be part of the spoken reviews we are collecting (described in the section on data collection). Although not all users will speak flawlessly complete sentences, we expect an underlying prosodic and language model to be present nonetheless. Automatic addition of periods and other punctuation has already been shown to be possible in speech and beneficial

to performance of automatic speech recognizers (Liu et al., 2006; Kolář et al., 2004). It has been further shown to help in identifying names in speech (Hillard et al., 2006). We anticipate that aspects of this technology will have to be applied to speech for content creation, and our data collection effort has been devised with this in mind, as well. In two separate styles of interaction, users will provide spoken reviews in ways that encourage both multiple utterance input and individual utterances.

Using Content

It is several months later. A friend of the original reviewer is now in Toronto, looking for a place to have lunch. He logs onto a shared website, navigates to the Toronto page, and sees a list of reviewed sites. He has shared information himself with his friends and the system has learned that he and our original reviewer have similar tastes. The Chinese restaurant fits perfectly into the

Figure 4. A representation of the user interface for displaying the results of content creation plus recommendation

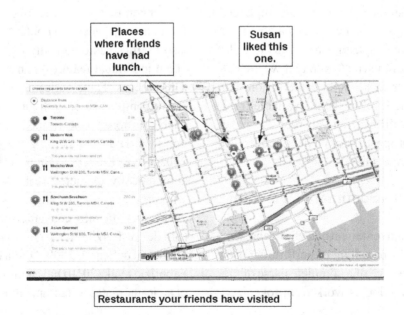

profile for both users, and is highlighted (Figure 4) in the results.

For building models of both trust and context, the content creation paradigm we envision would enable seamless and nonintrusive collection of the necessary data. Users' interaction can be time-stamped, at either the moment the speech is collected or in a later, post-processing stage. Because the information can be easily associated with specific users–in fact, many users may see that as the entire point of the interaction–sharing will be determined by who users feel would either benefit from or want to know that information. Existing social media and mobile phone contact lists can be leveraged to provide initial sharing networks.

In the remainder of this section, we examine how user models have been used effectively in dialogue systems, and then look more closely at enhancements to these systems via mobile technology.

Enabling Browsing/User Modelling in Dialogue Systems

Even outside a traditional dialogue, it is important for systems to process content to help users meet their information-seeking goals. Belkin et al. (1995) argue that browsing is a natural human activity and guidance is a necessary part of any information interface. In a study in the restaurant domain, it was shown that domain knowledge could be automatically summarized using a combination of machine learning and user modelling (Polifroni & Walker, 2008). The same automated technology for content selection was applied to a news corpus and shown to help journalists find background data for breaking news stories (Barker et al., 2009). By enabling a richer set of meta-data to associate with entities in a system, we anticipate more informative summaries (e.g., "Most people liked this restaurant but Victor didn't.").

User models have shown their utility in spoken dialogue systems for tailoring domain information to fit user preferences. Carenini and Moore (2001, 2006) applied a mathematical formalism to score, select and organize content for presentation to users in generated recommendations. This work showed that tailoring an evaluative argument to a user's preferences does increase its effectiveness. Carenini and Moore's work has been extended and applied to content selection for recommendations in a spoken dialogue system in a restaurant domain (Walker et al., 2004; Polifroni & Walker, 2008).

In deciding what to tell a user about options in a dialogue system, Demberg and Moore (2006) show the importance of pointing out trade-offs. They found that, for example, a user who prefers both flying on KLM and taking direct flights is more confident of a system that offers best-possible matching flights but also mentions a sub-optimal (e.g., connecting) flight if it is also on KLM. A follow-up study showed that these refinements to the options space, when volunteered by a system, help reduce dialogue turns (Paksima et al., 2009). Carenini and Rizoli (2009) specifically examine the presentation of opinion-based data in a multimedia setting and argue for the inclusion of dissimilar information and data on the degree of "controversiality" of opinions (i.e., how split the opinions were between positive and negative).

Recommender Systems

Regardless of the methodology or formalism, modelling preferences helps users process large amounts of data to find what is of specific interest to them. Both collaborative and content filtering are demonstrably useful and have become an expected aspect of any online shopping experience. Collaborative filtering, however, can be plagued by problems of sparsity, i.e., when few people have recommended only a handful of products, it's difficult to create and infer from communities of users. Content filtering, which can be more robust in the face of a limited population of users, still requires a stage of either online enrollment or a period of monitored usage in order to learn preferences. Hybrid systems have been proposed to address these issues (Balabanović & Shoham, 1997; Good et al., 1999), usually combining the two approaches for recommendations perceived as better overall by users.

Improvements to recommender systems build on computational models of trust (Abdul Rahman & Hailes, 1997). Systems make use of this notion as an adjunct to collaborative filtering, i.e., as a way of improving on recommendations by using this additional information (O'Donovan & Smyth, 2005; Massa & Avesani, 2004). If a friend has dined with you in the past and knows your tastes, a particular restaurant you reviewed and liked will be of interest, possibly defining interest. Metrics for trust can be gathered and refined via knowledge of the people users share reviews with, and the degree to which they agree or act upon those reviews.

Adomavicius et al. propose that the next generation of recommender systems make use of contextual information, both to address the sparsity problem, as well as to fill a specific and missing need in current systems (Adomavicius & Tuzhilin, 2005; Adomavicius et al., 2005). They cite a hypothetical example of making movie recommendations, where a system might offer a different choice for a Sunday afternoon matinee (a time when a given user might typically be seeing a movie with her children) than for a Saturday evening (when that same user may have established a pattern of seeing a more adult-oriented film with her partner). In an empirical study, it was shown that time and place had a significant effect on user ratings, in the case of movies (Adomavicius et al., 2005). Time-stamped reviews gathered from families with children could provide a prototype for another family's visit to the same city, including critical information implicitly, e.g., what were good morning activities.

DATA COLLECTION

As a first step in making the restaurant review scenario a reality, a data collection effort has been completed. The data were collected in a laboratory setting, with subjects recruited from among a self-reported population of people who frequently eat out at restaurants and who are familiar with on-line restaurant review sites. Each of these subjects spoke to a Nokia handset instrumented for the purpose of data collection. Users saw the questions shown in Table 1 on the handset and responded by clicking and holding to talk. All utterances were transcribed after recording.

Table 1. The two questionnaires used for the data collection effort in spoken restaurant reviews

1. What is the name of the restaurant?
2. Where is the restaurant located?
3. What type of cuisine does it serve?
4. What is its phone number?
5. Rate this restaurant on a scale of 1-5, where 1 is poor and 5 is excellent.
6. Rate the food quality on a scale of 1-5.
7. Rate the quality of service on a scale of 1-5.
8. Rate the atmosphere on a scale of 1-5.
9. Please review the restaurant and your experience in your own words.
1. What is the name of the restaurant?
2. Where is this restaurant located?
3. What type of cuisine does it serve?
4. What is its phone number?
5. Rate this restaurant on a scale of 1-5, where 1 is poor and 5 is excellent.
6. Rate the food quality on a scale of 1-5.
7. In words, please summarize the food quality.
8. Rate the quality of service on a scale of 1-5.
9. In words, please summarize the quality of service.
10. Rate the atmosphere on a scale of 1-5.
11. In words, please summarize the atmosphere.
12. Please review the restaurant and your experience. Repeating information is okay.

Subjects were randomly assigned to answer one of two questionnaires, both in the restaurant domain. In both questionnaires, the users were asked to rate the food quality, service, and atmosphere of each individual restaurant on a scale of 1-5. In one set of questions, shown at the top of Table 1, users were asked to simply assign a scalar value to the attributes and then rate the restaurant and their experience as a whole in a single, albeit lengthy turn. In the second set of questions, shown at the bottom of Table 1, users were asked to assign a scalar value and verbally describe each individual attribute. These users were also asked to provide an overall spoken review, in which they were told they could repeat information previously spoken.

This data collection effort was designed to give us flexibility in designing an initial application and also to provide insight into review data elicited under slightly different protocols. Both sets of questionnaires are designed primarily to collect data that can be used to associate users' spoken reviews with an automatically derived value representing their sentiment about the restaurant and its attributes. The first set of questions represents an ideal situation, i.e., where a user simply speaks in a free-form manner and we determine both features and polarity ratings. The latter set of data has been designed to capture specific information that might be useful to bootstrap training algorithms for automatically detecting specific features and assigning a graded sentiment representation for each. The latter set may also help us to train models for automatic assignment of utterance boundaries in free-form speech. Both sets of data have been used for language model training.

Preliminary experiments in extracting information from the spoken data have been promising and show that sentiment detection algorithms can be applied to spoken data (Polifroni et al., 2010). We plan to expand on the corpus already created to gain insight into the specific issues involved in using speech for creating review data. With a large enough corpus of spoken reviews, we can also begin looking into audio indexing as a way of finding specific information requested by users.

Speech data, along with ancillary information such as position and time, can be acquired via a relatively thin client. National and international consumer review sites, along with existing social media applications, have already accustomed people to the idea of expressing their opinions and sharing them with groups of friends. A convenient platform for expressing these opinions is the next logical step.

IN A BROADER CONTEXT

A user in a small village in the developing world critically depends on local, long-distance bus service to buy and sell goods. The bus comes irregularly and the bus stop is far away. However, the bus does follow a prescribed route. When it departs each stop, a user calls a central number and reports the bus's departure via speech. Users can call another number to quickly check on the position of the bus. Each user has to pay a small fee for the service, but users who call and report reliable departure information (e.g., determined by follow-up calls from farther along the route), receive credits for subsequent calls.

This last scenario shows our concept in a broader context. Speech for content creation should not be considered solely an idea for smartphone markets. We hope to make use of this idea to address a current need in developing countries: developing content and providing access to it (Marsden, 2003). Recent advances in the development of ASR for resource-poor languages indicate that "good enough" versions of ASR engines can be obtained for a relatively small cost (Novotney & Callison-Burch, 2010; Barnard et al., 2009; van Heerden, 2009; Barnard et al, 2010). These ASR engines can be used in basic spoken dialogue systems. If the system for creating content is also reduced to a more basic functionality, e.g., posting an alert that a bus is on its way, it should be feasible to use speech here, as well.

Sherwani et al. (2009a) showed that information access of a more restricted kind can be successful for low-literate users interacting with a speech interface. In their study, both high-literate and low-literate users achieved higher success rates using a speech interface than one that used touch-tone. This study highlighted the importance of training new users, along with the importance of local facilitators to help introduce new technology. If such an infrastructure is in place, speech interfaces can be profitably used. In general, it seems that an "orality-grounded" HCI is possible in the developing world; specific implementations must be designed carefully, with the exigencies of the developing world in mind (Sherwani et al., 2009b).

The information needs of farmers in Nigeria were the subject of a study from the Consultative Group on International Agricultural Research, based at the World Bank (Ozowa, 1997). Among the specific needs mentioned in this study were current prices, timing of crop planting, and information on group marketing. Infrastructure for information delivery is rudimentary for these farmers, who have limited access to television or even radios. A simple spoken interface could provide critical information in a relatively straightforward and easy-to-learn way.

In the simplest scenario, information could simply be recorded and a key sent via SMS to interested farmers. By calling a number and entering the key, users could hear product planning information. There would be no need for speech processing whatsoever. A simple small-vocabulary system, such as that developed in (Plauche et al., 2006), could provide access to the information via a menu-driven dialogue system instead of keypad input. Farmers who have information could follow the same sort of menu-driven dialogue to input information, e.g., speaking the name of the crop and the town where it was sold, followed by the price.

Economic interests are not the only driver for such systems. Another similar possibility would be a system that allows patients who take certain drugs to share their experiences with those drugs via a "living database." The system could provide access to both typed and spoken testimonials from contributors who share a common medical problem such as side effects from a particular drug. To access the database, people would speak a query such as "is there an association between Lipitor and shoulder pain?" and get back a display listing succinct summaries of all matching hits. Clicking on any one of them would launch an audio playback or open a window showing the complete text entry. The user could then enter their own new contribution to the database as well, if they so desired.

CONCLUSION

Spoken language systems should make people's lives easier. An ideal system would be viewed as a convenience, i.e., something users turn to when they are trying to simplify their lives. However, many studies have shown that speech actually increases user's cognitive load. Experiments in using speech in mobile environments have shown that multi-tasking and time pressure have measurable effects on users' speech patterns (Müller et al., 2001). Oviatt (2006) hypothesizes that less constrained speech systems increase cognitive load by imposing a demand for planning on the part of users. It would seem that a system deployed on a mobile device, eliciting unconstrained speech, could actually be an annoyance.

The key difference in the systems we propose, however, is that the interaction is managed the entire time by the user. Users can choose to devote attention to the system when it is convenient for them. In our most ambitious scenario, the system does not provide any immediate feedback at all. Users are, therefore, not held hostage to a system that cannot proceed until it has understood some part of the input. Whatever processing needs to be done on user input can be delayed indefinitely. When the user does choose to review the results, she can do so later, when she has the time to devote to the task and/or is in an environment where text-based corrections are supported.

Mobile devices make all these scenarios possible, for different reasons. Tourists carry them when they are sightseeing. People have them when they dine, see movies, or go to museums. In the developing world, they are more common than landlines or internet connections (United Nations International Telecommunications Union, 2009). Many of these devices can run thin clients that enable time-stamped speech capture. Higher end devices can associate geo-positioning data with speech. Used effectively, we feel this information can enrich the user experience with mobile devices.

ACKNOWLEDGMENT

The authors would like to thank Chao Wang, Janet Slifka, and Kalina Bontcheva for their valuable help and input on this paper.

REFERENCES

Abdul-Rahman, A., & Hailes, S. (1997). A distributed trust model. In *Proceedings of the Workshop on New Security Paradigms* (pp. 48-60).

Adomavicius, G., Sankaranarayanan, R., Sen, S., & Tuzhilin, A. (2005). Incorporating contextual information in recommender systems using a multidimensional approach. *ACM Transactions on Information Systems, 23*, 103–145. doi:10.1145/1055709.1055714

Adomavicius, G., & Tuzhilin, A. (2005). Toward the next generation of recommender systems: A survey of the state-of-the-art and possible extensions. *IEEE Transactions on Knowledge and Data Engineering, 17*(6), 734–749. doi:10.1109/TKDE.2005.99

Appelt, D., & Martin, D. (1999). Named entity extraction from speech: Approach and results using the textpro system. In *Proceedings of the DARPA Broadcast News Workshop* (pp. 51-54).

Balabanovič, M., & Shoham, Y. (1997). Content-based, collaborative recommendation. *Communications of the ACM, 40*(3), 66–72. doi:10.1145/245108.245124

Barker, E., Polifroni, J., Walker, M., & Gaizauskas, R. (2009). *Angle-seeking as a scenario for task-based evaluation of information access technology.* Paper presented at the International Workshop on Intelligent User Interfaces, Sanibel Island, FL.

Barnard, E., Davel, M., & van Heerden, C. (2009). ASR corpus design for resource-scarce languages. In *Proceedings of the 10th Annual Conference of the International Speech Communication Association* (pp. 2847-2850).

Barnard, E., Davel, M., & van Huyssteen, G. (2010). Speech technology for information access: A South African case study. In *Proceedings of the AAAI Symposium on Artificial Intelligence*, Palo Alto, CA (pp. 8-13).

Béchet, F., Gorin, A. L., Wright, J. H., & Hakkani-Tür, D. (2004). Detecting and extracting named entities from spontaneous speech in a mixed initiative spoken dialogue context: How may I help you? *Speech Communication, 42*(2), 207–225. doi:10.1016/j.specom.2003.07.003

Belkin, N., Cool, C., Stein, A., & Thiel, U. (1995). Cases, scripts, and information seeking strategies: On the design of interactive information retrieval systems. *Expert Systems with Applications, 9*(3), 379–395. doi:10.1016/0957-4174(95)00011-W

Bontcheva, K., Tablan, V., Maynard, D., & Cunningham, H. (2004). Evolving GATE to meet new challenges in language engineering. *Natural Language Engineering, 10*(3-4), 349–373. doi:10.1017/S1351324904003468

Branavan, S. R. K., Chen, H., Eisenstein, J., & Barzilay, R. (2009). Learning document-level semantic properties from free-text annotations. *Journal of Artificial Intelligence Research, 34*(1), 569–603.

Carenini, G., & Moore, J. D. (2001). A strategy for evaluating generative arguments. In *Proceedings of the First International Conference on Natural Language Generation* (pp. 1307–1314).

Carenini, G., & Moore, J. D. (2006). Generating and evaluating evaluative arguments. *Artificial Intelligence, 170*(11), 925–952. doi:10.1016/j.artint.2006.05.003

Carenini, G., Ng, R. T., & Zwart, E. (2005). Extracting knowledge from evaluative text. In *Proceedings of the 3rd International Conference on Knowledge Capture* (pp. 11-18).

Carenini, G., & Rizoli, L. (2009). A multimedia interface for facilitating comparisons of opinions. In *Proceedings of the International Conference on Intelligent User Interfaces* (pp. 325-334).

Dave, K., Lawrence, S., & Pennock, D. M. (2003). Mining the peanut gallery: Opinion extraction and semantic classification of product reviews. In *Proceedings of the 12th International Conference on World Wide Web* (pp. 519-528).

Demberg, V., & Moore, J. (2006). Information presentation in spoken dialogue systems. In *Proceedings of the 11ᵗʰ International Conference of the European Chapter of the Association for Computational Linguistics.*

Di Fabbrizio, G., Gupta, N., Besana, S., & Mani, P. (2010). Have2eat: A restaurant finder with review summarization for mobile phones. In *Proceedings of the 23ʳᵈ International Conference on Computational Linguistics: Demonstrations* (pp. 17-20).

Dowman, M., Tablan, V., Cunningham, H., Ursu, C., & Popov, B. (2005). *Semantically enhanced television news through web and video integration.* Paper presented at the Second European Semantic Web Conference Workshop, Crete, Greece.

Favre, B., Béchet, F., & Nocéra, P. (2005). Robust named entity extraction from large spoken archives. In *Proceedings of the Conference on Human Language Technology and Empirical Methods in Natural Language Processing* (pp. 491-498).

Feng, J., Bangalore, S., & Gilbert, M. (2009). Role of natural language understanding in voice local search. In *Proceedings of the 10ᵗʰ Annual Conference of the International Speech Communication Association* (pp. 1859-1862).

Glass, J., Hazen, T., Cyphers, S., Malioutov, I., Huynh, D., & Barzilay, R. (2007). Recent progress in the MIT spoken lecture processing project. In *Proceeding of the 8ᵗʰ Annual Conference of the International Communication Association* (pp. 2553-2556).

Goldberg, A. B., & Zhu, X. (2006). Seeing stars when there aren't many stars: Graph-based semi-supervised learning for sentiment categorization. In *Proceedings of TextGraphs: The First Workshop on Graph Based Methods for Natural Language Processing* (pp. 45-52)

Good, N., Schafer, J. B., Konstan, J. A., Borchers, A., Sarwar, B., Herlocker, J., et al. (1999). Combining collaborative filtering with personal agents for better recommendations. In *Proceedings of the Sixteenth National Conference on Artificial Intelligence* (pp. 439-446).

Gorin, A. L., Parker, B. A., Sachs, R. M., & Wilpon, J. G. (1997). How may I help you? *Speech Communication, 23*, 113–127. doi:10.1016/S0167-6393(97)00040-X

Gupta, N., Di Fabbrizio, G., & Haffner, P. (2010). Capturing the stars: Predicting rankings for service and product reviews. In *Proceedings of the NAACL HLT Workshop on Semantic Search* (pp. 36-43).

Hillard, D., Huang, Z., Ji, H., Grishman, R., Hakkani-Tür, D., Harper, M., et al. (2006). Impact of automatic comma prediction on POS/name tagging of speech. In *Proceedings of the IEEE/ACL Workshop on Spoken Language Technology* (pp. 58-61).

Horlock, J., & King, S. (2003). Discriminative methods for improving named entity extraction on speech data. In *Proceedings of the 8th European Conference on Speech Communication and Technology* (pp. 2765-2768).

Hu, M., & Liu, B. (2004). Mining opinion features in customer reviews. In *Proceedings of the 19th National Conference on Artificial Intelligence* (pp. 755-760).

Huang, J., Zweig, G., & Padmanabhan, M. (2001). Information extraction from voicemail. In *Proceedings of the Conference of the Association for Computational Linguistics* (pp. 290-297).

Hurst, M., & Nigam, K. (2004). Retrieving topical sentiments from online document collections. *Document Recognition and Retrieval 11, 5296*, 27-34.

International Telecommunications Union. (2009). *ICT statistics*. Retrieved from http://www.itu.int/ITU-D/ict/statistics/

Jansche, M., & Abney, S. P. (2002). Information extraction from voicemail transcripts. In *Proceedings of the ACL Conference on Empirical Methods in Natural Language Processing* (Vol. 10).

Kawahara, T., Lee, C.-H., & Juang, B.-H. (1997). Combining key-phrase detection and subword-based verification for flexible speech understanding. In . *Proceedings of the International Conference on Acoustics, Speech, and Signal Processing, 2*, 1159–1162.

Kolář, Ŝ., & Psutka, J. (2004). Automatic punctuation annotation in Czech broadcast news speech. In *Proceedings of the International Speech Communication Association* (pp. 319-325).

Liu, J., & Seneff, S. (2009). Review sentiment scoring via a parse-and-paraphrase paradigm. In *Proceedings of the Conference on Empirical Methods in Natural Language Processing* (pp. 161-169).

Liu, Y., Shriberg, E., Stolcke, A., Hillard, D., Ostendorf, M., & Harper, M. (2006). Enriching speech recognition with automatic detection of sentence boundaries and disfluencies. *IEEE Transactions on Audio, Speech, and Language Processing, 14*(5), 1526–1540. doi:10.1109/TASL.2006.878255

Malioutov, I., & Barzilay, R. (2006). Minimum cut model for spoken lecture segmentation. In *Proceedings of the 21st International Conference on Computational Linguistics and the 44th Annual Meeting of the Association for Computational Linguistics* (pp. 25-32).

Marsden, G. (2003). Using HCI to leverage communication technology. *Interaction, 10*(2), 48–55. doi:10.1145/637848.637862

Massa, P., & Avesani, P. (2004). Trust-aware collaborative filtering for recommender systems. In *Proceedings of the Federated International Conference on the Move to Meaningful Internet: CoopIS, DOA, ODBASE* (pp. 492-508).

Miller, D., Schwartz, R., Weischedel, R., & Stone, R. (1999). Named entity extraction from broadcast news. In *Proceedings of the DARPA Broadcast News Workshop* (pp. 37-40).

Müller, C., Grossmann-Hutter, B., Jameson, A., Rummer, R., & Wittig, F. (2001) Recognizing time pressure and cognitive load on the basis of speech: An experimental study. In *Proceedings of the 8th International Conference on User Modeling* (pp. 24-33).

Munteanu, C., Penn, G., & Zhu, X. (2009). Improving automatic speech recognition for lectures through transformation-based rules learned from minimal data. In *Proceedings of the Joint Conference of the 47th Annual Meeting of the ACL and the 4th International Joint Conference on Natural Language Processing of the AFNLP* (pp. 764-772).

Nigam, K., & Hurst, M. (2004). *Towards a robust metric of opinion.* Paper presented at the AAAI Spring Symposium on Exploring Attitude and Affect in Text, Stanford, CA.

Novotney, S., & Callison-Burch, C. (2010). Cheap, fast and good enough: Automatic speech recognition with non-expert transcription. In *Proceedings of the Annual NAACL Conference on Human Language Technologies* (pp. 207-215).

O'Donovan, J., & Smyth, B. (2005). Trust in recommender systems. In *Proceedings of the 10th International Conference on Intelligent User Interfaces* (pp. 167-174).

Oviatt, S. (2006). Human-centered design meets cognitive load theory: Designing interfaces that help people think. In *Proceedings of the 14th Annual ACM International Conference on Multimedia* (pp. 871-880).

Ozowa, V. N. (1997). Information needs of small scale farmers in Africa: The Nigerian example. *Consultative Group on International Agricultural Research News, 4*(3).

Paksima, T., Georgila, K., & Moore, J. D. (2009). Evaluating the effectiveness of information presentation in a full end-to-end dialogue system. In *Proceedings of the SIGDIAL Conference on the 10th Annual Meeting of the Special Interest Group on Discourse and Dialogue* (pp. 1-10).

Pang, B., & Lee, L. (2005). Seeing stars: Exploiting class relationships for sentiment categorization with respect to rating scales. In *Proceedings of the 43rd Annual Meeting on Association for Computational Linguistics* (pp. 115-124).

Pang, B., Lee, L., & Vaithyanathan, S. (2002). Thumbs up? Sentiment classification using machine learning techniques. In *Proceedings of the ACL Conference on Empirical Methods in Natural Language Processing* (pp. 79-86).

Plauche, M., Nallasamy, U., Pal, J., Wooters, C., & Ramachandran, D. (2006). Speech recognition for illiterate access to information and technology. In *Proceedings of the International Conference on Information and Communications Technologies and Development* (pp. 83-92).

Polifroni, J., & Seneff, S. (2010). Combining word based features, statistical language models, and parsing for named entity recognition. *In Proceedings of the 11th Annual Conference of the International Speech Communication Association* (pp. 1289-1292).

Polifroni, J., Seneff, S., Branavan, S. R. K., Wang, C., & Barzilay, R. (2010). *Good grief, I can speak it! Preliminary experiments in audio restaurant reviews.* Paper presented at the IEEE Workshop on Spoken Language Technology, Berkley, CA.

Polifroni, J., & Walker, M. (2008). Intentional summaries as cooperative responses in dialogue: Automation and evaluation. In *Proceedings of the ACL Conference on Human Language Technologies* (pp. 479-487).

Schapire, R. E., & Singer, Y. (2000). Boostexter: A boosting-based system for text categorization. *Machine Learning, 39*(2-3), 135–168. doi:10.1023/A:1007649029923

Seneff, S. (1992). TINA: A natural language system for spoken language applications. *Computational Linguistics, 18*(1), 61–86.

Sherwani, J., Ali, N., Rosé, C. P., & Rosenfeld, R. (2009). Orality-grounded HCI: Understanding the oral user. *Information Technologies & International Development, 5*(4).

Sherwani, J., Palijo, S., Mirza, S., Ahmed, T., Ali, N., & Rosenfeld, R. (2009). Speech vs. touch-tone: Telephony interfaces for information access by low literate users. In *Proceedings of Information and Communications Technologies and Development* (pp. 447-457).

Snyder, B., & Barzilay, R. (2007). Multiple aspect ranking using the good grief algorithm. In *Proceedings of the Human Language Technology Conference of the North American Chapter of the Association of Computational Linguistics* (pp. 300-307).

Suzuki, Y., Fukumoto, F., & Sekiguchi, Y. (1998). Keyword extraction using term-domain interdependence for dictation of radio news. In *Proceedings of the 17ʰ International Conference on Computational Linguistics* (pp. 1272-1276).

Thong, J.-M. D., Goddeau, D., Litvinova, A., Logan, B., Moreno, P., & Swain, M. (2000). Speechbot: A speech recognition based audio indexing system for the web. In *Proceedings of the 6th International Conference on Computer-Assisted Information Retrieval* (pp. 106-115).

Titov, I., & McDonald, R. (2008). Modeling online reviews with multi-grain topic models. In *Proceeding of the 17th International Conference on World Wide Web* (pp. 111-120).

van Heerden, C., Barnard, E., & Davel, M. (2009). Basic speech recognition for spoken dialogues. In *Proceedings of the 10ʰ Annual Conference of the International Speech Communication Association* (pp. 3003-3006).

Walker, M., Whittaker, S., Stent, A., Maloor, P., Moore, J., Johnston, M., & Vasireddy, G. (2004). Generation and evaluation of user tailored responses in multimodal dialogue. *Cognitive Science, 28*(5), 811–840. doi:10.1207/s15516709cog2805_8

Ward, W. (1989). Understanding spontaneous speech. In *Proceedings of the Workshop on Speech and Natural Language* (pp. 137-141).

Whitaker, S., Hirschberg, J., Amento, B., Stark, L., Bacchiani, M., Isenhour, P., et al. (2002). SCANMail: A voicemail interface that makes speech browsable, readable, and searchable. In *Proceedings of the Conference on Human Factors in Computing Systems* (pp. 275-282).

Zhai, L., Fung, P., Schwartz, R., Carpuat, M., & Wu, D. (2004). Using *N*-best lists for named entity recognition from Chinese speech. In *Proceedings of HLT-NAACL: Short Papers* (pp. 37-40).

This work was previously published in the International Journal of Mobile Human Computer Interaction, Volume 3, Issue 2, edited by Joanna Lumsden, pp.35-49, copyright 2011 by IGI Publishing (an imprint of IGI Global).

Chapter 8
3D Talking–Head Interface to Voice–Interactive Services on Mobile Phones

Jiri Danihelka
Czech Technical University in Prague, Czech Republic

Lukas Kencl
Czech Technical University in Prague, Czech Republic

Roman Hak
Czech Technical University in Prague, Czech Republic

Jiri Zara
Czech Technical University in Prague, Czech Republic

ABSTRACT

This paper presents a novel framework for easy creation of interactive, platform-independent voice-services with an animated 3D talking-head interface, on mobile phones. The Framework supports automated multi-modal interaction using speech and 3D graphics. The difficulty of synchronizing the audio stream to the animation is examined and alternatives for distributed network control of the animation and application logic is discussed. The ability of modern mobile devices to handle such applications is documented and it is shown that the power consumption trade-off of rendering on the mobile phone versus streaming from the server favors the phone. The presented tools will empower developers and researchers in future research and usability studies in the area of mobile talking-head applications (Figure 1). These may be used for example in entertainment, commerce, health care or education.

INTRODUCTION

Rapid proliferation of mobile devices over the past decade and their enormous improvements in terms of computing power and display quality opens new possibilities in using 3D representations for complementing voice-based user interaction. Their rendering power allows creation of new user interfaces that combine 3D graphics with speech recognition and synthesis. Likewise, powerful speech-recognition and synthesis tools are becoming widely available on mobile clients

DOI: 10.4018/978-1-4666-2068-1.ch008

or readily accessible over the network, using standardized protocols and APIs. The presented 3-dimensional talking head on a mobile phone display represents a promising alternative to the traditional menu/windows/icons interface for sophisticated applications, or a more complete and natural communication alternative to purely voice- or tone-based interaction. Such interface has proven many time to be useful as a virtual news reader (Alexa, Berner, Hellenschmidt, & Rieger, 2001), weather forecast (Kunc, & Kleindienst, 2007), healthcare communication assistant (Keskin, Balci, Aran, Sankur, & Akarun, 2007) blog enhancement (Kunc, Slavik, & Kleindienst, 2008) and can be very useful especially in developing regions where people often cannot read and write.

So far, talking-head interfaces have been used mostly on desktop PCs. Existing frameworks for talking-head development on desktop PCs (Wang, Emmi, & Faloutsos, 2007; Balci, 2005) have inspired our work. Emerging electronics such as mobile phones, pocket computers or embedded devices now possess enough power to enable a talking-head interface, but lack tools for creating such applications. In this paper we propose an effective architecture for interactive, fully-automated 3D-talking-head applications on a mobile client (Figure 1) and implement a framework for easy creation of such applications.

The main contributions of this work are:

- We document that performance limits of contemporary mobile devices are sufficient for running a 3D+audio interface by practical experiments and benchmarks;
- We describe practical techniques of synchronizing the audio stream and visual animation to deliver convincing talking-head interaction on the mobile device;
- We present a platform-independent prototype implementation of a distributed framework for creating and generating the 3D-talking-head applications.

Figure 1. Talking-head application on a Windows Mobile 6.1 device (HTC Touch Pro). It is able to articulate speech phonemes and show facial expressions (anger, disgust, fear, sadness, smile, surprise)

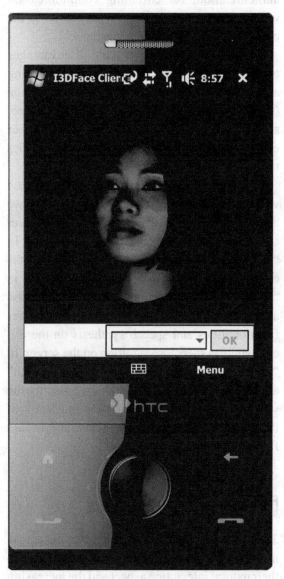

By providing a general tool for creating interactive talking-head applications on mobile platforms, we aim to spark future research in this area. It may open up space for many useful applications, such as interactive mobile virtual assistants, coaches or customer-care, e-government platforms, interactive assistants for the handicapped, elderly or illiterate, 3D gaming or navigation, quiz competitions or education (Wagner, Billinghurst, & Schmalstieg, 2006). It may be used for secure authentication, for enriching communication with emotional aspects or for customizing the communicating-partner's appearance.

3D talking-heads have their disadvantages too – consuming a lot of resources and not being appropriate for all types of information exchange (such as complex lists or maps). The first aspect should take care of itself by computing power evolution, the second by adding further modalities to the interactive environment.

The paper is organized as follows: in "Related work", relevant prior art is surveyed, and then we discuss components distribution between server and client in "Distributed design analysis". In "Performance measurements" we perform power-consumption and graphics benchmarks, and in "Architecture discussion and selection" we discuss architecture implications for performing graphics functionalities and speech synthesis on the client, whereas speech recognition on the server. In "Synchronization of face animation with speech" and "Framework implementation" we describe the voice and graphics synchronization and details of the software framework. We conclude and discuss future outlook in "Conclusion".

RELATED WORK

3D user interfaces are a general trend across multiple disciplines (Bowman et al., 2008), due to their natural interaction aspect and the increasing availability of relevant technology. In the domain of desktop computing, with large displays and multimedia support, use of multi-modal interaction and 3D virtual characters has been on the rise. Virtual characters improve telepresence (the notion of customer and seller sharing the same space) in e-commerce (Qiu, & Benbasat, 2005)

or interaction with technology for elderly people (Ortiz et al., 2007). Learning exercises with virtual characters (Wagner, Billinghurst, & Schmalstieg, 2006) have shown that audio components improve their perception and that 3D virtual characters are much better perceived than 2D ones. Much effort has also concentrated on building multi-modal mobile interaction platforms (Deng et al., 2004).

Research into Embodied Conversational Agents (ECA), agents with a human shape using verbal and non-verbal communications (Dryer, 1999), shows that people prefer human-like agents over caricatures, abstract shapes or animals, and, moreover, agents with similar personality to their own (Dryer, 1999; Nass, Moon, Fogg, Reeves, & Dryer, 1995).

Natural interaction with the resources of the global network (especially using voice), is a growing field of interest. Recent works for example develop the idea of the World Wide Telecom Web (WWTW) (Kumar, Rajput, Chakraborty, Agarwal, & Nanavati, 2007; Agarwal, Kumar, Nanavati, & Rajput, 2008; Agarwal, Chakraborty, Kumar, Nanavati, & Rajput, 2007), a voice-driven ecosystem parallel to the existing WWW. It consists of interconnected voice-driven applications hosted in the network (Kumar, Rajput, Chakraborty, Agarwal, & Nanavati, 2007), a *Voice Browser* providing access to the many voice sites (Agarwal, Kumar, Nanavati, & Rajput, 2008) and the Hyperspeech Transfer Protocol (HSTP) (Agarwal, Chakraborty, Kumar, Nanavati, & Rajput, 2007) allowing for their seamless interconnection. Developing regions with large proliferation of phones but little Internet literacy are set to benefit.

Similarly, mobile platforms would benefit from improved interaction. For example, mobile Web browsing has been shown to be less convenient than desktop browsing (Shrestha, 2007). Augmenting the interaction with voice and graphics assistance ought to improve it. Conversely, pure voice-response systems have been shown to benefit from augmenting with a visual interface (Yin &

Zhai, 2006). This motivates adding more modalities into the user-mobile-client-Web interaction.

Research in assistive technologies has focused on Web interaction by voice and its applicability for the handicapped or elderly. For example the HearSay audio Web browser (Ramakrishnan, Stent, & Yang, 2004; Sun, Stent, & Ramakrishnan, 2006) allows to automatically creating voice applications from web documents. An even larger group of the handicapped may be reached if more modalities are used for the interaction, allowing the use of animations or sign-language.

Synchronizing voice (speech) with animation (lip movement) has been addressed before, yet on desktop platforms. The BEAT animation toolkit (Cassell, Vilhjalmsson, & Bickmore, 2001) (based on language tagging) allows animators to input text to be spoken by an animated head, and to obtain synchronized nonverbal behaviors and synthesized speech that can be input to a variety of animation systems. The DECface toolkit (Waters & Levergood, 1993) focuses on correctly synchronizing synthesized speech with lip animation of virtual characters. A physics-based model (Albrecht, I., Haber, J., & Seidel, H. 2002) (relying on co-articulation – coloring of a speech segment by surrounding segments) and a distributed model (Poller & Muller, 2002) (based on phoneme timestamps, as our framework) for synchronizing facial animations with speech have also been presented.

Detailed 3D-face rendering has so far avoided the domain of mobile clients, due to limited computing capacity, display quality and battery lifetime. Previous attempts to render an avatar face on a mobile client have still used nonphotorealistic rendering (NPR), such as the cartoon shading (Choi, Kim, Lee, Lee, & Park, 2004). The platform in (Choi, Kim, Lee, Lee, & Park, 2004) also has ambitions for strong interactivity, allowing for visual interaction based on video capture and server-based face-expression recognition. However, the character is not automated, but merely conveying the visual expression of the person at the other end of the communication channel.

Previous mobile frameworks for easy application creation (Pandzic, 2002; Pandzic, Ahlberg, Wzorek, Rudol, & Mosmondor, 2003; Kadous & Sammut, 2002) were restricted to a particular mobile platform, yet currently there exist many mobile operating systems. Our proposed framework is not only platform independent, but also compatible with desktop facial-modelling tools.

Several languages convenient for talking-head scripting are available. We exploit the SMIL-Agent (Synchronized Multichannel Integration Language for Synthetic Agents) (Balci, Not, Zancanaro, & Pianesi, 2007) scripting language, based on XML. Related languages developed for talking head scripting are AML (Avatar Markup Language) (Kshirsagar, Magnenat-Thalmann, Guye-Vuilleme, Thalmann, Kamyab, Mamdani, 2002) and ECAF (Authoring Language for Embodied Conversational Agents) (Kunc & Kleindienst, 2007).

An open modular facial-animation system has been described in (Waters & Levergood, 1993). Commercial systems such as FaceGen (Singular Inversion, 2010) can be used for creating face meshes, and the Xface (Balci, 2005) represents an open toolkit for facial animations. We take inspiration from these tools, targeted for PC platform, and extend them with the network connection functionality, taking the features of mobile clients and their power-consumption limitations into a consideration.

DISTRIBUTED DESIGN ANALYSIS

During the design process of our framework we considered several possible architectures for talking-head-enhanced applications. For a natural conversation between the (real) user and the (virtual) head we need components for 3D rendering, speech recognition, speech synthesis,

Figure 2. Video-streaming architecture is convenient for less powerful mobile phones with fast Internet connection, because it delegates most of the application work to a remote server. It can be easily implemented as platform-independent. We did not include such architecture in our framework, because it is energetically inefficient.

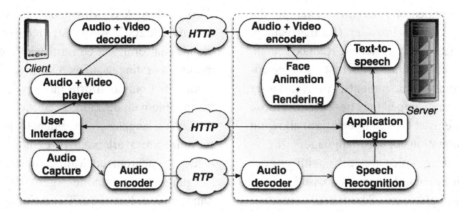

and application logic. Each of these components can reside either on the client or server side. This section discusses possible architecture alternatives (Figures 2, 3 and 4).

Speech Synthesis

Speech can be synthesized either on the mobile device or on a remote server. In the past the components for speech synthesis (also called Text-to-Speech engines) on mobile devices used

to have somewhat lower quality than components for synthesis on desktop/server PCs, which possess more resources. However, the computational power and available memory of present mobile devices allows generating voice output with a quality which satisfies the needs for computer-human dialogue. So the impact in quality is almost unrecognizable.

It is a challenging task to synchronize speech and face animation (lips movement). We address the synchronization problem by using phoneme/

Figure 3. Client-server configuration that uses the server for application-logic processing and for speech recognition. Results of the recognition process are directly provided to the application-logic module. The client side is used for text-to-speech processing, face animation and their synchronization. This architecture is supported by our framework.

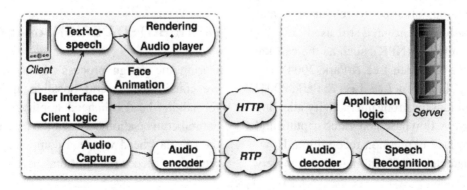

Figure 4. Client-server configuration that uses the server side for application-logic processing only. Our framework supports this type of configuration. It is suitable only for mobile devices with great computational power. Also this configuration is convenient in situations where only low bandwidth is available.

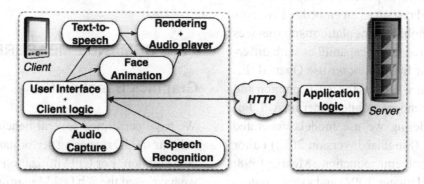

viseme timestamps (Poller, & Muller, 2002) (for details of the complete solution see "Synchronization of face animation with speech"). For this type of synchronization, it is necessary to have speech synthesis and animation component co-located together. That is why we only support speech synthesis on the client. Nevertheless, as discussed in "Performance measurements", the client-side synthesis is more energy-efficient anyway and therefore should be preferred over the server-side variant.

Speech Recognition

Speech recognition is significantly more CPU- and memory-intensive than speech synthesis. Suitable mobile speech-recognition solutions are available for scenarios when the set of recognized words is greatly limited (e.g. yes/no answers, one-of-N options or voice dial). Without such constraints (e.g. dictating an arbitrary letter), available mobile solutions are quite error-prone. In such case, for speech recognition it is better to send the recorded voice to a remote server.

Unlike in the case of speech synthesis, our framework supports both server- and client-side speech recognition. However, client-side speech recognition is limited to very small dictionaries

(about 50 words) with a simple acoustic and language model.

Graphics Rendering and Streaming

The visual application content can be rendered either on the mobile phone or on a remote server followed by video streaming to the phone. The second approach can be easily implemented as platform-independent because there is little code on the client side, but it has also many disadvantages.

Video streaming needs a lot of bandwidth that is often limited in mobile networks. Such architecture moves most components to the server side (Figure 2). The server renders video and synthesizes speech. Both are then streamed over the network to the client. The entire application logic resides on the server side.

We have tried the video-streaming approach and our experiments show that latency of up to 400 ms, caused by video compression and network latency, may occur between the user input and a response from the server. Such latency may make voice interface unpleasant, especially if the user expects an immediate response (e.g. using buttons to move a camera within a virtual world). Video-streaming on mobile phones is usually also more power-demanding.

Client-side graphics rendering is less power-demanding, however, it is far more challenging to be implemented as platform-independent and with the limited resources an embedded systems has. Different mobile phone platforms and devices have different rendering capabilities with different APIs. In our framework we use OpenGL ES (Kronous Groups, n. d.) as the most common and platform-independent mobile rendering API. For head/face rendering we use models generated from FaceGen (Singular Inversion 2010) editor with applied polygon reduction (Melax, 1998; Reddy, 2003; Hamann, 1994) and viseme reduction techniques (Danihelka, Kencl, & Zara, 2010) to reduce the model complexity.

Connection Requirements

According to our experiments, at least a 100 kbps connection throughput is needed for video streaming; otherwise the video quality is not acceptable for a user on a mobile client screen with resolution 320x240. For audio streaming architectures (Figure 3), 12 kbps data connection is enough.

Common usual throughput on connections for mobile phones is: GPRS 40 kbps, EDGE 100 kbps, UMTS 300 kbps, Wi-Fi on mobiles

600 kbps. While audio streaming works over all of the above, video streaming requires a higher-bandwidth connection.

PERFORMANCE MEASUREMENTS

Graphics Benchmarks

We have performed several benchmark tests to validate the 3D rendering performance and power consumption. For CPU utilization measurement we have used the acbTaskMan utility (Acbpocketsoft, n. d.). All measurements and tests were performed on the HTC Touch Pro mobile device with Qualcomm 528 MHz processor and Windows Mobile 6.1 operating system. Qualcomm chipsets are the most common in current Windows Mobile phones. For demonstration and testing we have developed an OpenGL ES rendering application called GLESBenchmark (Figure 5), inspired by (Kishonti Informatics, 2003), which renders a 3D head in a real-time. Selected performance test results are summarized in Table 1.

We conclude that the phone is able to render up to 8000 triangles illuminated by one directional light at 15 frames per second but the speed

Figure 5. Snapshots of the created GLESBenchmark application. The head is animated during performance measurements.

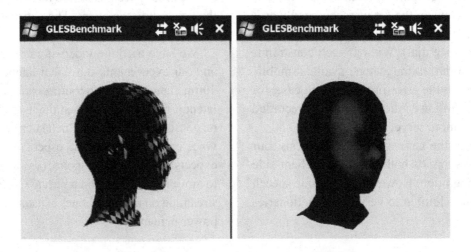

Table 1. Face rendering - Frames per second (FPS) depending on lighting, shading and texturing settings

	Female face	Male face
Triangles	8864	6352
Flat Shading	23.70	33.32
Smooth Shading	23.69	33.42
Flat, Directional Light	12.56	15.77
Smooth, Directional Light	12.58	15.76
Smooth, Point Light	3.76	5.77
Smooth, Directional, Textures	12.42	15.55
Flat, Directional, Textures	12.45	15.69

Table 2. HTC Touch Pro power consumption

	Consumption
OpenGL rendering (8192 triangles), WiFi on	899 mW
Video streaming WiFi (100 kb/s)	1144 mW
Video streaming EDGE (100 kb/s), WiFi off	2252 mW
Playing predownloaded video, WiFi on	752 mW
Display on, WiFi on	402 mW
Client voice recognition (PocketSphinx)	433 mW
Server voice recognition using WiFi	1659 mW

drops considerably when using a point light. Surprisingly, the rendering speed does not depend on choice of shading method (flat or smooth shading). According to GLBenchmark (Kishonti Informatics, 2003), some other phones (iPhone, Symbian phones) do not have difficulties with rendering of 3D objects illuminated by point light (the rendering speed is nearly the same as in the case of directional light). Textures affects the rendering performance only a little. We used a 512x512 pixel texture in our experiments. Maximum texture size in OpenGL ES is limited to 1024x1024 pixels or less on most mobile platforms.

Power Consumption

We have made estimates and rough measurements of power consumption for each of the architectures discussed. During the tests the Wi-Fi module with audio streaming was on, the display backlight was set to the minimum value and the automatic turn-off of the display (phone sleep mode) was disabled. Our rendering and Wi-Fi consumption values closely reflect those published at (Mochocki, Lahiri, & Cadambi, 2006; Acquaviva, Lattanzi, & Bogliolo, 2004; Devevey,

Lorenzon, & Tambary, 2005). Our own measurements (Table 2) show lower power consumption than estimated in these works, but have the same relative correspondence. This is probably due to lower per-instruction power consumption budget of novel mobile devices.

For video streaming, bandwidth and power consumption do not depend on number of rendered triangles, because we assume them to be processed at the sufficiently fast server. However, highly textured models can negatively affect the video-compression rate. In case that the 3D model is rendered on the client at stable FPS, power consumption rises with the number of triangles because every triangle needs some CPU instructions to be processed. Although we have performed measurements with only three different sizes of models, results show that we can expect power consumption to grow linearly with the number of rendered triangles.

The measurements demonstrate that the video-streaming power consumption is about twice that of the rendering power consumption. A typical 1340 mAh/3.7 V battery can supply 260 minutes of video streaming or 460 minutes of rendering of a high-detail (2000 triangles) scene.

Mobile device energy-efficiency computational tradeoff is set to have a continuously improving trend, as reported in (Koommey, Berard, Sanchez, & Wong, 2009), number of computations per kWh is doubling approximately every 1.6 years, which is the long-term industry trend. Therefore,

Figure 6. A synthetized word "Recently" contains three syllabes (down) and it is visually represented by seven visemes (up). Viseme position in the timeline is set by the speech synthetiser. During the animation process the model mesh blends between adjacent visemes

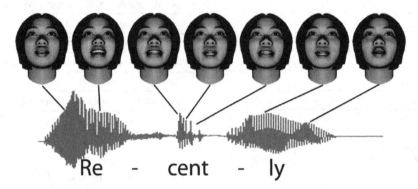

the power needed to perform a task requiring a fixed number of computations will halve every 1.6 years, or the performance of mobile devices will continue to double every 1.6 years while maintaining the same battery lifetime. Mobile wireless interfaces rather follow the same trend due to the vast processing required (Pandzic, & Forchheimer, 2002; Silven & Jyrkka, 2007) and are therefore unlikely to change the above balance favoring more computing on the mobile client instead of network data streaming (see Figure 6).

ARCHITECTURE DISCUSSION AND SELECTION

Different applications and mobile phones have different needs. Hardware performance of mobile devices varies greatly. That is why we decide to support both server and client speech recognition. We prefer server-side speech recognition over the client-side due to the limitation of memory and computational power of present mobile devices. Solutions for speech recognition on mobile phones have lower quality than on servers, which possess more resources and produce more natural speech dialog. Speech recognition is also memory- and CPU-intensive and these resources are required for rendering. However, with future increases of

computing power of mobile devices, we expect this to change in favor of client-side recognition.

Our video-streaming experiments have shown that latency of up to 400 ms may occur between user input and a response from the server. According to this and the power consumption estimates and tests in "Performance measurements", architecture with graphics rendered on the mobile phone appears more convenient and efficient than one with the video streamed.

We prefer and support 3D-rendering and speech synthesis to be performed on the client only. It reduces client power consumption and connection-bandwidth needs, and it is also more flexible in terms of user interaction and animation synchronization. Speech synthesis can be performed with sufficient quality on the more powerful mobile phones.

Therefore, we recommend creating applications with server speech recognition and application logic and client synthesis and graphics rendering (Figure 3).

SYNCHRONIZATION OF FACE ANIMATION WITH SPEECH

The synchronization process is presented in Figure 7. Text is sent to the Text-to-Speech module

Figure 7. Process for generating face animation based on phonemes durations

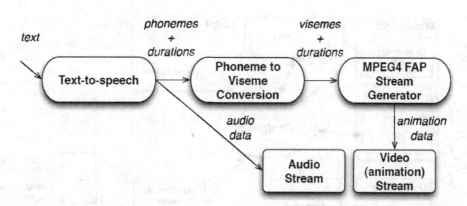

where the synthesis is performed. During the speech-synthesis process information about each generated phoneme and its duration is logged. While the audio wave data, created during the process, do not require any further processing and are directly saved into the audio stream, the logged phonemes and durations are passed to the conversion (Phoneme to Viseme Conversion). This conversion translates every phoneme to the relevant viseme (basic unit of speech in visual domain). Finally, based on the visemes and the timing information (durations), MPEG4 Facial Animation Parameters (FAPs) are generated and saved as animation stream. The synchronization of face and voice is then guaranteed when both streams are played simultaneously.

FRAMEWORK IMPLEMENTATION

On the basis of the above findings we have designed and implemented a platform-independent framework for creating talking-head applications for mobile devices. We have chosen the Qt library (Qt Software, 2008) for the user interface development and as a base for the entire framework for its flexibility and cross-platform portability. The framework is divided into several software modules and components (see Figure 8).

The modules User Interface, 3D Renderer and Multimedia are responsible for interaction with user in both visual and acoustic domain. Rendering of 3D contents is performed by OpenGL ES (Kronous Groups, n. d.) as discussed in "Graphics rendering and streaming".

Face animation is generated and processed by the Face Animation module. We decided to use MPEG4 Facial Animation standard (Pandzic, & Forchheimer, 2002) (MPEG4 FA) for the animation of talking head and for a face-model features description. For that purpose we have modified and optimized the Xface (Balci, K. 2005) library to be able to run on mobile devices and platforms. This library provides an API for the MPEG4-FA-based animation and the tools for the face-model description. The Xface library also contains a parser of the SMIL-Agent (Balci, Not, Zancanaro, & Pianesi, 2007) (Synchronized Multichannel Integration Language for Synthetic Agents) scripting language. It is an XML-based scripting language for creating and animating embodied conversational agents. We use this language for creating dialogues between the user and the talking head. The application is then created by connecting SMIL-Agent scripts into a graph, where the nodes correspond to SMIL-Agent scripts and edges to user decisions (Figure 7).

Speech recognition and synthesis is provided by the Automated Speech Recognition (ASR)

Figure 8. Framework architecture

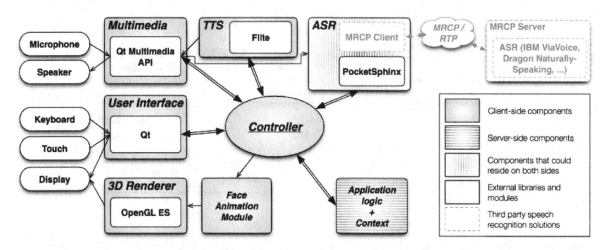

and Text-to-Speech (TTS) components. Both components have universal interfaces so that support for different engines is available via plugins. Our framework has a built-in support for the Flite TTS engine (Black & Lenzo, 2001) and the PocketSphinx ASR engine (Huggins-Daines, Kumar, Chan, Black, Ravishankar, & Rudnicky, 2006). However, support for other engines is feasible with only a little effort. Moreover, the framework also

contains MRCP client for speech recognition, so any existing MRCP server with ASR media support can be used for speech recognition. While the ASR component may reside either on the client or server side, TTS must reside on the client side only, due to necessity of synchronization of face animation and voice.

Application logic and context (e.g. user's session) is handled on the server side. The client

Figure 9. An example of created application – Virtual mobile phone operator. A snippet of our server application logic scripting – decision tree map (up) and corresponding script using XML based SMILAgent (Balci, Not, Zancanaro, & Pianesi, 2007) scripting language (down, simplified)

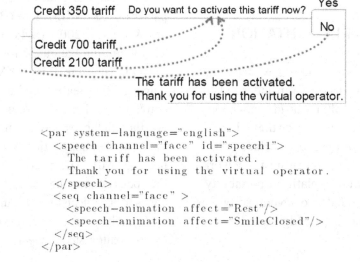

```
<par system-language="english">
    <speech channel="face" id="speech1">
       The tariff has been activated.
       Thank you for using the virtual operator.
    </speech>
    <seq channel="face" >
       <speech-animation affect="Rest"/>
       <speech-animation affect="SmileClosed"/>
    </seq>
</par>
```

communicates with the server using standard HTTP requests and responses. A standard web server is used for that purposes, but instead of HTML output the SMIL-Agent script is used as a response.

Applications created by our framework can run on Windows Mobile, Symbian platforms, desktop Windows, Linux and Mac OS (separate source code compilation for each of the platforms is required). We are currently working on support for the Android, iPhone and MeeGo platforms.

Using our framework we have created two example crossplatform applications. The first is a virtual customer care center and the second is a virtual shop (Figure 7). The applications use talking heads generated by FaceGen and they are capable to render an animated head model with 1466 triangles (Figure 1). The rendering speed of the applications is above 15 FPS (usual mobile video capturing framerate). (Reduction of Animated Models for Embedded Devices, 2010)

CONCLUSION

We demonstrate that as mobile clients are becoming more powerful, real-time rendering of a voice-interactive talking head is within their reach and we may expect a boom in voice-interactive 3D mobile applications in fields like entertainment, commerce, education or virtual assistance. The client-server architecture, rendering and synchronizing the 3D and audio components locally and controlling the logic and speech processing remotely, allows applications to be less power-hungry and improves the quality of virtual-character interaction.

By providing a framework for easy creating of virtual-character based application on mobile phones, we would like to spark future research and application development in the area. It is our intention to make the entire platform openly available in the near future.

Currently mobile-phone speech-application developers have to deal with many platform-depended interfaces. Speech application development can be facilitated by integrating synthesis and recognition libraries to the mobile operating systems. (Currently only Apple iPhone OS and Google Android OS supports native speech synthesis.)

In our future work we plan to do some usability testing of performance, voice recognition accuracy and user emotional response. We would also like to focus on the upcoming Windows Phone 7 operation system that supports both speech synthesis and speech recognition through classes that are also part of .NET Compact Framework 4.0. In the area of distributed architectures we intend to enable easy provisioning of mobile talking-head applications using cloud services. We see the future in such applications because they offer reduced server cost (paid incrementally as utility), better reliability (automated server duplicating), flexibility in computation power and storage space, highly automated server maintenance, scalability and allows software developers to focus more on their core work. The main challenge will likely be portability, as cloud applications have to be in a special form (e.g. .NET managed code for Microsoft Azure) and we expect many difficulties in porting current server applications to the cloud.

ACKNOWLEDGMENT

This research has been partially supported by the Grant Agency of the Czech Technical University in Prague, grant No. SGS10/291/OHK3/3T/13, the research program LC-06008 (Center for Computer Graphics) and by Vodafone Foundation Czech Republic.

REFERENCES

acbPocketSoft (n. d.). *acbTaskMan for PocketPC*. Retrieved from http://www.acbpocketsoft.com

Acquaviva, A., Lattanzi, E., & Bogliolo, A. (2004). Power-aware network swapping for wireless palmtop PCs. *IEEE Transactions on Mobile Computing, 5*(5), 571–582. doi:10.1109/TMC.2006.71

Agarwal, S. K., Chakraborty, D., Kumar, A., Nanavati, A. A., & Rajput, N. (2007). HSTP: Hyperspeech transfer protocol. In *Proceedings of the Eighteenth Conference on Hypertext and Hypermedia* (pp. 67-76). New York, NY: ACM Press.

Agarwal, S. K., Kumar, A., Nanavati, A., & Rajput, N. (2008). The world wide telecom web browser. In *Proceeding of the 17th International Conference on World Wide Web* (pp. 1121-1122). New York, NY: ACM Press.

Albrecht, I., Haber, J., & Seidel, H. (2002). Speech synchronization for physicsbased facial animation. In *Proceedings of the 10th International Conference on Computer Graphics, Visualization, and Computer Vision* (pp. 9-16).

Alexa, M., Berner, U., Hellenschmidt, M., & Rieger, T. (2001). An animation system for user interface agents. In *Proceedings of the 6th International European Conference on Computer Graphics, Visualization, and Computer Vision*.

Balci, K. (2005). Xface: Open source toolkit for creating 3d faces of an embodied conversational agent. In *Proceedings of the International Conference on Smart Graphics* (pp. 263-266).

Balci, K., Not, E., Zancanaro, M., & Pianesi, F. (2007). Xface: Open source project and smil-agent scripting language for creating and animating embodied conversational agents. In *Proceedings of the 15th International Conference on Multimedia* (pp. 1013-1016). New York, NY: ACM Press.

Black, A., & Lenzo, K. (2001). Flite: A small fast run-time synthesis engine. In *Proceedings of the 4th ISCA Tutorial and Research Workshop on Speech Synthesis* (pp. 20-24).

Bowman, D., Coquillart, S., Froehlich, B., Hirose, M., Kitamura, Y., & Kiyokawa, K. (2008). 3D user interfaces: New directions and perspectives. *IEEE Computer Graphics and Applications, 28*(6), 20–36. doi:10.1109/MCG.2008.109

Cassell, J., Vilhjalmsson, H. H., & Bickmore, T. (2001). Beat: The behavior expression animation toolkit. In *Proceedings of the 28th Annual Conference on Computer Graphics and Interactive Techniques* (pp. 477-486). New York, NY: ACM Press.

Choi, S.-M., Kim, Y.-G., Lee, D.-S., Lee, S.-O., & Park, G.-T. (2004). Nonphotorealistic 3-d facial animation on the PDA based on facial expression recognition. In A. Butz, A. Kruger, & P. Olivier (Eds.), *Proceedings of the 4th International Symposium on Smart Graphics* (LNCS 3031, pp. 11-20).

Danihelka, J., Kencl, L., & Zara, J. (2010). Reduction of animated models for embedded devices. In *Proceedings of the 18th International European Conference on Computer Graphics, Visualization, and Computer Vision* (pp. 89-95).

Deng, L., Wang, Y., Wang, K., Acero, A., Hon, H., & Droppo, J. (2004). Speech and language processing for multimodal human-computer interaction. *The Journal of VLSI Signal Processing, 36*(2), 161–187. doi:10.1023/B:VLSI.0000015095.19623.73

Devevey, P., Lorenzon, N., & Tambary, C. (2005). *Measuring wireless energy consumption on PDAs and on laptops* (Tech. Rep. No. 2005). Genoa, Italy: University of Genoa.

Dryer, D. C. (1999). Getting personal with computers: How to design personalities for agents. *Applied Artificial Intelligence, 13*(3), 273–295. doi:10.1080/088395199117423

Hamann, B. (1994). A data reduction scheme for triangulated surfaces. *Computer Aided Geometric Design, 11*(2), 197–214. doi:10.1016/0167-8396(94)90032-9

Huggins-Daines, D., Kumar, M., Chan, A., Black, A., Ravishankar, M., & Rudnicky, A. (2006). Pocketsphinx: A free, real-time continuous speech recognition system for hand-held devices. In *Proceedings of the IEEE International Conference on Acoustics, Speech and Signal Processing* (p. 1). Washington, DC: IEEE Computer Society.

Kadous, M., & Sammut, C. (2002). *Mobile conversational characters. Virtual conversational characters: Applications, methods, and research challenge.* Paper presented at the Joint HF/OZ-CHI Workshop on Human Factors and Human-Computer Interaction, Melbourne, Australia.

Keskin, C., Balci, K., Aran, O., Sankur, B., & Akarun, L. (2007). A multimodal 3D healthcare communication system. In [Washington, DC: IEEE Computer Society.]. *Proceedings of the Conference on, 3DTV*, 1–4.

Kishonti Informatics. (2003). *GL benchmark.* Retrieved from http://glbenchmark.com

Koommey, J. G., Berard, S., Sanchez, M., & Wong, H. (2009). Assessing trends in the electrical efficiency of computation over time. *IEEE Annals of the History of Computing.*

Kronous Groups. (n. d.). *OpenGL ES - the standard for embedded accelerated 3D graphics.* Retrieved from http://www.khronos.org/opengles/

Kshirsagar, S., Magnenat-Thalmann, N., Guye-Vuilleme, A., Thalmann, D., Kamyab, K., & Mamdani, E. (2002). Avatar markup language. In *Proceedings of the Workshop on Virtual Environments*, Aire-la-Ville, Switzerland (pp. 169-177).

Kumar, A., Rajput, N., Chakraborty, D., Agarwal, S. K., & Nanavati, A. A. (2007). WWTW: The world wide telecom web. In *Proceedings of the Workshop on Networked Systems for Developing Regions* (pp. 1-6). New York, NY: ACM Press.

Kunc, L., & Kleindienst, J. (2007). ECAF: Authoring language for embodied conversational agents. In V. Matousek & P. Mautner (Eds.), *Proceedings of the 10th International Conference on Text, Speech, and Dialogue* (LNCS 4629, pp. 206-213).

Kunc, L., Slavik, P., & Kleindienst, J. (2008). Talking head as life blog. In P. Sojka, A. Horak, I. Kopecek, & K. Pala (Eds.), *Proceedings of the 11th International Conference on Text, Speech, and Dialogue* (LNCS 5246, pp. 365-372).

Melax, S. (1998). A simple, fast, and effective polygon reduction algorithm. *Game Developer, 11*, 44–49.

Mochocki, B., Lahiri, K., & Cadambi, S. (2006). Power analysis of mobile 3d graphics. In *Proceedings of the Conference on Design, Automation and Test in Europe*, Leuven, Belgium (pp. 502-507).

Nass, C., Moon, Y., Fogg, B. J., Reeves, B., & Dryer, C. (1995). Can computer personalities be human personalities? In *Proceedings of the Conference Companion on Human Factors in Computing Systems* (pp. 228-229). New York, NY: ACM Press.

Ortiz, A., del Puy Carretero, M., Oyarzun, D., Yanguas, J., Buiza, C., Gonzalez, M., et al. (2007). Elderly users in ambient intelligence: Does an avatar improve the interaction? In C. Stephanidis & M. Pieper (Eds.), *Proceedings of the 9th Conference on User Interfaces for All* (LNCS 4397, pp. 99-114).

Pandzic, I. S. (2002). Facial animation framework for the web and mobile platforms. In *Proceedings of the Seventh International Conference on 3D Web Technology* (pp. 27-34). New York, NY: ACM Press.

Pandzic, I. S., Ahlberg, J., Wzorek, M., Rudol, P., & Mosmondor, M. (2003). *Faces everywhere: Towards ubiquitous production and delivery of face animation.* Paper presented at the 2nd International Conference on Mobile and Ubiquitous Multimedia, Norrkoping, Sweden.

Pandzic, I. S., & Forchheimer, R. (2002). *Mpeg-4 facial animation: The standard, implementation and applications* (pp. 15–61). Chichester, UK: John Wiley & Sons. doi:10.1002/0470854626. part2

Pentikousis, K. (2010). In search of energy-effcient mobile networking. *IEEE Communications Magazine*, *48*(1), 95–103. doi:10.1109/MCOM.2010.5394036

Poller, P., & Muller, J. (2002). Distributed audio-visual speech synchronization. In *Proceedings of the Seventh International Conference on Spoken Language Processing* (pp. 205-208).

Qiu, L., & Benbasat, I. (2005). An investigation into the effects of text-to-speech voice and 3D avatars on the perception of presence and flow of live help in electronic commerce. *ACM Transactions on Computer-Human Interaction*, *12*(4), 329–355. doi:10.1145/1121112.1121113

Qt Software. (2008). *Qt cross-platform application framework.* Retrieved from http://qt.nokia.com/products

Ramakrishnan, I. V., Stent, A., & Yang, G. (2004). Hearsay: Enabling audio browsing on hypertext content. In *Proceedings of the 13th International Conference on World Wide* Web (pp. 80-89). New York, NY: ACM Press.

Reddy, M. (2003). SCROOGE: Perceptually-driven polygon reduction. *Computer Graphics Forum*, *15*(4), 191–203. doi:10.1111/1467-8659.1540191

Shrestha, S. (2007). Mobile web browsing: Usability study. In *Proceedings of the 4th International Conference on Mobile Technology, Applications, and Systems and the 1st International Symposium on Computer Human Interaction in Mobile Technology* (pp. 187-194). New York, NY: ACM Press.

Silven, O., & Jyrkka, K. (2007). Observations on power-effciency trends in mobile communication devices. *EURASIP Journal on Embedded Systems*, (1): 17.

Singular Inversion. (2010). *FaceGen.* Retrieved from http://www.facegen.com

Sun, Z., Stent, A., & Ramakrishnan, I. V. (2006). Dialog generation for voice browsing. In *Proceedings of the International Crossdisciplinary Workshop on Web Accessibility* (pp. 49-56). New York, NY: ACM Press.

Wagner, D., Billinghurst, M., & Schmalstieg, D. (2006). How real should virtual characters be? In *Proceedings of the ACM SIGCHI International Conference on Advances in Computer Entertainment Technology* (p. 57). New York, NY: ACM Press.

Wang, A., Emmi, M., & Faloutsos, P. (2007). Assembling an expressive facial animation system. In *Proceedings of the ACM SIGGRAPH Symposium on Video* Games (pp. 21-26). New York, NY: ACM Press.

Waters, K., & Levergood, T. (1993). *DECface: An automatic lipsynchronization algorithm for synthetic faces* (Tech. Rep. No. 93/4). Cambridge, MA: Cambridge Research Laboratory.

Yin, M., & Zhai, S. (2006). The benefits of augmenting telephone voice menu navigation with visual browsing and search. In *Proceedings of the SIGCHI Conference on Human Factors in Computing Systems* (pp. 319-328). New York, NY: ACM Press.

This work was previously published in the International Journal of Mobile Human Computer Interaction, Volume 3, Issue 2, edited by Joanna Lumsden, pp.50-64, copyright 2011 by IGI Publishing (an imprint of IGI Global).

Chapter 9
feelabuzz:
Direct Tactile Communication
with Mobile Phones

Christian Leichsenring
Bielefeld University, Germany

René Tünnermann
Bielefeld University, Germany

Thomas Hermann
Bielefeld University, Germany

ABSTRACT

Touch can create a feeling of intimacy and connectedness. This work proposes feelabuzz, a system to transmit movements of one mobile phone to the vibration actuator of another one. This is done in a direct, non-abstract way, without the use of pattern recognition techniques in order not to destroy the feel for the other. The tactile channel enables direct communication, i. e. what another person explicitly signals, as well as implicit context communication, the complex movements any activity consists of or even those that are produced by the environment. This paper explores the potential of this approach, presents the mapping use and discusses further possible development beyond the existing prototype to enable a large-scale user study.

INTRODUCTION

Touch is arguably the most immediate, the most affective, and – when it comes to media – one of the most overlooked modalities used for human communication. It can convey emotions and feelings on a direct and primordial level (Eichhorn,

Wettach, & Hornecker, 2008; Heikkinen, Olsson, & Vaananen-Vainio-Mattila, 2009; Vetere et al., 2005).

We propose feelabuzz – a system to directly transform one user's motion into the vibrotactile output of another, typically remote device. Unlike previous work on tactile communication (Chang,

DOI: 10.4018/978-1-4666-2068-1.ch009

O'Modhrain, Jacob, Gunther, & Ishii, 2002), we do so using only mobile phones without any additional gear. Nowadays most mobile phones universally have both accelerometers for the sensing and vibration motors for the actuation of the interaction. Mobile phones have the key advantages of not only being widespread to the point of omnipresence but also to usually be in the direct vicinity of their users. Not having to buy and more importantly to carry around an extra piece of hardware is a property whose importance cannot be overstated. Using phones also makes it easy to integrate the new haptic channel with existing auditory, visual and maybe textual channels, thereby extending the phone's capabilities as a communication device. As we have our phones with us or nearby most of the time, they are well suited not only for direct communication but also for implicit context communication (e. g. walking or riding the bus) as well. The choice of vibration as an output modality not merely stems from its prevalence on the chosen platform and its availability and unobtrusiveness when carrying the phone in a pocket but also from the fact that movement naturally transforms into vibration and similar tactile feedback in the real world (e. g. footsteps on the floor, multiple persons using one stair rail, someone stirring on a sofa or even the feedback to one's own hand when stroking something).

Related Work

Similar approaches have been followed by others. The work of Heikkinen, Olsson, and Vaananen-Vainio-Mattila (2009) provides insights on the expectations of users regarding haptic interaction with mobile devices. Their results underline our design considerations. The participants brought up poking and knocking metaphors as well as the idea of a constantly open "hotline" between two participants. Their participants even saw the possibility of the emergence of a haptic symbolism or primitive language, which have been developed during the evolution of the interaction.

O'Brien and Mueller (2006) created special devices of various forms to examine the needs of couples when "holding hands over a distance". A main critique of their participants was concerned about the cumbersome and unfashionable design of their devices: "The participants stressed how they wanted a device that was more personal and easy to carry. They desired it to be small enough to fit it in their pocket. One participant noted that she wanted something she could relate to personally". Furthermore, their users disliked that the special device draw to much public attention.

Eichhorn et al. build a pair of stroking devices for separated couples. Each device has a sensor and a servo which expresses the stroke initiated by the remote device. The device functions as a proxy object to stroke each other over a distance. A lot of the work already conducted on vibrotactile interaction is focused either on the recognition of haptic gestures or on mapping different cues to haptic stimuli (Murray-Smith, Ramsay, Garrod, Jackson, & Musizza, 2007; Rovers & van Essen, 2004; Brewster & Brown, 2004; Mathew 2005; Enriquez & MacLean, 2003).

To our knowledge there is no practical work on direct mapping between the accelerometer readings and the vibration motor of a mobile phone. With feelabuzz we aim at creating a personal, lightweight and always ready-to-hand haptic communication channel. In this work we will first discuss aspects of haptic communication and then introduce the feelabuzz system.

The Challenge

Albeit the vibrations today's mobile phones can make are a poor substitute for the actual touch of another person, we believe that the knowledge that it is the very movement the other person is doing just now that makes a user's phone vibrate in a certain way can give them a real feeling of presence and intimacy. Imagine how a piece of clothes our loved one once wore or a letter or the place he or she used to sit can make us feel just because he or she touched it. Often, the fewer images there are, the more powerful the images our mind will conjure up. Instead of transmitting reality with as much sensory bandwidth as possible, we intend to give people something to build upon and depend on their minds to add in all the details.

Still, how much is there to really hook on to? Figure 1 shows accelerometer data for different activities. It is not necessary to be an expert to distinguish these four sample activities. We are optimistic though that people will become experts in the sense that they will learn even to pick up the comparatively subtle cues that separate the way of movement of close persons from everyone else's way. Provided that a strong social tie is a profound enough motivation to train their sense of touch to achieve this, we think the actual sensitivity of touch is often underestimated (TED, 2007).

We also included Figure 1d, showing a tapping with the palm on a mobile phone resting in one's pocket, because we think that people will become quite creative once given such a straightforward tool as feelabuzz and might for example develop their own signals to quickly inform someone about things without even taking their phones out of their pockets. In this example, you can clearly see that the sensor had been tapped on first four times, then three times, then twice and finally once.

The challenge now is to find a mapping from acceleration data to vibration output that makes it as easy for the users to discern these patterns with their skin as it is to tell them apart visually in the graphs and to deduce the underlying activity in an intuitive way, relying as much on pre-existing world knowledge on part of the user as much as possible.

Concepts

The information conveyed by feelabuzz can be split into two parts that we call direct communication and implicit context communication.

Direct Communication

Providing users with the possibility to intuitively induce tactile feedback in another person's mobile phone presents a new communication channel that can be used in many ways. The channel's possibilities for readily understood signals are limited though. Apart from knocking to do simple things such as requesting attention, synchronizing or timing pre-decided behavior, or giving short binary feedback, few intentional tactile communication events will be understood by the naive user. Although there are sophisticated means of communication through such narrow channels, most notably Morse code, we expect that to be employed only by experts and not to become widespread. Instead, we rely on people's ability to develop their own adapted communication strategies using a mixture of implicit and explicit negotiation. Quite complex and effective communication systems can emerge via such mechanisms (Galantucci, 2005; Healey, Swoboda, Umata & King, 2007; Goldin-Meadow & Mylander, 1998; Senghas, Kita, & Ozyurek, 2004; Kegel, 1994; Bickerton, 1981).

Implicit Context Communication

The other large sector of information that is conveyed by feelabuzz are the unintentional and implicit movements of the device. These can either originate from the users or from the environment, as already proposed by (Murray-Smith, Ramsay, Garrod, Jackson, & Musizza, 2007)

The time-series data in Figure 1 show that different kinds of activities by the users themselves lead to very different acceleration profiles. Like-wise, sitting in a driving vehicle will lead to an acceleration pattern that is notably different from those caused by human movements.

Note that none of this has to be detected by pattern recognition software. There are no predefined classes. Instead, the interpretation of many movement patterns is expected to come quite naturally

Figure 1. Accelerometer data of different movements recorded with 100 Hz (a) running (b) relaxing (c) swinging(d) hand tapping on the phone (Figures1a through 1d)

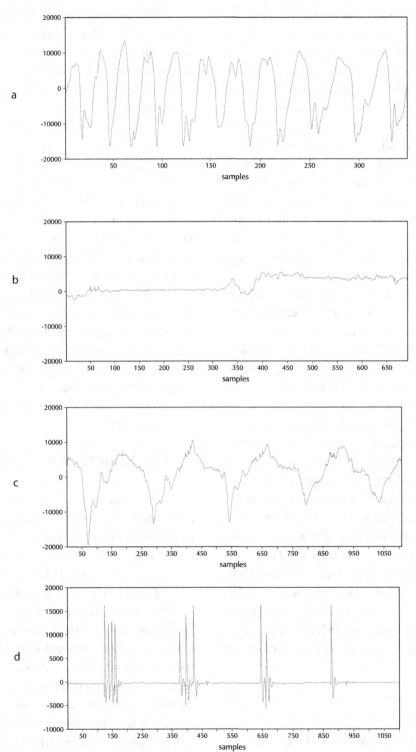

and involve all the rich context information and world knowledge humans have. Additionally, the sophistication of the interpretations can fluently increase with the user experience. As there are rarely clear class boundaries in the real world, transitions between different types of movement can be perceived in all their ambiguity and fuzziness in a near analogue fashion without the need to make clear distinctions. While regression models could do so as well, the subsequent mapping back to artificial vibrotactile stimuli in a way that allows intuitive access as well as in-depth learning of subtle features would be a major challenge to say the least. Actually one would have to know and reliably detect any such subtlety in advance before playing it back to a user in an alienated way. Relying on the human's long-evolved ability to interpret rich real-world data streams seems to be a more promising way in terms of effectiveness and a much more interesting way in terms of unintended uses and exploration by future users.

Implementation

Signal Processing and Vibrotactile Mapping

To map the S accelerometer readings s(t) with $s_i(t) \in [0, smax]$, $1 \leq i \leq S$ to the vibration module input value $y(t) \in [0, ymax]$ we perform a couple of steps.[1] First we compute the magnitude of the vector of sensor values:

$$m(t) = \rho \|s(t)\| = \rho \sqrt{\sum_{i=1}^{S} s_i(t)^2} \qquad (1)$$

with ρ being a normalization factor:

$$\rho = \frac{y\max}{\sqrt{S s_{\max}^2}} \qquad (2)$$

Now an RC high-pass filter is applied to the sensor values with the decay constant $\alpha h = 0.99$

$$b_h(t) = \alpha_h(b_h(t-1) + (m(t) - m(t-1))) \qquad (3)$$

which gets rid of the gravitational acceleration and other constant or long-term acceleration influences without losing as much inertia as a simple derivation would.

Subsequently, an exponential smoothing is applied with smoothing factor $\alpha l = 0.05$:

$$b_l(t) = \alpha_l \left| b_h(t) \right| + (1 - \alpha_l) b_l(t-1) \qquad (4)$$

It is important to give more inertia to the system in a controlled way so that a lot of activity from the sender will add up to give an increasingly strong signal on the receiving end (Figure 3). This turned out to be what best matched our intuitive a-priori expectations of how the system should behave.

It has the drawback of leveling out all of the more impulse-like parts of the signal which are a salient feature and also quite important for signaling. To preserve these impulse components as well, we add them back in with a simple kind of spike detection.

For this we compute the moving average over the last n time steps, defined for any function x(t) as

$$MA_n(x, t) = \frac{1}{n} \sum_{i=0}^{n-1} x(t-i) \qquad (5)$$

and check if the high-pass-filtered signal $b_h(t)$ exceeds a certain threshold of $\beta_a = 3$ times the moving average. If this is the case we perform an exponential mapping of the spike signal and add it back to the low-pass-filtered signal with the adjusting coefficients $\beta_{bh} = 2$ and $\beta_{bl} = 3$:

Figure 2. Step response to the rectangular signal m(t)

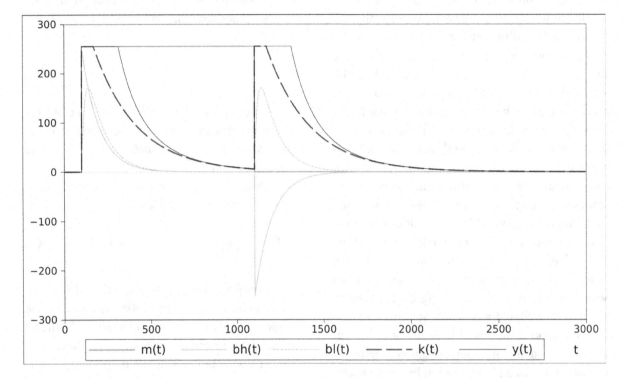

$$k(t) = \begin{cases} y\max\left(\dfrac{\beta_{b_h}b_h(t)}{y\max}\right)^{\alpha_e} & \text{if } b_h(t) > \beta_a MA_n(b_h, t) \\ 0 & \text{else.} \end{cases}$$

(6)

$$y(t) = \min(k(t) + \beta_{b_l}b_l(t), y\max)$$

(7)

with n = 5 and αe =0.4. Finally, the output is cropped to ymax. For our prototype platform, the Neo FreeRunner (see Section 5.2), we noticed that values of y(t) < 30 are not noticeable so we set them to 0 to not unnecessarily strain the battery (not included in Equation 7 and Figure 3).

Figures 2 and 3 show the behaviour of these steps combined. The step response in Figure 2 is shown with the intermediary steps $b_h(t)$, $b_l(t)$ and k(t). The high-and low-pass characteristics of $b_h(t)$ and $b_l(t)$ can clearly be seen. The output signal y(t) subsequently shows an immediate response

as well as a strong inertia that can be configured independently from $b_h(t)$. y(t) and k(t) are clipped to ymax which distorts their shape.

In Figure 3 a burst of delta pulses increasingly excites the system and this excitation takes a comparatively long time to wear off. At the same time, the pulses themselves are perfectly preserved and amplified. They are also clearly high-pass filtered as made apparent by the downward spikes that help them stand out in noisier signals.

Technology

The feelabuzz prototype hardware consists basically of two paired mobile phones. On the phones we gather the accelerometer data which are then preprocessed, transmitted and mapped to the vibrotactile actuator. Our prototype system was developed using the Neo FreeRunner devices (GTA2) (Moss-Pultz et al., 2010) running Openmoko SHR (Moss-Pultz & Chen, 2010). The communication is transmitted over two direct Open

Figure 3. Filter response y(t) to a burst of delta pulses m(t)

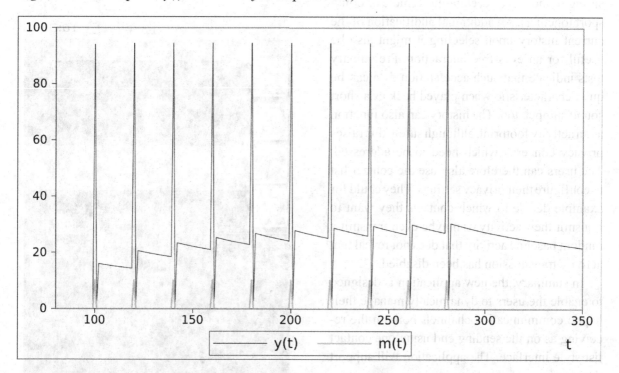

Sound Control (OSC) (Wright & Freed, 1997) connections between the paired devices. OSC is a UDP-based simple push protocol which is widely available in common programming languages. On the device itself we are using the Python programming language to acquire the sensor data and preprocess it, connect the devices over the network and then excite the vibration motor. The prototype showed that different activities as illustrated in Figure 1 could be distinguished. The tapping was recognizable as well as other activities that could well be separated from each other. To see whether these findings hold true for a wider range of users with different backgrounds, we are working on a large-scale user study. For conducting this evaluation we need to distribute feelabuzz as a conveniently downloadable application on a common mobile platform. The Neo Freerunner devices served well for the prototype system as they provide a very accessible platform, but they are already quite out-dated with respects to their hardware and not very common these days and

therefore not well-suited for the evaluation. The Android (Google, Inc., 2010) platform on the other hand seems to be the perfect match for our needs: it is widespread, it features multitasking and an Application Store (Google, Inc., 2010) for the easy distribution of the software.

We designed a concept for a new application which we are working on right now.

Application Concept

A sketch of the future interface can be seen in Figure 4. The application shows a short activity level history from all available contacts. The contact overview serves as a visual representation of the users' accelerations. Thereby a user can quickly see which contacts show interesting activity patterns at any time. The visual overview also serves as a chat room in which all users can simultaneously see and react to each other's activity patterns. It would also be easy to discern similar acceleration patterns arising from joint activities (e. g.

sitting in the same accelerating vehicle or doing sport together). An additional audification of the current history upon selecting it might also be useful for an eyes-free interaction. Preliminary tests indicate that such acceleration data can be quite characteristic when played back as a short sound snippet, too. The history can also function as an activity footprint, although surely this raises privacy concerns which need to be addressed. The users can therefore also use the contact list to configure their privacy settings. They could for example decide to which contacts they want to transmit their activity or maybe even transmit a random baseline activity that does not reveal that activity transmission has been disabled.

In summary, the new application is designed to enable the users to dynamically manage their haptic communication channels both on the receiving as on the sending end using one contact list style interface. The application will support the user in:

- Manage contacts and permissions
- Provide an activity overview
- Provide a haptic chat room

This among other things calls for a new network infrastructure. XMPP/TCP (Saint-Andre, 2004) is used as the transport layer for the communication between the devices. XMPP provides most of the needed features (such as dynamically connecting the users' phones, managing their contacts and sessions). More functionality is provided by XMPP's publish/subscribe extension (Millard & Saint-Andre, 2010). XMPP also has other advantages over OSC. The current implementation depends on a network connection without too much fluctuation in the amount of delay because each sensor reading is immediately transmitted and haptically displayed. Any jitter in the amount of lag will thus translate into stretching and contracting of the activity curve along the time axis. A fixed delay probably won't be that much of a problem as it only becomes noticeable in a decreased

Figure 4. Concept of a feelabuzz Android app

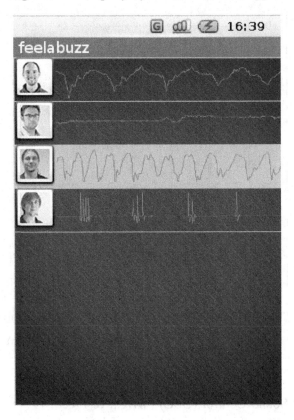

reaction time or when synchronizing with other communication channels. The obvious fix for the distortion problem will therefore be a buffering of a certain amount of data before sending it in packets. This will introduce a certain amount of extra lag but on the other hand will compensate for any fluctuations up to this deliberate delay. It will also have the effect of reducing the network load by vastly reducing the amount of data packets that have to be sent.

CONCLUSION

We presented the concept and a prototype of a near-analogue coupling of the accelerometers built into modern mobile phones to the likewise included vibration motors of a remote device to create a feeling of connectedness over a distance.

We described a mapping that can recognizably transmit such acceleration data and implemented it on a pair of prototype phones. By adapting the sensitivity of the system to the signal's energy level, even minute movements can be transmitted without saturating the output too quickly. Our prototype algorithm can even transmit small rotational movements of the device. Finding a good adaptive algorithm that can fade between the sensitive and the current mode is a very interesting perspective. Furthermore we presented the concept of the future application which will serve as a basis for our large scale evaluation. Enhancing the usability of the interface and enabling a user's whole social network to be haptically interconnected is another important step. To support users in the setup of the haptic communication channel we designed a user interface which provides a short visual history of the motion of their contacts (Figure 4). Contacts can be tuned into by just selecting them from the list. The design envisions either this one-to-one or a many-to-many mapping similar to a group chat. The above-mentioned move to the XMPP infrastructure will support the dynamic management of the haptic communication channels. Furthermore, XMPP's central routing is needed to connect users in different networks and behind NAT routers which are quite common at home. The aim is to get to run feelabuzz on many users' own phones by providing an improved application for download. This will not only make it possible to put the evaluation of our method on a broad basis but also to collect experiences with haptic communication channels in general with an omnipresent device to which the subjects can personally relate and which accompanies them in their daily life.

REFERENCES

Bickerton, D. (1981). *Roots of language*. Ann Arbor, MI: Karoma Publishers.

Brewster, S., & Brown, L. M. (2004) Tactons: Structured tactile messages for non-visual information display. In *Proceedings of the Fifth Conference on Australasian User Interface*, Darlinghurst, Australia (pp. 15-23).

Chang, A., O'Modhrain, S., Jacob, R., Gunther, E., & Ishii, H. (2002). Comtouch: Design of a vibrotactile communication device. In *Proceedings of the 4th Conference on Designing Interactive Systems* (pp. 312-320). New York, NY: ACM Press.

Eichhorn, E., Wettach, R., & Hornecker, E. (2008). A stroking device for spatially separated couples. In *Proceedings of the 10th International Conference on Human Computer Interaction with Mobile Devices and Services* (pp. 303-306). New York, NY: ACM Press.

Enriquez, M. J., & MacLean, K. E. (2003). The hapticon editor: A tool in support of haptic communication research. In *Proceedings of the 11th International Symposium on Haptic Interfaces for Virtual Environment and Teleoperator Systems* (pp. 356-362). Washington, DC: IEEE Computer Society.

Galantucci, B. (2005). An experimental study of the emergence of human communication systems. *Cognitive Science: A Multidisciplinary Journal, 29*(5), 737-767.

Goldin-Meadow, S., & Mylander, C. (1998). Spontaneous sign systems created by deaf children in two cultures. *Nature, 391*, 279–280. doi:10.1038/34646

Google, Inc. (2010) *Android*. Retrieved from http://www.android.com

Google, Inc. (2010) *Android market*. Retrieved from http://www.android.com/market

Healey, P., Swoboda, N., Umata, I., & King, J. (2007). Graphical language games: Interactional constraints on representational form. *Cognitive Science: A Multidisciplinary Journal, 31*(2), 285-309.

Heikkinen, J., Olsson, T., & Vaananen-Vainio-Mattila, K. (2009). Expectations for user experience in haptic communication with mobile devices. In *Proceedings of the 11th International Conference on Human-Computer Interaction with Mobile Devices and Services* (pp. 1-10). New York, NY: ACM Press.

Kegl, J. (1994). The Nicaraguan sign language project: An overview. *Signpost, 7*(1), 24–31.

Mathew, D. (2005) *vSmileys: Imaging emotions through vibration patterns.* Paper presented at the Forum on Alternative Access: Feelings & Games.

Millard, P., & Saint-Andre, P. (2010). *XEP-060: Publish-subscribe.* Retrieved from http://xmpp.org/extensions/xep-0060.html

Moss-Pultz, S., & Chen, T. (2010). *Openmoko.* Retrieved from http://www.openmoko.com/

Moss-Pultz, S., Welte, H., Lauer, M., Almesberg, W., Hung, T., Lai, W., et al. (2010). *Freerunner.* Retrieved from http://www.openmoko.com/freerunner.html

Murray-Smith, R., Ramsay, A., Garrod, S., Jackson, M., & Musizza, B. (2007). Gait alignment in mobile phone conversations. In *Proceedings of the 9th International Conference on Human Computer Interaction with Mobile Devices and Services* (pp. 214-221). New York, NY: ACM Press.

O'Brien, S., & Mueller, F. F. (2006). Holding hands over a distance: Technology probes in an intimate, mobile context. In *Proceedings of the 18th Australia Conference on Computer-Human Interaction* (pp. 293-296). New York, NY: ACM Press.

Rovers, A., & van Essen, H. (2004). HIM: A framework for haptic instant messaging. In *Proceedings of the International Conference on Extended Abstracts on Human factors in Computing Systems* (pp. 1313-1316). New York, NY: ACM Press.

Saint-Andre, P. (2004). *Extensible messaging and presence protocol (XMPP): Core.* Retrieved from http://datatracker.ietf.org/doc/rfc3920/

Senghas, A., Kita, S., & Ozyurek, A. (2004). Children creating core properties of language: Evidence from an emerging sign language in Nicaragua. *Science, 305*(5691), 1779–1782. doi:10.1126/science.1100199

TED. (2007). *Evelyn Glennie shows how to listen.* Retrieved from http://www.ted.com/

Vetere, F., Gibbs, M. R., Kjeldskov, J., Howard, S., Mueller, F. F., Pedell, S., et al. (2005). Mediating intimacy: Designing technologies to support strong-tie relationships. In *Proceedings of the SIGCHI Conference on Human factors in Computing Systems* (pp. 471-480). New York, NY: ACM Press.

Wright, M., & Freed, A. (1997). Open sound control: A new protocol for communicating with sound synthesizers. In *Proceedings of the International Computer Music Conference* (pp. 101-104).

ENDNOTE

[1] For the Neo FreeRunner, our prototype hardware, the number of sensors S can be either 3 or 6, smax is 2268 and ymax = 255. The sensor sampling rate was set to 100 Hz.

This work was previously published in the International Journal of Mobile Human Computer Interaction, Volume 3, Issue 2, edited by Joanna Lumsden, pp.65-74, copyright 2011 by IGI Publishing (an imprint of IGI Global).

Chapter 10
Human–Centered Design for Development

Hendrik Knoche
EPFL IC LDM, Switzerland

PR Sheshagiri Rao
CK Trust, India

Jeffrey Huang
EPFL IC LDM, Switzerland

ABSTRACT

This paper describes the challenges faced in ICTD by reviewing the lessons learned from a project geared at improving the livelihood of marginal farmers in India through wireless sensor networks. Insufficient user participation, lack of attention to user needs, and a primary focus on technology in the design process led to unconvinced target users who were not interested in the new technology. The authors discuss benefits that ICTD can reap from incorporating human-centered design (HCD) principles such as holistic user involvement and prototypes to get buy-in from target users and foster support from other stakeholders and NGOs. The study's findings suggest that HCD artifacts can act as boundary objects for the different internal and external actors in development projects.

INTRODUCTION

Rain-fed farming provides the bulk of the world's food supply and has tremendous potential to increase its productivity to meet the 2015 hunger reduction target of the Millenium Development Goal (MDG) (Trisorio-Liuzzi & Hamdy, 2008). Changes and innovations are needed in land, water and crop management but the efforts required

achieving this need to focus on increasing human and institutional capacity, build knowledge and improve management and infrastructure (Trisorio-Liuzzi & Hamdy, 2008). These changes should also improve the livelihoods of rainfed farmers' who have not participated in the economic booms of the last two decades. How will farmers adopt innovations and which role can ICT play in this?

DOI: 10.4018/978-1-4666-2068-1.ch010

With the mobile phone as ready-at-hand platform a number of commercial agricultural information services have been recently launched but it is unclear whether these services will be adopted and whether they will foster further adoption of innovations in land, water and crop management. Decision support systems (DSS) that can help farmers make decisions on the poorly understood complex interactions of soil moisture, seeds, fertilizers and pesticides. Crop-soil simulation model based DSS have not seen much uptake by farmers in developed countries (Stephens & Middleton, 2002) let alone developed countries (Matthews & Stephens, 2002). They seem a poor fit for the problems that farmers need to solve (Stone & Hochman, 2004). However, most reports point to the absence of user involvement in the different stages of designing these services, which calls for research and development guided by human-centered design. Wireless sensor networks (WSN) that could help reduce the effort required to gather environmental data from the field to feed DSS are slowly maturing but have so far not proven their usefulness in this context. This makes the adoption of DSS in agriculture by resource-poor farmers in developing countries a challenge on various levels.

ICTD efforts that are part of a development projects have many goals. Funding organizations demand measurable results such as sustainability, i.e. continued benefits to the target population after the projects end, to better justify allocation of funds. The scope is typically much larger than in HCI studies or in HCD. Uptake and continued use is usually outside the scope of these research areas and left to industrial players. The same goes for standard desirable development outcomes such as local empowerment and capacity building along with policy implementation. Involving target users in participatory workshops or design activities represents another challenge as scientists do not have the trust of the rural population and trusted intermediaries such as NGOs can be skeptical whether engaging in joint activities with

the researchers is worthwhile. We will present the lessons learned from the case study of an ICTD project, which aimed at improving the livelihood of resource-poor farmers but failed to interest them. The follow-up project relies on an HCD approach and focuses on getting buy-in from target users through iterative prototyping of applications. This will enable the farmers to envision using novel agricultural services on affordable mobile phones and in longitudinal studies they will be able to experience the value of these services themselves.

In the following section we provide background on innovation diffusion with a focus on agricultural contexts, the modes of farmer involvement in research, a wave of novel agricultural information services and the up-to-now disappointing adoption of DSS in agricultural contexts. Section 3 reviews the approach taken in an ICTD project developing a WSN-based DSS and describes the problems encountered. We then discuss the value of user evaluation of early prototypes and the use of mock-ups, storyboards, human subject consent forms and other HCD artifacts as boundary objects that provide can convince stakeholders, NGOs and other actors and present some wider ranging conclusions for ICT4D. The outlook section presents our revised approach to the second phase of the project.

BACKGROUND

Rain-fed farming produced the bulk of the world's food and generated 62% of the world's staple food (FAO, 2005). In 2009 agriculture employed over 240 million people in India – 52% of the workforce (CIA, 2009) – many on small landholdings. In the province of Karnataka the size of the farms of 87% of farming families was less than four hectares (Barker & Molle, 2004). The share of these small farms accounts for 50% of the total cultivated area, that of marginal farmers (less than one hectare) 39%. Marginal farmers in India have profited comparatively little from the economic boom and poverty reduction of the last two decades (Basu & Srivastava, 2005). This is mainly due to their reliance on rain, which is uncertain and occurs with great variance. However, the main gains in agricultural productivity lie in rain-fed farming and therefore solutions are sought as to how help bring these about.

Below we present how agricultural development through the extension model was a large success partly due to its reliance on an established trust relationship with farmers. Trust is time consuming to establish and many projects rely on established and trusted intermediaries such as NGOs or on the long-time involvement of researchers (Bentley, 1994). We present how the modes of farmer participation in research process are varied and the scope of participatory approaches is wider in development studies than in HCD. The section finishes with an overview of current agricultural ICTD initiatives and the dismal adoption of agricultural decision support systems due, to a large extent, to poor user involvement.

Development

Diffusion of agricultural innovations has a long tradition that predates HCI research by half a century. The US introduced extension services in 1914 to "*relay useful and practical information on subjects relating to agriculture and home economics*" and has been encouraging people to apply it. The success of the US model of agricultural extension and its innovation-development process was largely attributed to the fact that 50% of the funding was targeted at diffusion activities. The local-level extension agents not only brought innovations tested by research universities to the farmers but also gathered feedback and tried to understand their needs. This information was fed back to the agricultural agency, thereby fostering organizational learning and change. For example, the initial focus on increasing production was extended to include farmers' information and entertainment needs. In spite of mass media pro-

liferation a ratio of one extension worker to 100 farmers was maintained and, according to Rogers, much of the uptake of innovations could be attributed to the trust relationship they had with the farmers.

In terms of farmers' inclusion in the research process Biggs classified four modes of participation: contractual, consultative, collaborative and collegiate (Biggs, 1989). In the contractual mode the farmer's involvement is similar to that of a paid participant in a typical HCI study. The farmer acts and gets remunerated as a service provider of land, resources or services to the research project. The consultative mode follows a doctor-patient relationship in which the researchers try to elicit problems and suggest possible solutions to the farmer. In the collaborative mode the role of the farmers is more emancipated as they engage in continuous collaborations with the researchers as partners in the research process. This goes beyond typical user involvement in participatory designs in which participants rarely have ownership of the object of research and its insights and therefore obtain no direct benefit from the on-going research. In the collegial mode the researchers actively encourage the farmers to pursue research and development in rural areas.

In development parlance the *activities* carried out within the timeframe of the project produce *outputs* which ideally should result in *outcomes* in the mid-term (after the project has ended) and have long-lasting *impact* such as structural changes (Swiss Agency for Development and Cooperation SDC, 2008). During the 1980s, participatory approaches such as participatory rural appraisal (PRA) became popular with donors and organization active in the development domain as a response to the ineffectiveness of externally imposed, top-down, expert-oriented approaches (Cooke & Kothari, 2001). Research in development distinguishes between efficiency and empowerment arguments as justifications for the use of participatory approaches. For the former, participatory approaches represent a tool to achieve better project outcomes – identical to their purpose in HCD - whereas the latter views participation as a process that increases the capacities of participants from marginal groups to improve their livelihoods (Cleaver, 1999), which is outside the scope of HCD. Among development scholars, there are calls for a critical review of the benefits and limits of participatory approaches based on the limitations of its methods and tools and its theoretical, political and conceptual limitations (Cooke & Kothari, 2001).

ICTD

A growing number of information services have been recently trialed or launched to satisfy the information needs of farmers in India: aAqua mini (Bahuman & Kirthi, 2007), mKrishi (Horvath, 2008), Reuters Market Light (RML) (Mehra, 2007), IFFCO Kisan Sanchar (Awasthi, 2008), and Nokia's Life Tools (Nokia Siemens Networks, Nokia and Commonwealth Telecommunications Organization (CTO), 2008) and CERES (Anurag, Vivek, & Sasank, 2008) - see Rao and Sonar (2009) for a detailed comparison of these services. The provisioning of market prices for crops and agricultural inputs and the availability of the latter is a focus of many of these services. So far, research has found that the provisioning of market prices and news through mobile devices is of little interest to the farmers (Blattman, Jensen, & Roman, 2003; Bahuman & Ramamritham, 2010). Advice for the prevention, diagnosis and control of pests and diseases features high in farmers' needs but is not easy to implement in an automated way. Current services such as mKrishi and aAqua are human expert based and might not scale well to large numbers of users. Agrocom's (2008) real-time disease alert service through local weather stations in farmers' orchards proved financially unsustainable due to the high cost of the weather stations.

A number of ICTD projects have aimed at improving rural communication and knowledge building through solutions for illiterate users, e.g. audio wikis (Kotkar, Thies, & Amarasinghe, 2008), discussion forums that extend existing mass media coverage such as community radios (Patel, Chittamuru, Jain, Dave, & Parikh, 2010), and spoken web interfaces for user generated content (Kumar, Agarwal, & Manwani, 2010). Information uptake can benefit from the fact that speech can provide proof of relevance and trustworthiness of the information; for example, if the voice is that of a known expert (Patel et al., 2010) or is attributed expert status by a trusted entity (Sherwani et al., 2009).

Decision Support Systems in Agriculture

For a number of crop varieties, agricultural research has come up with simulation models that can be harnessed to predict growth and yield. Their accuracy depends on the availability of environmental parameters such as, e.g. soil type, soil moisture and temperature, ideally over the lifetime of the crop. Crop simulation models have been used in some agricultural decision support systems (DSS) but this research community was largely unfamiliar with diffusion theory and participatory methods commonly found in HCI and HCD research. Reviews of decision support systems in agriculture in general (Cox, 1996; Newman, Lynch, & Plummer, 2000) and in particular in development contexts (Stephens & Middleton, 2002) read like textbook motivations for human centered design. Poor adoption abounds and, according to authors unfamiliar with HCD, is linked to unclear target users, non-inclusion of end-users prior or during development, mismatches between solution and end-users problems, poor user interfaces, users' distrust in the technology, lack of field testing, and insufficient training and support. Most DSS evolved from research tools and assumed very different tasks than those a farmer faces.

COMMON SENSE NET FOR DSS

The initial project set out in 2004 to improve marginal farmers' livelihoods in India through wireless sensor networks (WSN). Project partners included two technical experts (one in India and one in Switzerland), who were working in the field of WSN, an atmospheric research institute, and a local NGO in Chennakeshava Pura, which is a small village in the Karnataka province. Draughts there are common and rain-fed farmers are faced with uncertainty about their harvest yield each year due to inconsistent and unpredictable rainfall and crop pests and diseases.

A wireless sensor network (WSN) consists of a set of nodes, also called motes. Each mote contains sensors and a radio component to communicate with other motes in a networked multi-hop fashion. This allows for timely delivery of information and has demonstrated its value in a range of environmental monitoring contexts such as forest fires and avalanche detection. Commercial WSN solutions exist for home and building automation. In an agricultural context WSNs can be used to collect data relevant to plant physiology e.g., soil moisture and ambient temperature. Fed into crop simulation models the WSN data along with the weather forecast can be turned into forecasts of crop yield and help improve water management. With Moore's law as a guiding light its proponents deem it only a matter of time until prices of WSN equipment become cheap enough such that marginal farmers in developing countries will be able to be supported by WSN deployments in conjunction with crop, pest and disease simulation models.

Within the project the technical setup in the field included two WSN clusters. The collected data from each cluster of battery-driven motes was forwarded first through a base station node, which consisted of a single board computer (SBC) connected to the power grid and an uninterrupted power source (UPS). The SBC acted as a *gateway* from the cluster through its wireless LAN access point to a local server in the village. The server in turn forwarded the data to a central server at the Indian research institute. This wireless LAN gateway approach was abandoned due to connectivity problems. It was replaced by a GPRS based gateway as the mobile phone network in the area improved. From then on the base station forwarded the data directly to the central server.

Methods and Approach

The project started with a series of parallel activities: reviews on appropriate sensor board solutions, sensors for environmental monitoring and crop models. In collaboration with another research institute, the local NGO gathered general information needs on farming and livestock management in the following manner. First Rao et al. (2004) identified the livelihood activities of the rural community through a survey of each neighborhood or caste group in the village. The survey included all members of the community since the researchers deemed that the introduction of a new technology might affect all of them. People took part in group meetings and focus groups, according to their major livelihood activities, e.g., rain-fed farming, irrigated farming and shepherding. The group meetings centered on the information needs of participants' livelihood activities and the participants identified problems and prioritized them by consensus resulting in a ranked list of information needs. Subsequently, between three to six interested individuals from each livelihood group participated in a focus group hosted in the home of one of them to provide details about each identified problem in two to four hour facilitated discussions. High school teachers were then instructed to follow the same approach in 14 other localities. The information needs in the community were diverse and demonstrated that environmental data could be valuable to the farmers. Due to later disputes about ownership the user needs data was not further analyzed and remains unpublished.

Although further user involvement in the project was mainly contractual, no formal contract was established nor was remuneration offered. However, the prospect of a better future through the technology was deemed enough of an incentive for the farmers to:

1. Protect the hardware if it was put into their plot for testing purposes.
2. Report on the conditions in the field.
3. Provide feedback on the value of the technology.

Technical research on WSN components and software was coupled with trials of casings that could be deployed in the field and withstand the climatic conditions. The main problems perceived by the researchers were due to the deployment of the hardware in the field. Energy was by far the biggest constraint. The lifetime of the WSN motes powered with two 3.6V lithium Ion batteries was typically in the range of weeks with a 5% duty cycle due to the synchronization overhead between the nodes (from wake-up to full operation) and networking protocol overhead. Excessive heat in the casings, which were exposed to sunlight and lacked cooling, reduced battery lifetime. The SBC required uninterrupted power for reception and forwarding of data, which was a challenge due to frequent power cuts and the limited duration covered by the UPS. The reliability of hardware was another concern; there was frequent/occasional malfunctioning and failure of node hardware due to unknown causes, lightning strikes and theft. This created the need for theft and dead motes detection algorithms.

The deployment of soil moisture probes proved difficult in the dry compacted soil. The soil around probes that were placed in a dug hole and covered afterwards was not representative of the rest of the field. The untouched compacted soil resulted in much higher run-off during rains than the soil in which a probe resided. Under very dry conditions the soil moisture probes often got detached from the enclosing soil and provided erroneous data.

Non-cooperating farmers who did not allow the placement of motes on their plots for propagation purposes required longer range radio connections. Radio wave propagation changed during the year for example due to crops growth and required higher margins during deployment.

Overall, the wealth of technological problems encountered did not leave room for understanding the needs of the marginal farmers and how the overall system could be designed in order to support those best in their decision-making processes. Perhaps the most interesting question – how the collected data would be presented to the prospective users – was left unaddressed. One notable exception was a controlled lab experiment geared at assisting novices in deploying a WSN. The metaphor of radio reception was employed for people to gauge signal strength and connectivity of motes in a hands- and eyes-free way when deploying the motes (Costanza et al., 2010). However, the participants in the study were technical students in Switzerland rather than farmers and extension workers in India.

During a local festivity one of the technical researchers gave a presentation to an unmoved community about the value of the research and the project's vision of the future for agriculture with WSN. This was not the only occasion on which it became clear that the target users were not interested in the technology, which they had experienced only through the presence of grey boxes installed by technical personnel in some fields. Informal discussions with marginal farmers revealed that they were not interested in any technology that did not bring them direct benefit in terms of rain, a perennial bore well, a road to the village, or monetary advantages e.g. through loans or subsidies (H. Jamadagni, private communication, 2007). Another researcher noted that *"access to marginal farmers is not easy as they are very cautious; also, there is a significant danger in raising their expectations when approaching them."* A representative from the local NGO explained that marginal farmers:

1. Generally felt left behind in the existing innovation processes.
2. Felt they furthered the careers of the scientists more than the scientists helped them improve their livelihoods.
3. Did not understand the scientific agricultural jargon used by the personnel deploying and maintaining the WSN (e.g. soil moisture and evapotranspiration).
4. Felt uncertain about the cost-benefit ratio of the WSN technology.

According to the representative, in order for the farmers to adopt an innovation the benefits of the innovation have to be clearly demonstrated and be substantial. Marginal farmers in the area would not consider improvement in harvest in the range of 30% given the uncertainty, risk and assumed added cost and effort. It should be noted that the marginal farmers never experienced any interaction with the data from the WSN.

The project then refocused on scientists as the target user. A web-based application that allowed for monitoring environmental variables was developed and deployed. Each participating scientist completed a survey prior to a two-week test run. Data logging showed that only six of the thirty participants had used the system. The researchers were asked to identify possible use cases for WSN in the context of agriculture in individual debrief interviews. The four use cases that emerged focused on soil science, entomology, crop physiology and water management.

DISCUSSION

Since the inception of the original project wireless sensor networks hardware has not seen the expected price drops common in mass produced computer hardware. Because WSN currently represents a niche product economies of scale might still render them cheaper once they become more popular and mass-produced. An even bigger impediment to the WSN approach was the unavailability of power for the required infrastructural backbone to forward data to storage centers or off-site servers and the dismal battery life in the motes.

The project was confronted with theft of sensor nodes and very little interest on the farmers' part. Theft could partly be attributed to the fact that one of the participating farmers was quite rich but overall the lack of communication with the local population about what the technology was achieving and how the community would benefit from it seemed to be the main problem. Local outreach activities that improve people's livelihoods in the area where the research is conducted would have been one way to secure benevolent attitudes from the population about the research activities.

The bigger cultural differences appeared to be between technology savvy scientists and rural farmers than between technical partners from different cultures (India and Switzerland). One of the values of HCD lies in getting buy-in from the farmers by letting them experience potential benefits through interaction with early prototypes or envision them through mock-ups. If the benefits of an innovation cannot be reaped in the near future participatory approaches at least need to provide the participants and other facilitators such as NGOs with value for their time and involvement. In the context of sensor data made available to citizens and community groups for environmental activism Aoki *et al.* (2009) reported similarly disenfranchised opinions on the involvement of researchers developing new instruments or conducting studies from concerned citizens, environmental activists, employees of governmental regulatory agencies and environmental action organizations: "... researchers [...] who were perceived as promising a great deal, requiring significant effort to educate and support, and ultimately delivering nothing of relevance." This problem is systemic in part because of the relatively short project durations of ICT projects in comparison to the slow process through which technology adoption occurs and community groups foster societal change.

Despite extensive research on the topic of adoption and diffusion of innovations (Rogers, 1995), many ICTD projects devote most of their attention to research activities unrelated to diffusion. For adoption that involves risk trust lies at the heart of adoption. Few if any development projects can rely on an organizational infrastructure and locally trusted diffusion agents that made agricultural extension such a big success in the US, especially when the time frames for these funded activities are short. NGOs have become popular partners to provide access to participants in research studies since they often have earned the trust of locals due to their long-term commitment and proven interest in the community. The introduction and maintenance of commercial services for rural populations face the same challenges. The Reuters Market Light service, for example, was introduced and partly maintained by the local post office network (Mehra, 2007). An entrepreneurial approach through the introduction of product or services is one approach to making development projects sustainable. But due to a lack of incentives researchers in the ICT domain are often uninterested in investing time and effort in the entrepreneurial side or diffusion activities.

Human subjects' approval and consent forms – often dreaded as a bureaucratic burden by researchers in HCI – might help to put the involvement of farmers on solid footing and serve as basic guidelines for technical personnel unfamiliar with user based research. Participant consent forms, along with the description of standard protocols for interaction with participants, could be seen as *boundary objects* that provides different angles of understanding for partners from diverse scientific backgrounds.

REVISED APPROACH: COMMON SENSE NET 2.0

For the follow-up project the focus shifted to the development of an application that would directly benefit the farmers; thus, a different approach was devised. The consortium was restructured to include an HCI partner working in collaboration with a technical implementation partner and an agricultural expert working on the design and implementation of the application, while the NGO provides access to the local community and serves as a village base for participatory design activities, pilot and field tests.

Technology

The mobile phone provides the only feasible information platform when dealing with unreliable availability of electricity. It is one of the biggest successes in rural ICT development - the poster child of a sustainable technology. The cell towers are independent of the energy grid and ordinary mobile phone use is possible with available windows of opportunity for recharging the battery. Most importantly, many mobile phone models serve as programmable platforms.

To circumvent the power constraints we are integrating off-the-shelf components to create a low power sensor box, which continuously collects and stores data on soil moisture, humidity, soil temperature, ambient temperature and atmospheric pressure in the farmer's field. The farmer uses a mobile phone application to retrieve the sensor data from this field diary via low-power communication protocols, e.g. Bluetooth. The phone can transfer this information along with manually entered parameters about, e.g. the crop being farmed, on what soil, and the sowing date, to a server via the GSM network. This removes the need for inter-mote communication

Applications

The target application(s) running on mobile phones should be able to convey the following information and procedural knowledge to potentially illiterate but numerate users:

1. Farming strategies (such as choice of crops, choice of mono and multiple crops), price, expected yields and risk scenarios based on predictions by the weather board, see (Knoche et al., 2010).
2. A schedule for farming practices.
3. Crop simulation model based predictions of worst, average, and best case yields in financial terms.
4. Probabilities of and control strategies for pest and disease incidences for crops.
5. Guidelines for harvesting.
6. Economic aspects of water management of existing bore well use or procurement of water in relation to c).
7. Local water levels and their fluctuations in bore wells.

For example, the crop simulation model takes the data sent to the server along with weather forecast information and computes yield scenarios (in kg/hectare) that are easy to understand for the farmer, e.g. best, worst and most probable case. Further effects on the yield if fertilizers and pesticides were applied or if more water could be supplied to the crop, are also included. Depending on the farmer's literacy rate the scenarios could be forwarded to the mobile phone through text messages (SMS), multimedia messages (MMS) or voice mail. Another option to produce the spoken word content is a phone based text-to-speech application. Any of these approaches will allow the farmer to repeatedly review and ponder the scenarios during his decision-making process. For this to come through, however, the system has to be designed such that the potentially illiterate farmers with little technical background can perform the following user interactions: pairing the

sensor box with the mobile phone, installing and calibrating the sensor box, entering data manually, retrieving data from the sensor box, receive and reviewing outcome scenarios, re-installing the sensor box on a different plot, and understanding when and how to change the batteries in the sensor box (Knoche et al., 2010).

As a first step towards these applications, we have developed first mobile application prototypes, which allow for easy distribution of information through SMS. We are using the open source software FrontlineSMS (Banks & Hersman, 2009) to provide members of the local NGO with an interface to disseminate and collect information transparently via SMS. Users can enter data through mobile phones forms that arrive via SMS at the FrontlineSMS server application. The data is kept in a database, which a script can automatically query for newly arrived messages and disseminate them through SMS GupShup – an Indian SMS gateway – to eligible receivers. Currently the technical work focuses on providing support on mobile phones for Kannada, the script and language spoken in the province of Karnataka, India. To support illiterate people the application requires a text-to-speech component as well as support for rendering the script. Our planned research on illiterate user interfaces will extend the existing research on mobile accessibility such as (Lalji & Good, 2008).

Although WSN might not currently be viable as a technology in rural India, we have started using it for scientific research that will benefit rain-fed farmers. We are using the motes developed for the first project for research on agro-forestry, specifically to understand which trees and shrubs to plant to improve the stability of bunds and trenches for soil and rain water conservation in a given soil composition given and soil moisture content profile. Bunds and trenches represent one of the most viable and promising water management strategies but the effort to build these is substantial and farmers might need to see the benefits before they invest in them.

Method

The current project follows an HCD approach adapted to the development arena. Field trips have been undertaken to better understand the context and conditions under which the farmers live. These included visits to the next bigger town in which the farmers sell their goods, buy their inputs and implements, obtain credit, and access government and agricultural extension services and health care. Most importantly, a great deal of attention is given to how to involve the target users in participatory design activities and longitudinal tests. The farmers will be remunerated to participate in interviews, focus groups and design workshops and hopefully will recognize the benefits of participating in a longitudinal field study that provides new services.

The information needs of the rain-fed farmers have been revisited as has the trust they place in a range of information sources including their peers in the same location, similar farmers in different locations, government extension agents, inputs retailers, researchers and new findings from participatory research. Our work aims at furthering the understanding of the larger context – the ecosystem in which the farmer acts and makes his decisions based on available information both sensed and via other means. How can other actors and stakeholders make use of this information and trust it? Currently, the rain-fed farmers place almost no trust in information provided by scientific sources; they prefer information provided by local or regional peers even though these might have no sound basis in reality, e.g. in the case of weather forecasting.

Through initial interviews with rain-fed farmers and extension workers we will try to understand how they currently obtain the relevant information for the above topics and which role mobile phones play in their daily activities. After having understood whether the farmers themselves or extension workers will be the primary users of the application we will create personas, usage scenarios and storyboards to communicate the user requirements to the development team and to help in the iterative design process with the farmers. In participatory design sessions we will try to elicit suitable ways to convey the desired information to the target users and find appropriate interaction designs. We will then conduct a longitudinal field study in which participants can use a working prototype system in real life over the course of six months during a cropping season.

We will use a bootstrapping approach, which will provide participants in a longitudinal field test with a growing set of information services that are not necessarily related to sensor or agricultural data. One of the first ways to provide a tangible benefit to participating farmers could be day labor opportunities, availability of government programs and weather forecasts on their mobile phones. In the spirit of action research the project will support a member of the local NGO to provide this information to the local population throughout one year. We aim to make this service self-sufficient through an advertising supported or subscription payment model. This entrepreneurial approach could provide the foundation for continued provision of benefits to the local community and possibly grow to encompass other adjacent communities. Another incentive for the NGO is that they can use this mobile phone based group communication tool for other purposes in the future.

As a part of this strategy we are planning to rope in further actors and stakeholders that could use the data from the sensors and entered by farmers for other purposes. Local agricultural offices and extension agents should be able to better serve their community if they knew more about which crops are currently planted, what fertilizers might be in demand, and which pest and diseases are rampant and have a means to convey information about remedies, subsidies and their programs to them. Retailers of agricultural inputs and their wholesalers might improve their supply chains if more information on potential local demand were available. The sensor driven field diary could be used in the determination of rain insurance claims. Proving eligibility of claims has been problematic since the available rainfall data has been too coarse in its granularity. Shepherds or goat herders whom farmers have to pay for grazing on their land to naturally fertilize it could use the field diary to estimate the risk of parasites on the plot with an appropriate application. This information would have high value due to the high cost for parasite treatments. The farmers could learn more from each other if the information about their farming strategies and approaches and resulting yield was shared amongst them. HCD artifacts such as scenarios, storyboards and prototypes of any kind can help in communicating with these target groups. Especially prototypes that help envision actual use and its benefits can help in getting buy-in from them and keep discussions with them focused on concrete and achievable goals. Furthermore, they increase the conveyed credibility and determination on part of the researcher and that the interaction with him is worthwhile.

One of the problems encountered with the HCD approach in a project of this kind is the upfront time required to understand the contexts of use, the target user and which information they need in the context of use and through what kind of interface. Technical partners unfamiliar with the modus operandi and value of HCD and under pressure having to deliver the front- and back-end solutions may be eager to start with their own vision of what needs to be built without fully understanding what will be required.

CONCLUSION

The cost-benefit of ICT solutions and participation in research projects need to be clearly communicated to potential target users. This was particularly problematic during the first phase of this project during which farmer involvement was contractual but involved no remuneration. It also did not provide any opportunity for farmers to envision the use of the technology or experience benefits first hand as no user interface was ever made available. Lo-fi prototyping and other participatory techniques that are standard in the tool set of HCD provide many opportunities to envision use of technology and get a glimpse of potential benefits. Both prototyping artifacts and consent forms can act as boundary objects that help scientists from different backgrounds to design better ICT solutions. Especially working prototypes that include user interfaces can fill this role and help convince NGOs, stakeholders and other external actors of the value of the project and that it will actually deliver on its visions.

On a more strategic level the HCI community needs to further raise funding organizations' awareness of the value of HCD in delivering research that addresses the needs of people in ICTD projects. The iterative approach of HCD also makes for a good transparent way to monitor progress and therefore aid in project evaluations. In order to not fall short on the expectations created the HCD community should look into potential shortcomings of participatory approaches in the larger scope in the development domain.

REFERENCES

Agrocom. (2008). *Agrocom software technologies private limited - achievements.* Retrieved from http://agrocom.co.in/achievements.php

Anurag, P., Vivek, B., & Sasank, T. (2008). *CERES information services.* Retrieved from http://ceres.co.in

Aoki, P. M., Honicky, R. J., Mainwaring, A., Myers, C., Paulos, E., Subramanian, S., & Woodruff, A. (2009). A vehicle for research: Using street sweepers to explore the landscape of environmental community action. In *Proceedings of the 27th International Conference on Human Factors in Computing Systems* (pp. 375-384).

Awasthi, U. (2008). *IFFCO Kisan Sanchar Ltd.* Retrieved from http://www.iffco.nic.in/applications/iffcowebr5.nsf/?Open

Bahuman, A., & Kirthi, R. (2007). *aAqua mini.* Retrieved from www.agrocom.co.in

Bahuman, A., & Ramamritham, K. (2010). *aAQUA mobile - almost all questions answered.* Retrieved from http://www.slideshare.net/bahuman/aaqua-mobile-pilot-to-advise-50000-farmers-over-the-telephone

Banks, K., & Hersman, E. (2009). FrontlineSMS and ushahidi - a demo. In *Proceedings of the 3rd International Conference on Information and Communication Technologies and Development*, Doha, Qatar (pp. 484-484).

Barker, R., & Molle, F. (2004). *Evolution of irrigation in South and Southeast Asia.* Colombo, Sri Lanka: Comprehensive Assessment Secretariat.

Basu, P., & Srivastava, P. (2005). *Scaling-up microfinance for India's rural poor.* Washington, DC: World Bank. doi:10.1596/1813-9450-3646

Bentley, J. W. (1994). Facts, fantasies, and failures of farmer participatory research. *Agriculture and Human Values*, *11*(2), 140–150. doi:10.1007/BF01530454

Biggs, S. D. (1989). Resource-poor farmer participation in research: A synthesis of experiences from nine national agricultural research systems. *OFCOR Comparative Study Paper*, *3*, 1–4.

Blattman, C., Jensen, R., & Roman, R. (2003). Assessing the need and potential of community networking for development in rural India. *The Information Society*, *19*(5), 349–364. doi:10.1080/714044683

CIA. (2009). *The world factbook.* Retrieved from https://www.cia.gov/library/publications/the-world-factbook/geos/in.html

Cleaver, F. (1999). Paradoxes of participation: Questioning participatory approaches to development. *Journal of International Development*, *11*(4), 597–612. doi:10.1002/(SICI)1099-1328(199906)11:4<597::AID-JID610>3.0.CO;2-Q

Cooke, B., & Kothari, U. (2001). *Participation: The new tyranny?* London, UK: Zed Books.

Costanza, E., Panchard, J., Zufferey, G., Nembrini, J., Freudiger, J., Huang, J., et al. (2010). SensorTune: A mobile auditory interface for DIY wireless sensor networks. In *Proceedings of the 28th International Conference on Human Factors in Computing Systems*, Atlanta, GA. (pp. 2317-2326).

FAO. (2005). *Database.* Retrieved from http://faostat.fao.org/

Horvath, R. (2008). Innovation - mobile services. In *Proceedings of the CII/GIS Conference.*

Knoche, H., Prabhakar, T., Jamadagni, H., Pittet, A., Sheshagiri Rao, P., et al. (2010). Common sense net 2.0 - minimizing uncertainty of rain-fed farmers in semi-arid India with sensor networks. In *UNESCO Technologies for Development.* Lausanne, Switzerland.

Kotkar, P., Thies, W., & Amarasinghe, S. (2008). An audio wiki for publishing user-generated content in the developing world. In *Proceedings of the HCI Workshop for Community and International Development*, Florence, Italy.

Kumar, A., Agarwal, S. K., & Manwani, P. (2010). The spoken web application framework: User generated content and service creation through low-end mobiles. In *Proceedings of the International Cross Disciplinary Conference on Web Accessibility*, Raleigh, NC (pp. 1-10).

Lalji, Z., & Good, J. (2008). Designing new technologies for illiterate populations: A study in mobile phone interface design. *Interacting with Computers*, *20*(6), 574–586. doi:10.1016/j.intcom.2008.09.002

Matthews, R. B., & Stephens, W. (2002). *Crop-soil simulation models: Applications in developing countries*. New York, NY: CABI. doi:10.1079/9780851995632.0000

Mehra, A. (2007). *Reuters market light now available in local post offices across Maharastra*. Retrieved from http://news.thomasnet.com/companystory/Reuters-Market-Light-Now-Available-in-Local-Post-Offices-across-Maharashtra-808401

Nokia Siemens Networks, Nokia and Commonwealth Telecommunications Organization (CTO). (2008). *Towards effective e-governance: The delivery of public services through local e-content*. Retrieved from http://www.e-agriculture.org/en/news/towards-effective-e-governance-delivery-public-services-through-local-e-content

Patel, N., Chittamuru, D., Jain, A., Dave, P., & Parikh, T. S. (2010). Avaaj Otalo—a field study of an interactive voice forum for small farmers in rural India. In *Proceedings of the 28th International Conference on Human Factors in computing systems*, Atlanta, GA (pp. 733-742).

Rao, K. V., & Sonar, R. M. (2009). M4D applications in agriculture: Some developments and perspectives in India. *Defining the 'D' in ICT4D*, 104-111.

Rao, S., Gadgil, M., Krishnapura, R., Krishna, A., Gangadhar, M., & Gadgil, S. (2004). *Information needs for farming and livestock management in semi-arid tracts of Southern India (Tech. Rep. No. AS 2)*. Bangalore, India: CAOS.

Rogers, E. M. (1995). *Diffusion of innovations* (4th ed.). New York, NY: Free Press.

Sherwani, J., Palijo, S., Mirza, S., Ahmed, T., Ali, N., & Rosenfeld, R. (2009). Speech vs. touch-tone: Telephony interfaces for information access by low literate users. In *Proceedings of the IEEE/ACM International Conference on Information and Communication Technologies and Development* (pp. 447-457).

Stephens, W., & Middleton, T. (2002). Why has the uptake of decision support systems been so poor. *Crop-Soil Simulation Models*, 129-147.

Stone, P., & Hochman, Z. (2004). If interactive decision support systems are the answer, have we been asking the right questions? In *Proceedings of the 4th International Crop Science Congress on New Directions for a Diverse Planet*, Brisbane, Australia.

Swiss Agency for Development and Cooperation SDC. (2008). *Annual report 2008 Switzerland's international cooperation*. Retrieved from http://www.cosude.ch/de/Home/Dokumentation/ressources/resource_en_181617.pdf

Trisorio-Liuzzi, G., & Hamdy, A. (2008). Rainfed agriculture improvement: Water management is the key challenge. Paper presented at the 13th IWRA World Water Congress, Montpellier, France.

This work was previously published in the International Journal of Mobile Human Computer Interaction, Volume 3, Issue 3, edited by Joanna Lumsden, pp.1-13, copyright 2011 by IGI Publishing (an imprint of IGI Global).

Chapter 11
A Festival–Wide Social Network Using 2D Barcodes, Mobile Phones and Situated Displays

Jakob Eg Larsen
Technical University of Denmark, Denmark

Arkadiusz Stopczynski
Technical University of Denmark, Denmark

ABSTRACT

This paper reports on the authors' experiences with an exploratory prototype festival-wide social network. Unique 2D barcodes were applied to wristbands and mobile phones to uniquely identify the festival participants at the CO2PENHAGEN music festival in Denmark. The authors describe experiences from initial use of a set of social network applications involving participant profiles, a microblog and images shared on situated displays, and competitions created for the festival. The pilot study included 73 participants, each creating a unique profile. The novel approach had potential to enable anyone at the festival to participate in the festival-wide social network, as participants did not need any special hardware or mobile client application to be involved. The 2D barcodes was found to be a feasible low-cost approach for unique participant identification and social network interaction. Implications for the design of future systems of this nature are discussed.

INTRODUCTION

In the Web 2.0 area social networks have played a pivotal role. They connect people throughout the world, regardless of their actual localization, enabling participants to communicate and easily share experiences and content. Recently, social networks that take advantage of the physical and spatial proximity of the participants have gained attention. Festivals are creating websites that allow people to register, exchange information, opinions, ratings, etc. In addition festivals

DOI: 10.4018/978-1-4666-2068-1.ch011

increasingly use popular social networks, such as Facebook and Twitter as communication means for participants. These social networks are more focused on a particular event and can have some unique features related to the fact that participants will be (at some point) in the same location at the same time. Music festivals are interesting in the context of urban computing (Kindberg, Chalmers, & Paulos, 2007) as they typically form small city-in-a-city environments with short lasting communities.

In this paper we present a prototype system that uses 2D barcodes on festival participant wristbands as means to participant identification in the system, and enable linking the festival participants to a personal profile. This profile can be used in several different applications in a festival-wide social network. The social network applications are built on top of the simple social network framework we created for the CO2PENHAGEN music festival event where our initial experiments were carried out.

RELATED WORK

Barcodes on wristbands have been applied in health care typically to promote patient safety, such as patient identification to eliminate medical errors and medication mistakes (Mun, Kantrowitz, Carmel, Mason, & Engels, 2007). However, in the area of mobile social applications (Smith, 2005; Thom-Santelli, 2007) barcodes have mainly been applied to link physical objects in the environment to available information (Hansen & Grønbæk, 2008). Other applications of barcodes include games (Schmidmayr, Ebner, & Kappe, 2008), situated learning (Kurti, Milrad, & Spikol, 2007), tourist applications (O'Hara & Kindberg, 2007), focusing on the barcode augmenting a physical object with information typically presented through a mobile device. Thus the use of 2D barcodes for festival participant identification in this study is a novel approach. Swedberg (2009)

reports on the use of RFID as identification technique used on tickets at the Ohio Music Festival, however, the experiment did not include other applications of RFID in the festival context. Zeni, Kiyavitskaya, Barbera, Oztaysi, and Mich (2009) employed RFID for crowd tracking at different festival events.

In the present study the 2D barcodes allow us to use low-cost off-the-shelf solutions as the means to support social network interaction, and thereby enable quick and cheap prototyping of different social network applications, such as interaction by means of situated displays in a festival setting. Prior work on large situated displays has focused on applications in CSCW research and groupware systems (Brignull & Rogers, 2003; Churchill, Girgensohn, Nelson, & Lee, 2004; Greenberg & Rounding, 2001). Tuulos, Scheible, and Nyholm (2007) describe how mobile phones and large public displays were used in a large-scale game involving collaborative story writing in an urban environment. An experiment with social interaction on a situated display in a festival setting has been tested using a collaborative story writing game (a WAP based solution) by Coulton, Bamford, and Edwards (2008) and Peltonen et al. (2007) experimented with additional touch-based interaction on large displays. Jacucci, Oulasvirta, Ilmonen, Evans, and Salovaara (2007) carried out field trials with CoMedia as an approach to support spectators in event coordination and sharing of media and presence at large-scale events (such as a music festival) using mobile phones.

Our focus is to study the use and feasibility of participant wristbands with 2D barcodes for unique identification of participants to enable interaction in festival-wide social network services that we created for the CO2PENHAGEN festival. The focus is on combining the physical presence in a festival environment with the virtual one in terms of a number of social network applications, where the interaction in enabled by the 2D barcodes linking individuals to the social network.

The CO2PENHAGEN Festival

The CO2PENHAGEN festival was a two-day music festival and arts event with 40+ bands playing, lounge areas, and activity areas with a set of installations with an emphasis on environmental aspects. Thus one of the aims of the festival was making it run entirely on renewable energy produced locally at the festival site situated just north of Copenhagen in Denmark. The grounds were divided into several areas (zones) including:

- **Music Zone:** With two music stages.
- **Activity Zone:** With a DJ-stage.
- **Lounge Zone:** Where participants could relax, listen to music, or watch a film.
- **Explore and Future Now Zone:** Showcasing environmental-friendly technologies (e.g. electric cars).
- **Food and Fuel:** With restaurants and bars.

An overview of the festival area with the different zones is shown in Figure 1.

CO2PENHAGEN was a new festival and therefore a relatively "small" event, compared to other music and arts festivals in Denmark. That made it possible for us to propose experimenting with the novel systems and technologies as presented in this study. The organizers were quite open to testing new technologies even of an experimental nature. The festival was highly dependant on volunteer workers helping out on all aspects of running the festival. This meant that there was some freedom for volunteers suggesting contributions to the festival, such as, in our case creating a novel festival-wide social network enabling the participants to share experiences at the festival site and thereby augment the festival experience. However, as our festival-wide social network had not been tested on a larger scale before the CO2PENHAGEN festival, it was not broadly announced to the participants. Instead people were asked if they wanted to join the network and the experiment on a voluntary basis.

Figure 1. Sketch of the festival area with zones, stages, entrance, and the Information area in the center

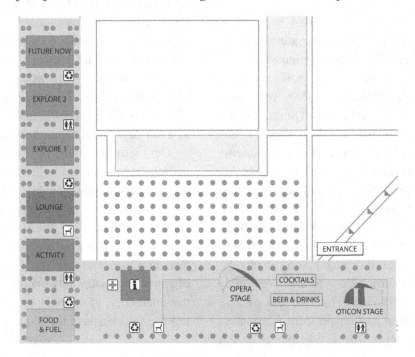

Social Network System Architecture and Applications

In order to build and experiment with the festival-wide social network as outlined above, a set of off-the-shelf components were used, including a set of standard software components. The system architecture also contained a set of software components that was written by us to create our social network. That included the back-end system with a database and a front-end application for mobile phones containing a number of different applications and a component for scanning the 2D barcodes.

System Architecture

We created a prototype system with an architecture around the notion of identification of the participants in the system with 2D barcodes as the means to application interaction. The emphasis was on prototyping and therefore making it extensible with new applications. The overall conceptual system architecture is outlined in Figure 2.

The details of the technical implementation of the architecture and mobile phone applications is beyond the scope of this paper. Instead we describe the key components and the four different domains in which they are used: participants' domain, volunteers' domain, backend domain and presentation domain. The key components in the system architecture include:

- **2D Barcodes:** Containing encoded information
- **Mobile Phones:** Used for barcodes scanning, picture taking, and sending text messages.
- **Computers/Laptops:** Used for data input, and connected to the large displays.
- **SMS Gateway:** For receiving sms messages sent from participants' mobile phones.
- **Server:** Backend system for the social network applications containing a database with profiles, pictures, and messages.
- **Large Displays:** Used for presentation.

Figure 2. Conceptual system architecture with indication of the four different domains of use

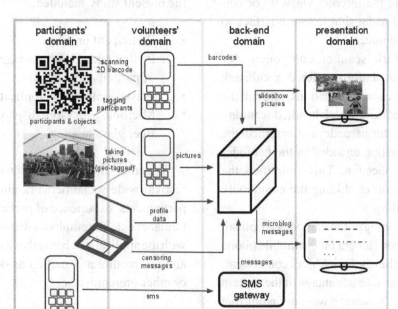

Festival participants operated only in participants' domain, where no special equipment was required. In the volunteers' domain, special equipment (for scanning barcodes and taking pictures) was required, but no advanced technical skills were necessary. The complexity of the system was hidden in the backend domain.

Participant Identification

The identification of participants in the system was based on the barcodes on the participant wristbands. The barcodes were scanned with a system built for Nokia mobile phones (Nokia N95 8GB) including a software component from QuickMark for the barcode scanning using the embedded mobile phone camera. We found the Quickmark application to be among the best barcode scanners for the particular Nokia N95 8GB device at our disposal for the experiments. We built a custom Symbian mobile application, which handled the results of scanning acting as a thin client with the server backend. The mobile client application contained several other modules: camera, GPS, networking, which were customized for the particular purpose. The mobile phones were connected to the Internet via WiFi or cellular network and had a simple user interface to be used by the volunteers.

The QuickMark scanner component for the Symbian S60 platform offered a callback mechanism: our custom application could call the QuickMark component, which handled scanning and decoding of the barcode and returned the result (the information encoded in the barcode) to the calling application. This simplified the implementation, entirely hiding the complexity of barcodes handling.

The actual scanning of the barcodes is optical, using the embedded camera on the mobile phone and controlled within the QuickMark component. The component can take advantage of the camera auto-focus feature, however it was our finding that for the purpose of scanning, turning auto-focus off would speed up the process: scanning of participants' wristbands was usually performed from a fixed distance that could also be easily adjusted with movement of the phone.

We conducted several tests of the feasibility that included varying barcodes size, rotation, distance and lighting conditions. Those tests proved that a barcode of a size fitting on the wristband (1.8 x 1.8 cm), from a comfortable distance (15-20 cm) could be scanned quickly (below one second for the actual scanning) regardless of its rotation until about 9 p.m. The result was either a successful scan (decoding of the barcode) or failing to decode the barcode, in which case the scanner keeps trying until interrupted by the user. For technical reasons we were not able to use camera flash to increase performance in poor light conditions (such as in the late evening). Our only concern was network availability, as the application depended entirely of the communication with the server backend.

Social Network Applications

The specific social network applications that were built on top of the described platform for the present study included:

- Participant profiles.
- Picture taking with geotagging and tagging of participants.
- Waste competition application.
- Microblog for large displays.
- Beer glass tagging.

Together these applications constitute the festival-wide social network aiming to enhance participants' experience of participation in a particular event. The emphasis was on the interaction with the applications being that it should be simple and not require any particular skills, equipment or other prerequisites.

Participant Profiles

Profile creation in our platform involved the process of linking a participant assigned code (embedded in a barcode) with personal information and optionally a photo of the participant. A first step to create a new profile was to scan the code of the participant (randomly generated 8 characters, encoded in 2D barcode). As the application running on the phone is set up to profile creation, it automatically notified the server that a profile with the given ID was being created. A profile photograph could then be taken by a volunteer (using the mobile phone application) and it would also be uploaded to the server as corresponding to given ID. Once this is done, profile creation procedure would move to a computer, where personal information of the participant can be put in. We decided to use computers for this information input primarily because of full-size keyboards, which would allow participants to input desired information by themselves. This approach is justified, especially in a multinational setting, where spelling the name to a volunteer can be problematic.

Picture Taking with Geotagging and Tagging of Participants

Pictures in our platform are taken within our application that supported autofocus and the full resolution available in the phone. The picture was geotagged and it was also possible to tag participants in it: scanning participants wristbands after taking the photo adds them to the picture metadata. This way pictures could be automatically added to the galleries of the participants. Photos can either be uploaded instantly after they are taken or stored and uploaded in batches with availability of a network connection.

The photos taken by volunteers were presented on large displays using a simple web application that shuffles them and created a slideshow. In addition, Google Maps was used to fetch a map where we overlaid the location where the picture was taken with a marker.

Microblog Application and Large Displays

Microblogging is a separate application for the large displays, where participants' messages (submitted by SMS) were presented in a storyline message stream. Those messages were handled by an external SMS Gateway provider that interfaces with our backend server. When the messages were added to our database, they first needed approval by a moderator (through a web interface), before they were displayed. Using SMS for microblogging still allows vast majority of people present to participate.

Waste Competition Application

As our platform allows for quick and unique participant identification, it was possible to build simple competitions using the platform as an infrastructure. In our prototype application, the logic is simple, as the server simply added a point to the participants' profile, every time the scanning was done. The logic of the actual competition was not implemented in the application, but had to be executed by the volunteers operating the scanners. It was however important to observe that because of using cellular network scanning points can easily be set up in any location or even be mobile, still having constant connectivity with the server.

Beer Glass Tagging

As mentioned previously one of the primary goals of the festival was the emphasis on environmental aspects and the purpose was to have an environmentally neutral festival, where aspects such as re-use were promoted. As one example, participants were encouraged to keep the glasses from the bars so they could be re-used (refilled) several times.

As unique barcodes could also be attached to objects (in this case glasses) it was possible to create accounts that was attached to objects. At the festival a glass of beer was tagged with a 2D barcode and initially three beers were added to the corresponding account. There was a discount compared to buying three separate glasses of beer, in order to promote the environmental aspect. When a participant wanted a refill, the barcode on the glass was scanned by the bartender, and a single credit was taken from the account. Once the account was empty, it was possible to buy additional credits for the same glass. Participant would pay the bartender, who would add another three credits to the account of the presented glass, by choosing that option in the mobile application. Beyond the environmental aspects an additional advantage was that it was easier for the crew at the bars to manage payment and sales.

Methodology and Experiment

Initial experiments with the above described system were carried out as an "in-the-wild" field experiments at the CO2PENHAGEN festival. In the center of the festival grounds was the Information area, where two 42" displays were situated, as shown in the overview of the festival area in Figure 1. A total of 10 volunteers took shifts over the two days to work in the Information area and other areas at the festival to help carry out the experiments. The volunteers were changing activities and there were always about six volunteer's active. These included recruiting and enrolling participants in the social network by creating profiles and making sure the applications were working properly. It also included taking pictures, managing the waste competition, and censoring microblog message (see the domains illustrated in Figure 2). But the volunteers also had duties in the information booth answering general questions from festival participants. The 10 volunteers were instructed on how to use the different parts of the mobile application mentioned above. This took place a few days before the festival started and again when their first shift started.

After entering the festival area and getting their wristbands, some participants would walk by the Information area located in the center of the festival area (Figure 1). At the Information area random festival participants were asked if they wanted to create a profile to be part of the CO2PENHAGEN social network. They were briefly explained about the social network and the above mentioned applications. A waterproof sticker with a unique 2D barcode was then attached to the participant wristband. A participant profile included a personal webpage at the festival website and was created by scanning the participant's wristband with a mobile phone (Figure 3a) along taking a profile picture of the participant (Figure 3b). The participants could add additional information to the profile, including a nickname and phone number.

A participant profile on the festival website contained a profile picture, nickname, status messages, personal photo gallery, competition point counters, and a questionnaire about the festival. The status could be updated by the participants by sending an SMS with the new status text to an SMS gateway (standard SMS charge). Pictures taken by the volunteers using the mobile phones were automatically uploaded to the server. After taking a photo, it was possible to scan the wristbands on the participants in order to tag the picture with those participants, adding the picture to the personal photo gallery in the personal profile page. The mobile phones used GPS to geotag the pictures taken by the volunteers. It was possible for the participants to upload their own photos from the festival to their personal page, having a single place to view and share photos from the event.

On the two displays a slideshow was presented (Figure 4a), showing random photos taken by the volunteers, including a small map segment (obtained from GoogleMaps) indicating where the photo was taken, similar to the approach described by Cheverst, Coulton, Bamford, and Taylor (2008).

Figure 3. Participant profile creation. a) Scanning of wristband barcode with mobile phone. b) Participant profile picture taken by a volunteer. On the left side is the large display showing pictures and the microblog.

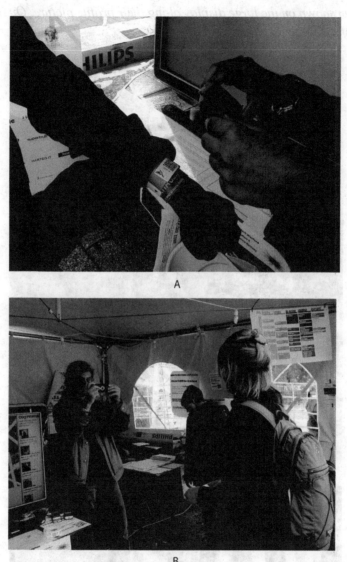

A

B

Moreover, the displays also showed a microblog containing the latest 5-8 status messages from different participants (sent via SMS). It was shown as a list of messages creating a short storyline. Those messages where first accepted by a moderator, to avoid offensive content (a requirement from the festival organizers). The participant profile name and picture was shown next to the status mes-

sage, and only the last three digits of the phone number were shown in case of a message from an unregistered participant. Participants were typically standing a few meters away from the large display interacting with microblog messages, as illustrated in Figure 4b.

Recycling was registered as part of festival competitions. The festival had several zones where

Figure 4. Interaction with the large displays. a) One of the large displays showing pictures and the microblog on the right hand side. The items on the microblog contained the profiles pictures of the participant. b) Reactions from participants interacting with the large displays submitting microblog messages and see them shown on the large display along their profile picture. On the right hand side of the picture the edge of the large display is visible.

A

B

people could throw out waste into different containers and our volunteers were scanning the wristbands of the participants who correctly sorted and threw out waste (Figure 5a). Collected points were added to the participant profile and prizes were given for the most active participants.

The moment a wristband was scanned the photo of this participant (if available) was shown on the large displays.

Finally, beer glasses were tagged with 2D barcodes, making it possible to buy refills for one glass (and to buy additional refills later). In the

Figure 5. Barcode-enabled applications. a) Scanning of the participant wristband during the waste competition, whereby the participant earns points on his profile. b) The bar with the 2D barcode payment area.

A   B

bar there was a special fast line for people with tagged glasses and a volunteer scanning them (Figure 5b). Thus, in the environmental spirit of the festival people were encouraged to keep the glasses instead of throwing them out.

During the entire festival we were logging all activities in the social network, which technically meant logging all activity on the backend system, as all activity went through the server backend. It also meant we could use the server as a "universal clock" for time stamping all activities in the log. The data logging included:

- Interaction.
- Pictures.
- Messages.
- Profiles.
- Weblog.

Moreover the authors took pictures to document the activities in addition to the (geotagged and participant tagged) pictures that were taken by volunteers. The pictures taken for documentation were not part of the information in the social network, and thus were not shown on the large displays during the festival.

A challenge in the collection of qualitative data during the festival was that we had to minimize the interference with the festival participants. The intention of the organizers was that the festival participants should experience the festival "as it was", without getting disturbed with questions, interviews, and questionnaires to fill out during the festival. Therefore a constraint in collecting this data was that we could only acquire informal feedback from the participants when they were interacting with the volunteers. This was typi-

Figure 6. Illustration of the social network activities during two days. Day 1: 2pm-1am. Day 2: 9am-10pm. The activites include: a) participants registered, b) microblog messages, c) beers sold, d) waste thrown out during competition.

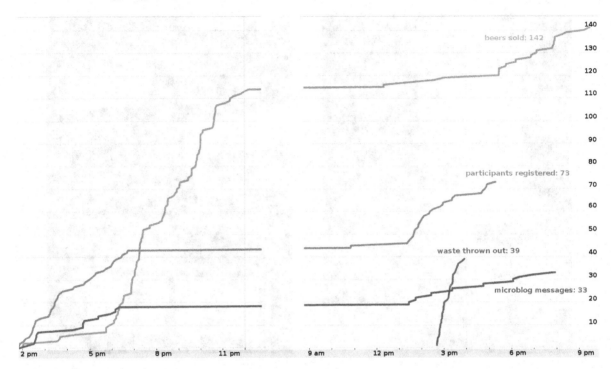

cally when they were told about the festival-wide social network, while setting up the profile, or if they had questions about the system. Thus the qualitative data was based on observations of the participants made by the volunteers and authors as they used the social network applications during the profile creation, while a picture was taken for the profile, and while they were texting messages to the microblog on the large displays.

Results

About six hundred people visited the festival in two days. Our pilot study was mainly done in the time slots 1 pm. – 1 am. the first day and 1 pm. - 10 pm. the second day (with most activity in the afternoons). 73 unique participant profiles were created, 67 (92%) of them were created with profile photo and 55 (75%) of the participants

registered their phone number. Four profiles (5%) were created as a couple profile (with two people in the profile picture). Participants sent a total of 33 messages. 30 messages came from people with a profile, and three messages were sent by unregistered participants. Our volunteers took 341 pictures and 89 (26%) had a valid geotag, enabling the applications to display the photo on a map where it was taken. There were five registered participants taking active part in the waste competition (children or young teenagers). Scanning of wristbands when people threw out waste was done 39 times, each time earning points to the participants. 55 glasses of beer, with the barcode attached, were sold. When purchased, they had three beers on the account. By the end of the festival, 43 of them had empty accounts indicating a successful use of the barcodes for this purpose.

Figure 6 provides an overview of the two day duration of the festival with cumulative statistics of the activities. The activities include:

- Participant profiles registered.
- Messages sent to the microblog.
- Beers sold (by means of tagged glasses).
- Waste thrown out during the competition.

From the figure we can see that the participant registration (profile creation) finished around 7 in the evening day one and 5 in the afternoon day two. The microblog was used and beers sold until the end in the evening both days. When analyzing the profile creation closer, we discover that it is common for the profiles to be created in pairs, one after another. We also observe larger groups that created profiles one-by-one. Slightly more profiles were created on the first day: 43 versus 30. On both days, profiles were created in pretty constant rate throughout the day, starting early afternoon. The festival participants typically registered when they arrived at the festival, which could explain the lower number of registrations on day two.

The beer application was not related to the participants' profiles, as the barcodes were attached to the glasses, not the wristbands. We can see the constant growth during both days; in the steep part on the first day almost 1 beer per minute was refilled (28 beers in 32 minutes). We can see that on the second day much fewer beers were sold, also due to the fact that the festival closed earlier (10 p.m. versus 12 p.m. on the first day). In general, this application was used with fairly constant rate, starting in early evening and finishing at night hours.

The waste competition was a short-time localized event with a scanning rate of 39 scans within 1 hour and 30 minutes. As we see in the chart, it was started on the second day around 2.30 pm and finished at 4 pm, and the participant with the most points was announced as the winner of the competition.

When people had their wristband scanned and photo taken to create the profile we had a chance to explain the idea and get their immediate reactions. Thus our results are partly based on qualitative observations and informal participant feedback. It was our observation that people liked the idea of the barcodes used for identification. Less than 10% of the participants that were asked declined to have a profile created, but we did not obtain information why they declined. For some participants it was the first time they had seen 2D barcodes. Some did know about their existence but had never actually used or seen them being used.

Another concern was lighting conditions: since 2D barcodes are scanned optically, it is necessary to have decent lighting for the scanner to work. The pilot study was running until 7 pm. and we did not observe any issues with scanning speed related to the lighting conditions. The beer line continued to operate well into night hours. Because of the lighting in the booths, there was no problem with scanning of the barcodes. A scan (including starting the QuickMark application) took up to ten seconds, which was used for waking up the phone from the standby mode (to preserve battery), choosing the right option of scanning from the application menu, adjusting the scanning position, decoding the barcode and handling the scanning results. Thus the actual reading of the 2D barcode was typically done in less than a second. This seemed to be an acceptable time when considering participants reactions, as their focus was on the scanner and right placing of the barcode. They were also using this time to ask questions about the technology and project in general. Together with photo taking and typing the name in, the whole profile creation process was taking about 30 seconds. We did not observe problems with scanning accuracy, after several test scans our volunteers were able to perform a scan in the first try. Participants did not express any dissatisfaction with the speed or accuracy of the whole process. On the contrary, when asked they described it as 'working well' and 'quick'.

The two large displays with photos and the microblog were attracting people's attention, as they were highly visible at the Information area. It was our observation that having both pictures and microblog on the screen was advantageous, as photos attract people's attention from a distance and the microblog was readable when people were closer and kept their attention. After the profile was created, people were expecting to see their photo on the screen. Some were sending a status message, waiting for it to show up (it took about 30 seconds). It seemed that messages were considered by the participants rather as a kind of "graffiti" message that was left behind, than the status message or feed for a live blog. This was related to the fact that the displays were seen from a limited distance and by a limited number of people at the Information area at any given moment. We would expect this behavior to change with larger displays seen by a larger number of people. The participants enjoyed that they could immediately share the experience with their friends with the message shown in the microblog on the displays. Some even checked back later to see if their message was still visible on the display. The microblog messages fall into three categories, as shown in Table 1.

It was observed that the interaction (including sending messages, creating profiles, taking pictures etc.) increased when participants were part of a group. Participants were treating such festival activities as a social activity, especially if it al-

lowed them to interact with other members of the group. For example messages sent to the microblog often address other members of the group. The waste competition was very appealing to children at the festival who were very eager to participate. They were not concerned about the prize, but were apparently participating for the sole reason of competition. When volunteers took pictures at the festival they had the option of tagging the people in the picture using their wristbands. When explaining this to volunteers and participants it was observed several times that people spontaneously uttered that this was just like tagging someone in a picture on Facebook, even though we did not describe it in this way. This clearly indicated that people immediately understood the concept of tagging them for the picture using the 2D barcode.

Discussion

We were satisfied that we managed to setup the system and that it was running smoothly. A social network in a festival environment was created and tested and our setup was suitable for a small-scale experiment. It enabled us to collect interesting qualitative data based on direct feedback from participants and volunteers. We were able to observe people interacting with our applications, discovering patterns (e.g. people interacting in groups) and ways the technology influenced the participants. The platform utilizing 2D barcodes turned out to be a suitable technology for the activities implemented, which demonstrated the potential in this setting.

Previous studies have mainly been carried out in laboratory settings (Mäkelä, Belt, Greenblatt, & Häkkilä, 2007), whereas we were able to study the use of barcodes "in-the-wild" in a real music festival setting. This required planning including briefing and instructing volunteers. It also made the qualitative data collection a challenge as more than 10 people were involved. Considering the nature of the highly dynamic setting that a music

Table 1. Categories of microblog messages

Message category	Examples
General environmental statements (due to the festival environmental theme)	"Keep the nature green" "Stop destroying the environment"
Statements about the festival itself, including the concerts, bands, food, and areas	"szhirley was cool" "BioM has good burgers!" "this is a really great concept!"
Weather statements (it was raining on the first day)	"having fun in the rain:-)" "Enjoying the sun today.."

festival is, the experiment turned out to provide some interesting indications on the interactions in such an environment.

Nevertheless, we were somewhat disappointed by the limited uptake of the microblog application in particular considering that we were able to get 73 participants to create profiles. We believe that the main reason was the displays being situated only in the Information area that people visited short-term and infrequently. Some participants registered a profile, but were then in a hurry to go to a concert and did not try the service later. Some tried out the microblog immediately with a positive attitude, as they could immediately see the result on the displays, as indicated by the participant reactions shown in Figure 4b.

As expected the large displays were given attention by participants visiting the Information area. The participants would watch the pictures and read the messages in the microblog. The original intention was to have displays situated at each stage and at the lounge area where people would spend longer time and thus it could stimulate interaction. However, the idea was abandoned by the festival organizers due to concerns about the total energy consumption at the festival, as the goal was to create an environmentally neutral festival. We had to accept this unfortunate decision, knowing it would weaken the deployment and experiment with the microblog and image sharing applications. It weakened the festival-wide social interaction aspect as people would have to be in the same place (the information area) to participate. On the other hand it made it easier from an experimental point of view to observe the uptake and reactions to the applications, as it was mostly happening in the information area. That would have been more difficult had there been several more displays in the festival area as originally planned, as the area from which people could have participated would then have been much larger, and direct observations would have been difficult to make. For instance it would have been difficult to obtain the qualitative data

observing those posting to the microblog when situated in a large crowd at a concert.

Interaction with the social network through the barcodes and mobile phones was new to most of the participants and our results confirmed the findings by Mäkelä et al. (2007) that participants were open and enthusiastic about the concept, but the majority was unfamiliar with the technology. We observed that it was easy for participants of all ages to grasp the idea of the social network and many participants were genuinely curious about various details of the concept, from the general motivation to technical details. All the technical details were hidden from the participants, so no special skills were required to participate. In general, people wanted to have their picture taken for the profile. They also were not hesitant to submit their phone number. We were only collecting first names (there was no reason to collect last names), but people were still typing in their full name. It seems that letting people type in their own name is a good solution as it made the whole process quicker and less error-prone. Also people liked the fact that they participated in the profile creation process in an active way. However, it was somewhat surprising that the majority of participants were not hesitant to provide full name and phone number when registering. For privacy reasons only the nickname was shown on the microblog on the situated displays (not the phone number).

Microblog messages usually followed profile creation, as some participants immediately wanted to try using their profile in the social network. Once the profile was created they interacted by sending a message to the microblog. However, on day one we observed that no participants came back later to send additional messages to the microblog. Again we believe that the way the displays were situated played a notable role, as discussed above. On day two however, we could see that even though the profile creation was finished, we still received several messages on the microblog. These messages containing statements about the festival ending, e.g. "We miss the festival already".

So some participants had taken up the microblog and posted more messages to it. In general, we can observe less uptake on the second day, as this was related to the festival program: roughly the same number of people appeared on the second day, but more concerts could have drawn people's attention away from our project.

Our pilot study has shown that it is feasible to use 2D barcodes on wristbands for unique identification of participants. The scanning process was acceptable given the necessary scanning frequency and the conditions at the festival. We have observed some interesting behavior patterns related to the activities in our project. Interaction of single participants with the system is limited to the time and place where they and other participants can immediately see the effect of their actions. For example, newly uploaded photos should appear instantly in the slideshow. Research in the area of urban computing has received critique in terms of applications not being designed for a broad audience (Thom-Santelli, 2007) as it may require special equipment or advanced mobile phones. No special equipment was required of the participants in this project as the scanners (mobile phones) and software was carried by the volunteers that took part in the experiment at the festival. This allowed anyone with a wristband and 2D barcode to become part of the social network. Adding to the microblog only required a mobile phone with SMS capabilities.

In the present study we have used 2D barcodes to uniquely identify participants, but an obvious alternative would be RFID tags (O'Neill, Thompson, Garzonis, & Warr, 2007). Scanning RFID tags require special hardware. Presently only a few mobile phones have such technology integrated that would allow a similar system to be built based on RFID for participant identification. Compared to the 2D barcodes, the biggest differences are that RFID scanning time is typically shorter and that it can be done from a larger distance. Light and weather conditions do not influence the scanning process as it does for the optically scanned 2D barcodes. In addition an RFID chip can encode more data and also data can be written to the chip. The largest difference from a participant perspective is that it is not transparent to the participant when his/her RFID chip is scanned. Some participants expressed their opinion that the 2D barcodes had a more "personal touch" than RFID tags, which some knew from work or school. The 2D barcodes allow the participants more privacy (Lee & Kim, 2006; Mäkelä et al., 2007), as it is ultimately the decision of the participant whether to show or to hide the barcode. This means that they are completely in control of their own level of participation.

In our test environment a scan of the 2D barcode takes about 10 seconds (including starting of the camera). If the camera is running constantly a scan can be performed in less than one second, but this requires more battery resources on the mobile phone. Using 2D barcodes gives many new opportunities (Schmidmayr et al., 2008), when compared to classical tickets or RFID cards. It is a low-cost solution and as demonstrated in this study, it is easy to integrate different applications on a mobile phone: barcode scanning, photo taking, GPS location, network communication, etc. The collection of software components and flexibility on a mobile phone device can facilitate easy prototyping of additional applications.

Future Work

In this study we have found 2D barcodes to be suitable for person and object identification when deploying a number of prototype applications with casual activities and a fairly low scanning frequency. However, we recognize that larger-scale deployment could benefit from different technologies for participant identification, such as RFID, or perhaps a combination of the two technologies. RFID supports an additional domain of applications, such as crowd tracking, as the RFID chips can be scanned from a larger distance. Analysis of movement patterns and activity level can address potential optimizations of the festival layout for enhanced social interaction by knowing how festival participants are distributed across the festival in terms of different profiles and interests. A potential downside of this approach is the transparency of the technology for the participants as mentioned above.

In our setup mobile phones were applied as 2D barcode scanners, meaning that the scanning process had to be carried out by a volunteer. This could potentially prevent some participants from interacting in the social network, as it requires interacting with the volunteer as a mediator. Future deployment could use fixed scanners and be extended to include special features for advanced mobile phone users, such as scanning the barcode using a personal mobile phone. Moreover the barcode could encode more information, such as a URL linking directly to the personal profile webpage or a Facebook application if integrated with an existing social network service. Having an RFID-based solution could perhaps benefit from the option of writing data to the RFID chip on the participant wristband, rather than the present one-way reading of a unique identifier embedded in a 2D barcode. In the present implementation we addressed this limitation of the 2D barcodes by writing information to the participant profile on the backend server instead, which then relied on network availability.

In the pilot study we have created our own special-purpose social network, with profile pages, photo sharing, and microblogs messages. In future deployment it would probably be beneficial to integrate those services with an existing social network, such as Facebook, to utilize already existing user profiles. This could also facilitate participant interaction taking place beyond the duration of the event. The architecture of our platform makes it possible to expand it and new applications can be added to the architecture application layer. The enabling technology is the unique identification

provided by the 2D barcodes, enabling people and objects to be linked to a system entity, such as a participant profile. We suggest that there is a domain of social networking applications which are relevant in the context of an event related to the spatial and temporal proximity of participants.

Finally, it could be interesting to deploy the platform and applications at larger festival events, with thousands or 10s of thousands participants who would participate in the activities created. It is important for such deployment that special equipment is operated by volunteers, where only a short training is required, as the applications are designed to be easy to operate. It is our experience that one of the largest challenges in deploying the system on a large scale (a large physical area) is providing a reliable network connection for all scanning points. Experiments at a larger scale event would provide further insights on the social interaction as well as the technical issues.

CONCLUSION

We have deployed a festival-wide social network with participant profiles with photo sharing and microblog for large displays at the CO2PENHA-GEN music and arts festival. Positive feedback was received from both festival participants and the organizers. Based on our observations and the experimental results we have initial indications that the approach taken using 2D barcodes to identify participants and linking them to a social network profile is a useful low-cost approach for events at this scale. Our novel approach enabled anyone at the festival to participate in the social network, as participants did not need any special hardware or mobile applications to participate. The pilot study has provided valuable insights how to improve the social network application for future experiments at larger scale events, and we find further studies relevant to obtain further insights.

ACKNOWLEDGMENT

We would like to thank the organizers of the CO2PENHAGEN festival for enabling us to carry out the experiments and to the volunteers and 73 participants that took part in the study. Also thanks to QuickMark for sponsoring the 2D barcode reader software that was used in the prototype application and to Forum Nokia for the mobile phones used in the experiment. Finally we would like to thank Kathy Smith.

REFERENCES

Brignull, H., & Rogers, Y. (2003). Enticing people to interact with large public displays in public places. In *Proceedings of INTERACT* (pp. 17-24).

Cheverst, K., Coulton, P., Bamford, W., & Taylor, N. (2008). Supporting (mobile) user experience at a rural village 'scarecrow festival': A formative study of a geo-located photo mashup utilising a situated display. In *Proceedings of the Mobile HCI Workshop on Mobile Interaction in the Real World* (pp. 27-31).

Churchill, E., Girgensohn, A., Nelson, L., & Lee, A. (2004). Blending digital and physical spaces for ubiquitous community participation. *Communications of the ACM, 47*(2), 38–44. doi:10.1145/966389.966413

Coulton, P., Bamford, W., & Edwards, R. (2008). Mud, mobiles and a large interactive display. In *Proceedings of the OzCHI Conference on Public and Situated Displays to Support Communities.*

Greenberg, S., & Rounding, M. (2001). The notification collage: Posting information to public and personal displays. In *Proceedings of the SIGCHI Conference on Human Factors in Computing Systems* (pp. 514-521).

Hansen, F. A., & Grønbæk, K. (2008). Social web applications in the city: A lightweight infrastructure for urban computing. In *Proceedings of the 19th ACM Conference on Hypertext and Hypermedia* (pp. 175-180).

Jacucci, G., Oulasvirta, A., Ilmonen, T., Evans, J., & Salovaara, A. (2007). CoMedia: Mobile group media for active spectatorship. In *Proceedings of the SIGCHI Conference on Human Factors in Computing Systems* (pp. 1273-1282).

Kindberg, T., Chalmers, M., & Paulos, E. (2007). Guest editors' introduction: Urban computing. *IEEE Pervasive Computing / IEEE Computer Society [and] IEEE Communications Society, 6*(3), 46–51. doi:10.1109/MPRV.2007.57

Kurti, A., Milrad, M., & Spikol, D. (2007). Designing innovative learning activities using ubiquitous computing. In *Proceedings of the Seventh IEEE International Conference on Advanced Learning Technologies* (pp. 386-390).

Lee, H., & Kim, J. (2006). Privacy threats and issues in mobile RFID. In *Proceedings of the First International Conference on Availability, Reliability and Security.*

Mäkelä, K., Belt, S., Greenblatt, D., & Häkkilä, J. (2007). Mobile interaction with visual and RFID tags: A field study on user perceptions. In *Proceedings of the SIGCHI Conference on Human Factors in Computing Systems* (pp. 991-994).

Mun, I. K., Kantrowitz, A. B., Carmel, P. W., Mason, K. P., & Engels, D. W. (2007). Active RFID system augmented with 2D barcode for asset management in a hospital setting. In *Proceedings of the IEEE International Conference on RFID* (pp. 205-211).

O'Hara, K., & Kindberg, T. (2007). Understanding user engagement with barcoded signs in the 'coast' location-based experience. *Journal of Location Based Services, 1*(4), 256–273. doi:10.1080/17489720802183423

O'Neill, E., Thompson, P., Garzonis, G., & Warr, A. (2007). Reach out and touch: Using NFC and 2D barcodes for service discovery and interaction with mobile devices. In A. LaMarca, M. Langheinrich, & K. N. Truong (Eds.), *Proceedings of the 5ᵗʰ International Conference on Pervasive Computing* (LNCS 4480, pp. 19-36).

Peltonen, P., Salovaara, A., Jacucci, G., Ilmonen, T., Ardito, C., Saarikko, P., et al. (2007). Extending large-scale event participation with user-created mobile media on a public display. In *Proceedings of the 6th International Conference on Mobile and Ubiquitous Multimedia* (pp. 131-138).

Schmidmayr, P., Ebner, M., & Kappe, F. (2008). What's the power behind 2D barcodes? Are they the foundation of the revival of print media? In *Proceeding of the International Conference on New Media Technology* (pp. 234-242).

Smith, I. (2005). Social-mobile applications. *IEEE Computer, 38*(4), 84–85.

Swedberg, C. (2009). *Ohio music festival sings RFID's praises.* Retrieved from http://www.rfid-journal.com/article/articleview/4985

Thom-Santelli, J. (2007). Mobile social software: Facilitating serendipity or encouraging homogeneity? *IEEE Pervasive Computing / IEEE Computer Society [and] IEEE Communications Society, 6*(3), 46–51. doi:10.1109/MPRV.2007.60

Tuulos, V. H., Scheible, J., & Nyholm, H. (2007). Combining web, mobile phones and public displays in large-scale: Manhattan story mashup. In A. LaMarca, M. Langheinrich, & K. N. Truong (Eds.), *Proceedings of the 5ᵗʰ International Conference on Pervasive Computing* (LNCS 4480, pp. 37-54).

Zeni, N., Kiyavitskaya, N., Barbera, S., Oztaysi, B., & Mich, L. (2009). RFID-based action tracking for measuring the impact of cultural events on tourism . In Höpken, W., Gretzel, U., & Law, R. (Eds.), *Information and communication technologies in tourism* (pp. 223–235). New York, NY: Springer.

This work was previously published in the International Journal of Mobile Human Computer Interaction, Volume 3, Issue 3, edited by Joanna Lumsden, pp.14-31, copyright 2011 by IGI Publishing (an imprint of IGI Global).

Chapter 12
Wearable Tactile Display of Landmarks and Direction for Pedestrian Navigation:
A User Survey and Evaluation

Mayuree Srikulwong
University of Bath, UK

Eamonn O'Neill
University of Bath, UK

ABSTRACT

This research investigates representation techniques for spatial and related information in the design of tactile displays for pedestrian navigation systems. The paper reports on a user survey that identified and categorized landmarks used in pedestrian navigation in the urban context. The results show commonalities of landmark use in urban spaces worldwide. The survey results were then used in an experimental study that compared two tactile techniques for landmark representation using one or two actuators. Techniques were compared on 4 measures: distinguishability, learnability, memorability, and user preferences. Results from the lab-based evaluation showed that users performed equally well using either technique to represent just landmarks alone. However, when landmark representations were presented together with directional signals, performance with the one-actuator technique was significantly reduced while performance with the two-actuator approach remained unchanged. The results of this ongoing research programme can be used to help guide design for presenting key landmark information on wearable tactile displays.

DOI: 10.4018/978-1-4666-2068-1.ch012

INTRODUCTION

Tactile navigation displays have the potential to be deployed as an alternative or complement to visual navigation displays. They have been reported to work effectively in environments where there are different forms of noise and environmental constraints and when users' attention, visibility and audibility may be limited (Tan et al., 2003; Van Erp et al., 2005; Duistermaat, 2005; Ross & Blasch, 2000).

Our eventual design goal is to create a spatial display that imposes fewer requirements for extensive transformations between frames of reference by a human operator (Millar & Al-Attar, 2004) and allows the user to achieve high task performance in challenging situations. Outstanding challenges with tactile displays for navigation systems include selection of spatial information types and their representation. In this paper, we describe two linked empirical studies. The first identifies contextually prioritised landmark categories important for different types of navigation. The second describes an experimental comparison of tactile representation techniques for such landmark categories on a wearable device for pedestrian navigation.

USER SURVEY OF LANDMARK USE

Several researchers, May et al. (2003), Burnett et al. (2001), Raubal and Winter (2002), and Klippel and Winter (2005), have suggested that a navigation system's value could be improved by providing landmark information in addition to the common use of directional information, however, there has been no reported use of landmark information in tactile navigation displays.

Landmarks for human navigation can be any objects or places that are stationary, distinct and salient (May et al., 2003; Grabler et al., 2008). Landmarks are identified by their salience (Raubal & Winter, 2002; Klippel & Winter 2005), subjec-

tively and depending on the mode of navigation (Allen, 1999). That is, landmarks are not objective and universal but are chosen subjectively by individuals, particularly in learning and recalling turning points along routes (Sorrows & Hirtle, 1999).

Landmarks play two major roles in navigation: as an organizing concept for space and as a navigation tool (Golledge, 1999). In organizing space, landmarks can represent a cluster of objects at a higher level of abstraction or scale and present an anchor for understanding local spatial relations (Golledge, 1999). For example, symbolic landmarks, such as the Eiffel Tower in Paris or the Statue of Liberty in New York, can come to represent an entire city. They serve as reference points; other landmarks or objects are recalled as being near them and not vice versa. These symbolic landmarks are defined by their visibility from a distance and, especially, their great cultural importance (Sorrows & Hirtle, 1999).

As a navigation tool, landmarks are used to identify decision, origin and destination points. They also provide confirmation of route progress and orientation cues for homing vectors (Golledge, 1999). Landmarks enable the human to construct spatial relationships between objects and routes for the development of her cognitive map of the space (Raubal & Winter, 2002; Millonig & Schechtner, 2005; Michon & Denis, 2001).

According to Allen (1999), human wayfinding can be categorized into three types: traveling to a familiar destination (*commuting*); traveling to an unknown destination (*questing*); and exploring the area, which may or may not involve visiting important landmarks (*exploring*).

Based on human perception and memory limitations, previous research has recommended that the number of tactile patterns to be presented should not exceed seven (Millonig & Schechtner, 2005; Chan et al., 2005; Gallace et al., 2006). These findings suggest an upper bound on the number of landmarks it may be useful to represent within a given navigation task and context. Given such a constraint, it is important to identify a small set of landmarks that are most likely to be useful. However, existing navigation systems typically present quite large sets of landmark information (Millonig & Schechtner, 2005; Nokia Maps™ 2.0; Garmin Nuvi™) (see the Appendix). Our first study empirically identified and classified a set of landmarks or landmark types appropriate for use in tactile navigation systems that support the three navigation purposes, commuting, questing and exploring.

Online and Face to Face Survey

Given our desire to include participants from different urban settings around the world, an online survey was an appropriate approach for this study. However, online surveys can be limited by their lack of direct interaction between interviewer and interviewee, therefore, we also conducted face-to-face interviews *in situ* with participants who had just been engaged in an urban pedestrian journey. The online and face-to-face surveys were intended to be different and complementary.

Choice of Landmarks

We gathered our reference set of landmarks by combining lists published in several research papers, Burnett (1998), Burnett et al. (2001), Millonig and Schechtner (2007), as well as ones that are presented in commercial pedestrian navigation systems, e.g. Nokia Maps™ 2.0; Garmin Nuvi™. Our set includes 50 kinds of landmarks such as traffic lights, monuments and markets. The full set is given in the Appendix.

Prior to our main questionnaire study, we ran several pilot interviewing sessions in an attempt to regroup and classify these landmarks at an appropriate level of abstraction. For example, using higher-level abstractions, e.g. grouping monument, museum, memorial and gallery into a tourist attraction category, might be less useful for navigation since tourist attraction is some-

times too generic to make identification easy on the ground. On the other hand, providing finer detail, e.g. identifying each individual landmark as specifically as possible, could make the set of landmarks unmanageably large, exacerbating the problems of using mobile navigation aids as described above. Feedback from these sessions confirmed the level of abstraction given in the Appendix as appropriate. To mitigate the forced choice nature of the resulting questionnaire, we provided free text areas where participants could report landmarks that were not included in our reference set.

Procedures and Rating Scales

For the online version, each participant answered three parts of the questionnaire, corresponding to questions about using landmarks in pedestrian navigation for three purposes: commuting, questing and exploring. In the face-to-face interviews, each participant first identified which of the three purposes they had just been engaged in and then answered the questions only with respect to that purpose of navigation.

For each journey with a particular navigation purpose, each participant first identified: (1) a navigated area, (2) if they used landmarks, and (3) if such landmarks were in the physical space or on any guidance system, e.g. a map. They then rated each of the 50 landmarks in our reference set by their importance as navigational aids for the journey. This was done on a 5-point scale, 1 being 'not use', 2 being 'use, not important' and 5 being 'use, very important'. We also collected data on the timing of their use of each landmark. Following May et al. (2003), we presented three choices of usage timing: before decision points, between points on the route, and both. Participants were given opportunities to specify other kinds of landmarks used that were not included in our set.

For any landmarks that received equal rating scores for a journey, the participant was asked to rank them by their descending importance for that particular navigational purpose.

Results

From the online participants, we collected 100 complete responses from different geographic locations, 40 males and 60 females. 61% of online responses were from Asia; 33% were from Europe; 5% were from Africa and 1% was from Australia. Navigated locations were urban areas within different sized cities, including for example London, Pisa, Bangkok, Aachen and San Francisco. There are of course differences in land use in the different geographical locations but for the routes described by our online participants, these urban spaces are similar with respect to the key characteristics we are interested in here. That is, they can be considered as dense (i.e. there are relatively large numbers of objects and cues in the space) and cluttered (i.e. the number of objects is so great that they may obscure important landmarks or cues) (Carter & Fourney, 2005).

We conducted 60 face-to-face interviews in one UK city, 32 males and 28 females. The city is relatively small, with an area of 11 square miles (28 km²). The population of the city is approximately 100,000 inhabitants. It is a major tourist centre of the region with over one million staying visitors and 3.8 million day visitors per year (B&NES Council, 2008). The city centre where the interviews were conducted is dense and cluttered.

Results From Online Participants

Of the 52 online participants who reported using landmarks to aid commuting, 48 (92%) used only landmarks in the physical spaces through which they were navigating while the other 4 stated that they used landmarks both in the physical spaces and on public map displays, i.e. large two-dimensional visual representations of the area. The fuller responses of these 4 explained

that the areas in which they commuted are large transportation hubs, such as a main train station, an airport and very large department stores where the interior components and structures look alike. They are crowded places and pedestrians needed an orientation aid to maintain their pace. This was achieved by glancing at the public maps provided on display stands along their routes.

Of the 75 online participants who reported using landmarks during a questing journey, 36 (48%) used landmarks only in the physical spaces, while the other 39 (52%) matched landmarks on maps with landmarks in the physical spaces to aid their navigation.

Of the 62 online participants who reported using landmarks during an exploring journey, 15 (24%) reported using landmarks only in physical spaces while the other 47 (76%) used landmarks both in the physical spaces and on maps.

As the results were not normally distributed, we ran non-parametric statistics. For the overall percentage of landmark use, Friedman's ANOVA found a significant difference in the number of landmarks used (both physical and in maps) across the three navigational purposes ($\chi2(2) = 12.03, p = .002$). Wilcoxon tests were used to follow up this finding. A Bonferroni correction was applied and all effects are reported at a .0167 level of significance. There were no significant differences in the number of landmarks (both physical and in maps) used between quest and explore ($T = 315, r = -.17$) and between commute and explore ($T = 891, r = -.18$). However, participants used significantly more landmarks (physical and in maps) when questing than when commuting ($T = 943.5, r = -.34$).

Prior to the study, we suspected that pedestrians would depend most on landmarks during their exploration trip, however, our results indicate that they used landmarks most while questing. The qualitative data reveal the reason for this to be that many explorers prefer 'getting lost in space' to truly appreciate the exploratory experience.

Results From Face-To-Face Participants

The face-to-face interviews yielded the following results. Of the 19 people who used landmarks during questing, 6 (31.5%) used landmarks only in the physical spaces, while the other 13 (68.5%) matched landmarks on maps and in the physical spaces to aid their navigation.

Of the 18 'explorers', 1 (5.5%) depended on landmarks only in physical space while the other 17 (94.5%) matched physical landmarks to landmarks on maps.

Of the 20 commuters, 1 (5%) stated that he always looks at one particular physical landmark during his (frequent) performance of his commuting journey. The other 19 (95%) commuters each reported that they used no landmarks to support commuting.

As the results were not normally distributed, we ran non-parametric statistics. In contrast to our online participants, participants across the three navigation purposes from our interview sessions were independent groups. Hence, we used the Kruskal-Wallis test. Results showed that the number of landmarks used by the face-to-face interviewees was significantly affected by navigation purpose ($H(2) = 43.33$, $p < .002$). Mann-Whitney tests were used to follow up this finding. A Bonferroni correction was applied and all effects are reported at a .0167 level of significance. There were no significant differences in the number of landmarks used between questing and exploring ($U = 190$, $r = -.09$). However, participants used significantly more landmarks when questing ($U = 20$, $r = -.88$) and when exploring ($U = 30$, $r = -.84$) than when commuting.

Detailed Results

Results from both online questionnaires and face-to-face interview were used to calculate: frequency (F), importance (I) and ranking (R) scores. The frequency (F) score is the number of times each landmark was used across respondents for a particular navigational purpose. The importance score is a summation of the weighted importance of each landmark across all respondents. The ranking score is a summation of weighted ranked scores of each landmark across all respondents.

Tables 1 and 2 show online questionnaire results and Table 3 presents results from the interview sessions.

In the online questionnaires, participants were asked to think about actual journeys they had made for each navigational purpose and described the landmarks they used in those journeys together with the landmarks' importance for that particular journey. These results are presented in Table 1. The frequency score (F) is the number of times each landmark was used across respondents for a particular navigational purpose. For example, for the exploring purpose, tourist attraction was used 41 times so its frequency score is 41.

Table 1. Common Landmarks Used in Specific Journeys with their Frequency (F) and Importance (I) for those Journeys (from Online Questionnaire)

Purpose	Common Landmarks	F Scores	I Scores
Commute	Mall and Market	32	96
	Traffic light	32	91
	Public transport	31	87
	Bridge	31	73
	Financial service	29	74
Quest	Mall and Market	40	113
	Bridge	38	110
	Railway stations	31	91
	Tourist attraction	31	89
	Religious place	32	85
	Traffic light	33	83
	Restaurant	33	81
Explore	Tourist attraction	41	131
	Hotels	30	96
	Mall and Market	32	93
	Bridge	33	83
	Monument and Memorial	28	89
	Religious place	29	81
	Public transport	25	73

Table 2. Most Important Landmarks with Their General Importance Ranking (R) Scores (from Online Questionnaire)

Purpose	Important Landmarks	R Scores
Commute	Well-known shops / business	111
	Mall and Market	107
	Traffic light	99
	Public transport	82
	ATM	75
	Educational institute	53
	Bridge	48
Quest	Mall and Market	147
	Well-known shops/business	134
	Bridge	103
	Tourist attraction	97
	Hotels	86
	Religious place	75
	Restaurant	67
Explore	Tourist attraction	145
	Hotels	87
	Mall and Market	83
	Other unique landmarks	74
	Monument and Memorial	66
	Railway station	62
	Religious place	52

Table 3. Top Landmarks in the city of Bath with their Frequency (F), Importance (I), and General Importance Ranking (R) Scores (from Face-to-Face Interviews)

Purpose	Top Landmarks	F	I	R
Commute	Monument and Memorial	1	2	7
Quest	Mall and Market	8	25	42
	Public transport	7	23	32
	River	7	22	36
	Religious place	7	20	35
	Bar and Pub	7	15	31
	Railway Station	5	18	30
	Monument and Memorial	5	15	15
Explore	Tourist attraction	18	64	114
	Railway station	16	58	66
	Museum and Gallery	17	52	76
	Monument and Memorial	16	50	44
	River	17	44	22
	Public transport	13	47	60
	Religious place	14	38	27

We derived the importance (I) scores as follows:

Importance score $= \Sigma$ ((Σ (not important rating) * 1), (Σ (slightly important rating) * 2), (Σ (important rating) * 3), (Σ (very important rating) * 4))

Thus, importance score is a summation of the weighted importance of each landmark across all respondents. For example, the frequency of 41 for tourist attraction was divided into frequencies of 2, 8, 11, and 20 respectively for landmarks being (1) used but not important, (2) slightly important, (3) important and (4) very important. Hence, the overall importance score of tourist attraction = (2*1)+(8*2)+(11*3)+(20)*4 = 131. Mall & market, and Bridge are amongst the most common landmarks that were important as navigation cues across all three navigation purposes.

In the online questionnaires participants were also asked to rank the importance in general of landmarks that they usually rely on when they embark on journeys, regardless of the area they are navigating. Respondents were asked to select the 7 most important of these generally used landmarks. Table 2 presents a list of the most common landmarks that participants (subjectively) considered important in general, rather than important for the particular journeys described in Table 1. We derived the list by calculating each landmark's general importance ranking (R) score as follows:

Ranking score $= \Sigma$ ((Rank1*7), ((Rank2*6), (Rank3*5), (Rank4*4), (Rank5*3), (Rank6*2), (Rank7*1))

Thus, for Table 2, a ranking score is a summation of weighted ranked scores of each landmark across all respondents. For example, 8, 7, 7, 1, 2, 1, and 0 respondents gave a rank of 1 to 7 respectively to tourist attraction. Hence, its ranking score equals (8*1) + (7*2) + (7*3) + (1*4) + (2*5) + (1*6) + (0) = 145.

Table 2 corroborates the finding in Table 1 that Mall & Market is very important as a navigation cue because it appears consistently across all three navigation purposes. The category 'Well-known shops/business' emerges in the commuting and the questing purposes because pedestrians refer to landmarks by their brands, e.g. McDonald's is one of the most frequently used landmarks. Similarly, the category 'Other unique landmarks' appeared fourth in the explore column because pedestrians did not refer to some symbolic landmarks as tourist attractions but rather by their unique names, e.g. the Eiffel Tower.

Table 3 shows similar ranking of landmarks in a single city from the face-to-face interviews. Journey specific frequency, journey specific importance and general importance were calculated in the same way as for the online questionnaire results presented in Tables 1 and 2. (Since the results from the face-to-face interviews were less diverse than the results from the online questionnaires it is possible to present the face-to-face interview results in a single table.)

The average number of landmarks used per journey for different navigation purposes are as follows: 6.5 (commuting), 12.47 (questing), and 11.04 (exploring) from the online survey; 1 (commuting), 4.42 (questing) and 10.4 (exploring) from the face-to-face interviews. Based on routes taken by each interviewee, we found that factors influencing these numbers include: differences in length of journey, i.e. a quest is normally shorter than an exploratory journey; and differences in the number of destinations, i.e. a quest normally involves one destination while an exploration may involve one or many more destinations. Nevertheless, further study on the factors influencing these patterns would be required.

Qualitative Results

Qualitative data from the face-to-face interviews revealed several interesting navigation patterns:

- For exploration, pedestrians generally have an intention to visit some culturally important landmarks. These landmarks serve as their destinations. Nevertheless, they have little idea what generic landmarks they would use to aid their navigation to reach such destinations. Hence, they decided not to use landmarks as navigation cues or confirmation that they were on the right path. Instead, to reach destination landmarks, they relied on directional and textual information, e.g. street names.

- Some explorers navigated blindly by following another explorer, e.g. a friend who excels in navigation. These pedestrians were not able to remember any landmarks along the route except destination landmarks. Although they remembered these destination landmarks, they were not able to associate their locations with the whole route.

- About 60% of explorers who use maps tend to use them continuously throughout their journey and depend entirely on them.

- Most of the 'questers' stated that they would first study the route and try to memorise directions and landmarks leading to the destination. Once they embarked on the journey, they would try to recall the route and associate landmarks seen in physical space with landmarks in their memory.

- The number of landmarks used per journey for the quest and exploratory purposes may vary depending on the nature of routes and areas. For example, some large cities contain a wide range of different landmarks that may be distant from each other, while other smaller urban areas contain fewer landmark categories that are more proximally located in the immediate vicinity.

The Preferred Set of Landmarks Based on Their Scores

Table 4 shows a side-by-side comparison of the overall ratings by both online and face-to-face participants. Based on the frequency of appearance across cells in Table 4, the most important landmark is *mall and market* since it scores highly in all but 2 of the cells (the commuting and exploring purposes in the city in which we ran the interview sessions).

As our findings reveal that pedestrians use landmarks primarily for questing and exploring, we will focus on the results for these two purposes. As illustrated in Table 4, the second and third most important landmarks are religious place and tourist attraction. These 2 categories of landmark were used extensively for these two navigation purposes. The next most important landmarks were railway station, monument *and* memorial, public transportation and bridge.

Thus, if we have to choose a small set of landmarks for use in mobile pedestrian navigation aids, according to results of our study, the most suitable landmarks should be: mall and market, religious place, tourist attraction, public transportation, bridge, monument and memorial, and railway station. These findings corroborate the results of previous research that at the top end of the scale there are some generic landmarks, which will be appropriate across different environments (Burnett, 1998).

In addition to generic landmarks, there is the 'other unique landmark' category that is crucial to navigation but is not generalisable. This category includes symbolic or iconic landmarks of the city or famous chain stores that are located in strategic areas or at important decision points on the route. For the symbolic landmarks type, they could be instances of the generic landmarks, e.g. *tourist attractions*.

Nonetheless, these top-ranked landmark categories might not be generalisable due to different

Table 4. Top Ranked Landmarks in Descending Order Based on a Summation of F, I and R Scores

Purpose	Top Landmarks (Global Rating) From Online Responses	Top Landmarks (Global Ranking) From Online Responses	Top Landmarks (of One City) From face-to-face Responses
Commute	Mall and Market Traffic light Public transport Bridge Financial service	Well-known shops / business Mall and Market Traffic light Public transport ATM Educational institute Bridge	Monument and Memorial
Quest	Mall and Market Bridge Railway stations Tourist attraction Religious place Traffic light Restaurant	Mall and Market Well-known shops / business Bridge Tourist attraction Hotels Religious place Restaurant	Mall and Market Public transport River Religious place Bar and Pub Railway Station Monument and Memorial
Explore	Tourist attraction Hotels Mall and Market Bridge Monument and Memorial Religious place Public transport	Tourist attraction Hotels Mall and Market *Other unique landmarks* Monument and Memorial Railway station Religious place	Tourist attraction Railway station Museum and Gallery Monument and Memorial River Public transport Religious place

morphologies of the surveyed cities. Religious places provide a good example. Some cities contain hundreds of landmarks of this category, e.g. temples in Bangkok that are highly visible and distinct from their environments, hence their high frequency of use as landmarks by our respondents. On the other hand, some other cities, e.g. London, contain hundreds of local churches that are not visually or structurally salient (Klippel & Winter, 2005) and our participants did not select such landmarks as cues for navigation in those cities.

The value of a particular category of landmarks varies from one situation to another. For example, tourist attractions offered little assistance to a quest journey while they were the main, if not sole, purpose of exploring. Thus, the same landmark may be used for different navigational purposes but have greater or lesser value for each of them.

It is worth noting that continuous objects, such as a river, were identified in our face-to-face interviews as crucial to navigation. While a visual navigation aid can readily represent such features, an auditory or tactile pedestrian navigation guide would struggle to indicate such landmarks clearly, more so even than the other kinds of landmarks considered. This issue will require further investigation to clarify its potential for real world use.

TACTILE REPRESENTATION OF LANDMARKS

Having empirically identified a small set of landmark types for supporting pedestrian navigation, we next investigated the tactile representation of these landmark types.

Tactile Representation Techniques

To create a distinguishable and learnable set of tactile stimuli, researchers have manipulated attributes such as frequency, amplitude and duration of vibration signals as in MacLean and Enriquez (2003), Tan et al. (2003), and Ternes and MacLean

(2008). Ternes and MacLean (2008) suggest that signal rhythms created by manipulating signal duration provide the most effective result. In this study, we closely followed the design of Ternes and MacLean's heuristic tactile rhythms for our tactile stimuli (Figure 1). Each signal in a set contains different note length and evenness. For example, a 2-second short stimulus contains a number of repetitions of 125 millisecond (ms) notes.

Human skin adapts to continued pressure stimulation resulting in a decrease in sensory experience (Schiffman, 1976). In a navigation system that provides both directional and landmark information, there is a possibility that a user might not be able to identify the differences between signals after her skin has been continually stimulated with similar vibrations. Schiffman (1976) suggests that introducing *discontinuity* can help stabilise sensory perception of different types of signals. This discontinuity can be achieved by increasing the number of contact points on the body, e.g. using a combination of two or more actuators to generate unique stimuli. Although suggested, this technique had not been investigated; therefore, we also examined this technique in our experiment.

To summarise, this study involved investigating the following tactile landmark representation techniques: (1) manipulating the signal rhythms,

Figure 1. One-actuator Technique. Each row represents one bar, represented 2 times as a 2-second stimulus. Each note contains vibration on-time (grey) and off-time (white) that separates it from the next note.

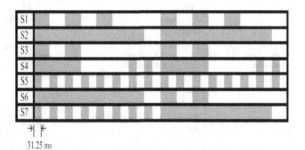

31.25 ms

and (2) increasing the number of body contact areas (i.e. increasing the number of actuators used to display information). In this paper, we refer to the two techniques as the one-actuator and two-actuator techniques respectively. Both techniques for tactile representation of landmarks were presented alone and together with tactile directional signals.

Experimental Evaluation

The prototype wearable device that we built consisted of 8 actuators mounted in a waist belt (Figure 2). Following previous research, Van Erp et al. (2005) and Tsukada (2004), the actuators were unequally spaced (ranging from 50 mm to 130 mm between adjacent actuators) to account for participants' varying body shape and size. The waist belt was worn over light clothing such as a T-shirt. Tasks involved the system generating tactile stimuli and participants identifying perceived directions (Srikulwong & O'Neill, 2010a, 2010b) or landmarks by choosing corresponding pictures of directions or landmarks on a touch screen tablet PC.

There were 20 participants: 10 males and 10 females with an average age of 29 (SD = 4.94, range 20 – 40 years of age). Participants' average waist size was 78 centimetres (cms) (SD = 9.93, range 62-99 cms). We established from pre-test

Figure 2. The Waist Belt Prototype (motor number 3 is the front centre actuator)

questionnaires that all participants understood the concept of "direction" and "landmark" and had no difficulties identifying them. Participants were given training prior to each session.

Each directional tactile stimulus involved actuation of one motor in the corresponding direction and consisted of 12 repetitions of the signal at 50-millisecond pulse and inter-pulse duration (Srikulwong & O'Neill, 2010a, 2010b). The eight directions represented included: east, west, north, south, southeast, southwest, northeast, and northwest. Each actuator represented a direction based on its location around the participant's waist, with north represented by the front centre actuator.

There were 5 experimental conditions in which directions, landmarks or both were represented by the tactile signals: (C1) direction; (C2) landmark with one actuator; (C3) landmark with one actuator + direction; (C4) landmark with two actuators; and (C5) landmark with two actuators + direction. Measurements included learnability, memorability, and distinguishability of landmark signals and their associations, and users' preferences (Table 5).

For the two-actuator conditions, the sets of actuator pairs were as follows (see Figure 2 for referents of actuator numbers):

- The 180° actuator pairs were 3-7, 2-6, 1-5 and 4-8.
- The 90° actuator pairs were 1-3, 2-4, 3-5, 4-6, 5-7, 6-8, 7-1 and 8-2.
- The 135° actuator pairs were 1-4, 2-5, 3-6, 4-7, 5-8, 6-1, 7-2, and 8-3.

Table 5. The study design

Study	Minute start	Training	Measurement	Conditions
Stage 1	1	Yes	Distinguishability, Learnability	C1 – C5
Stage 2	30	No	Memorability	C2 and C4

Experimental Procedures

Our empirical study had two independent variables: representation technique (one actuator or two) and the presence (or absence) of directional signals; dependent variables were reaction time (in milliseconds) and performance accuracy. Reaction time refers to the duration between the end of each stimulus and the participant's response to it.

For both techniques, perception of these arbitrary signals was expected to improve through explicit learning and remembering (Garzonis et al., 2009). Prior to carrying out the experiment for each of the tactile stimulus types, participants were given opportunities to learn the signal patterns and their associations in a four-step training process. The four steps included (1) the system displaying signals, (2) participants freely choosing stimuli to be displayed, (3) participants indicating the landmark corresponding to a given tactile stimulus, and (4) participants indicating a set of landmarks according to generated stimuli. Training stopped when participants scored over 71% accuracy or had been through 4 repetitions of the whole 4-step process.

Stage 1: Measuring Learnability and Distinguishability

At this stage, we investigated whether performance with the two tactile representation techniques for landmarks would differ in terms of learnability and distinguishability. Tasks involved the system generating tactile stimuli and participants identifying perceived directions or landmarks by choosing corresponding pictures on a touch screen tablet PC. We compared a range of performance measures: corrected perceived directions, corrected perceived landmarks and reaction time.

The order of the experimental conditions was counterbalanced amongst participants. Vibration signals in all conditions were generated in a pseudo-random order. Vibration signals and meaning associations were counterbalanced amongst

participants. In addition, actuator pairs used in C4 and C5 were counterbalanced.

In condition C1 (direction only), participants experienced 3 repetitions of 8 directional signals (24 signals in total). In C2, C3, C4, and C5, participants experienced 21 signals (i.e. 7 landmarks x 3 repetitions) for each condition. Repetitions were introduced to reduce the likelihood that participants might achieve correct answers by chance. In C3 and C5, the system generated a random directional signal, paused for 2 seconds, and then generated a landmark signal on one actuator (C3) or on a pair of actuators (C5). Once participants had finished all 5 conditions, they were asked to answer questions comparing the one-actuator and two-actuator techniques.

Stage 2: Measuring Memorability

Stage 2 took place 30 minutes after they had been exposed to each type of landmark vibration stimuli. During this time participants completed a distraction task (answering a questionnaire and discussing their experience of the experiment. Participants were then asked to repeat conditions 2 and 4 in the same counterbalanced order in which they had carried them out in stage 1.

Hypotheses

The vibration signals designed for directions are symbolically straightforward. They involve symbolic mapping of a limited set of cardinal and ordinal directions to their respective vibration signals on corresponding parts of the body; in this and other cases (Van Erp et al., 2005), an absolute point vibration for each designated direction on a distributed placement of actuators around the waist. The representation of landmarks is more challenging. In the navigation design domain, the large set of landmarks studied in research papers and used in commercial systems are not systematically classified, are highly diverse and often are poorly differentiated. As a result, signal patterns for landmarks and their meaning associations are effectively arbitrary. Hence, it was hypothesised that learning time required for landmark representations will be significantly longer than those for directions (**H1**) as participants have to learn the association between the signal and what it represents.

Previous research by Tan et al. (2003), Brown et al. (2006), and MacLean and Enriquez (2003), has suggested that humans can recognize 4-7 abstract tactile patterns and associate them with predefined meanings. We hypothesized that participants will be able to recognize 7 landmarks with at least 80% accuracy in at least 1 non-control condition, either in condition 2 or 4 (**H2**). Based on the same previous research, we predicted that participants will be able to distinguish landmark from directional signals in conditions 3 and 5 (**H3**).

However, in conditions 3 and 5 where we present directional signals together with landmark signals, we hypothesised that the presence of direction signals will reduce participants' performance in recognizing landmark patterns (**H4**). This is due to the human memory and limited attention capacity (Schiffman, 1976).

While many researchers like Ternes and MacLean (2008), Brown et al. (2006), and Tan et al. (2003), have concluded that using different signal rhythms will effectively make stimuli distinguishable but the combination of two simultaneous actuator vibrations might make the iconic stimuli

for landmarks more functionally unique (Schiffman, 1976; Loomis & Lederman, 1986) and clearly distinguishable from directional stimuli. Hence, we predicted that the two-actuator technique will produce better performance than the one-actuator technique when representing landmarks in a waist-belt tactile display that provides both directional and landmark information (**H5**).

Finally, both the one-actuator and the two-actuator representation techniques are abstract and are arbitrarily associated to landmarks. Therefore, we predicted that both techniques' forgetting rate would be equal (**H6**).

Results

For the two-actuator technique, we varied pairs of actuators used (explained in the section *Experimental Evaluation*; see Figure 2 for a reference layout). All the 180° actuator pairs were used by all participants. Other pairs were distributed evenly across all participants.

Participants performed well with the actuator pairs that were vertically or horizontally aligned with their body. These pairs included the 2-4, 1-5, 6-8, 2-8, 3-7 and 4-6 pairs. The next best pair among the rest was the 3-8 pair. The actuator pairs that produced the highest performances are shown in Figure 3.

We had expected that participants would have had better performance with the other two 180°

Figure 3. Best actuator pairs

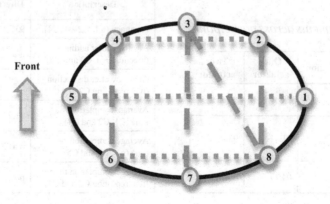

actuator pairs (i.e. the diagonals 4-8 and 2-6). However, results revealed that asymmetric or diagonal pairs did not produce good performance.

Learnability

During the 4-step training, participants were allowed to learn each of the types of vibration signals and their associated meanings. We recorded the number of signals tried, duration in each round and the number of rounds.

A repeated-measures ANOVA indicated a significant difference in overall training requirements for all representation techniques: number of training rounds $F(2, 38) = 16.93, p < .01$, training duration $F(2, 38) = 26.07, p < .01$ and number of signal trials $F(2, 38) = 20.91, p < .01$. Summary data for training requirements is shown in Table 6.

Post-hoc pairwise comparison results (Bonferroni adjustment) indicated that training requirements for both landmark representation techniques were significantly greater than for the direction technique: number of training rounds (both $ps <$ .01), training duration (both $ps < .01$), and number of training signals (both $ps < .01$). No significant difference was found in training duration, rounds and number of signal trials between the one-actuator and the two-actuator techniques (all three with $ps > .05$). Hence, we accepted *H1*.

Prior to the study, we predicted that participants would spend more time and effort in learning landmarks with the one-actuator technique (according to *H5*), as using one actuator might be less

distinguishable than using two actuators. However, our results indicated that participants spent just as much time and effort on either technique. In addition, performance scores for all techniques at the end of the training sessions showed no significant difference $F(2, 38) = 2.82, p > .05$.

Performance and Distinguishability

Table 7 shows accuracy performance across the conditions. A repeated-measures ANOVA showed that the time to complete each condition was not significantly affected by the type of representation technique, $F(2, 38) = 1.60, p > .05$. However, different techniques significantly affected accuracy performance, $F(2, 38) = 3.82, p < .05$. Post-hoc pairwise comparison (Bonferroni adjustment) revealed that participants performed significantly better with directional identification than with both landmark techniques, (both $ps < .05$). There was no difference in accuracy performance between the one-actuator and the two-actuator techniques ($p > .05$).

Results in Table 7 (1st row) showed that participants were able to recognize landmark signals with over 80% accuracy rate for both landmark techniques. Therefore, *H2* was accepted.

Table 6. Training requirements across participants

Training requirements	Direction	One-actuator	Two-actuator
Average number of rounds (round)	4.3	7.6	6.85
Average number of signals tried in all rounds	41	91	95
Average training duration (min:sec)	02:40	04:17	04:14

Table 7. Performance Measures: Accuracy in %, Time in mm:ss

Description	Direction	One actuator	Two actuator
Accuracy (C1, C2 and C4)	93.75	80	82.14
Accuracy after adding direction (n/a, C3 and C5)	n/a	68.1	81.43
Accuracy after distraction (n/a, repeating C2 and C4)	n/a	77.14	83.57
Average completion time (C1, C2 and C4)	01:22	01:40	01:37
Average completion time (n/a, C3 and C5)	n/a	03:08	02:33
Average completion time (n/a, repeating C2 and C4)	n/a	00:55	01:00

Based on the landmark accuracy performances, we concluded that all participants were able to distinguish landmarks from directional signals in both condition 3 and 5. Therefore, we accepted *H3*.

We predicted that the performance of landmark signal perception would be affected by the presence of directional signals. We ran the dependent *t*-test, which compared accuracy performance of C2-C3 (one-actuator technique) and C4-C5 (two-actuator technique). With the one-actuator technique, performance of landmark identification was significantly lower when directional information was presented than when it was absent, $t(19) = 2.65, p < .05$. In contrast, with the two-actuator technique, participants were able to identify landmarks equally well both with and without directions being presented $t(19) = 0.32, p > .05$. Hence, we rejected *H4* since the presence of directional signals affected only the one-actuator technique.

Based on performance in C3 and C5 (Table 7, 2nd row), participants were able to perform significantly better with the two-actuator technique than with the one-actuator technique $t(19) = -2.63$, p $< .05$. Therefore, *H5* was accepted.

Memorability

In order to measure memorability, we distracted participants with interviewing and questionnaire sessions before asking them to repeat conditions 2 and 4. Results are presented in Table 7, 3rd row. Paired-samples *t*-tests showed that there was no significant difference in forgetting rates between the two landmark representation techniques. Specifically, there was (1) no significant difference in accuracy performance between a repeated C2 and a repeated C4, $t(19) = -0.95, p > .05$; and (2) no significant difference in performance between C2 and a repeated C2, $t(19) = 0.64, p > .05$ or between C4 and a repeated C4, $t(19) = -0.51, p > .05$. Therefore, *H6* was accepted.

Table 8. Average Scores of Subjective Measures: n of 5 on a 1-5 likert scale, 1 being low and 5 being high

Subjective Measurements	One actuator	Two actuator
Distinguishable among themselves	3.55	3.55
Memorable	2.75	3.2
Associable with landmarks	2.6	2.65
Distinguishable from directions	4.25	4.5
Interference with directions	2.55	2.3

Subjective Data and Preferences

Post questionnaires were applied at the end of each experimental condition. We gathered user's subjective data on the two landmark representation techniques on several measures. They included: distinguishability of landmark signals from direction signals, distinguishability amongst landmarks themselves, memorability, ease of meaning association, and the level of direction signals' interference with landmark signals. Participants were asked to rate on a 1-5 likert scale, 1 being low and 5 being high. The one-actuator technique scored lower than the two-actuator technique in all subjective measures except for distinguishability amongst landmarks, in which it scored equal marks with the two-actuator technique (Table 8).

As for subjective preferences between the two landmark representation techniques, 12 participants (60%) preferred the two-actuator to the one-actuator technique.

Summary

Our results established that the one-actuator and two-actuator techniques were similar and offered almost equal support for landmark representation alone. To be precise, they required equal amount of training effort, and users performed equally well in the experimental conditions in which solely landmark signals were presented.

However, the two-actuator technique provided better performance than the one-actuator technique when landmarks were presented together with directional signals. This is crucial to the development of a tactile pedestrian navigation system that provides both directional and landmark information.

CONCLUSION

The empirical studies reported in this paper form part of an ongoing research programme investigating the use of tactile displays to support pedestrian navigation. The user-based survey was used to identify and classify the use of various types of landmarks for different navigation purposes (Table 4). Following on from this study, the experimental study implemented and evaluated a prototype wearable tactile interface for indicating direction and types of landmarks. Results suggest that using a two-actuator approach to representing landmarks and directions on a wearable device for pedestrian navigation may be fruitful.

Participants had an average response time of 4 seconds per signal across all conditions. This value is probably just about satisfactory for our intended use. Nevertheless, if these signals were to be used in outdoor urban environments, performance levels might drop since there are several other factors such as different levels of users' cognitive load and levels of noise in those environments. We anticipate that further training might help decrease response time in the lab setting, which might in turn reduce response time in applied environments. Further study is necessary to investigate whether extensive training can better the performance and the extent to which external factors such as noise might affect the results, especially in the field.

Users perceived vibration signals quite well and they were able to recognize a signal's meaning. Our findings suggest that the two-actuator technique was better than the one-actuator technique in several respects, especially as it afforded better performance when landmark signals were presented together with directional signals.

Our experimental study assessed short-term learnability and memorability and we have obtained some promising results. However, based on these results, it is not possible to determine the effects of longer term use. A longitudinal study is required to address such issues.

Our next steps are to refine the tactile navigation prototype for use in field trials in an urban area. Through these investigations, we will evaluate and improve the design and address performance-related benefits and challenges of a wearable tactile pedestrian navigation system.

ACKNOWLEDGMENT

Mayuree Srikulwong's research is supported by the University of Thai Chamber of Commerce, Thailand. Eamonn O'Neill's research is supported by a Royal Society Industry Fellowship at Vodafone Group R&D.

REFERENCES

Allen, G. L. (1999). Cognitive abilities in the service of wayfinding: A functional approach. *The Professional Geographer, 51*(4), 554–561. doi:10.1111/0033-0124.00192

B&NES Council. (2008). *Bath in focus.* Retrieved from http://www.business-matters.biz/site.aspx?i=pg64

Brown, L. M., Brewster, S. A., & Purchase, H. C. (2006). Multidimensional tactons for nonvisual information display in mobile devices. In *Proceedings of the 8th International Symposium on Human Computer Interaction with Mobile Devices and Services* (pp. 231-238). New York, NY: ACM Press.

Burnett, G. E. (1998). *'Turn right at the King's Head' driver's requirements for route guidance information*. Unpublished doctoral dissertation, Loughborough University, Leicestershire, UK.

Burnett, G. E., Smith, D., & May, A. (2001). Supporting the navigation task: Characteristics of 'good' landmarks . In Hanson, M. A. (Ed.), *The annual conference of the ergonomics society* (pp. 441–446). London, UK: Taylor & Francis.

Carter, J., & Fourney, D. (2005). Research based tactile and haptic interaction guidelines . In Carter, J., & Fourney, D. (Eds.), *Guidelines on tactile and haptics interaction* (pp. 84–92). Saskatoon, SK, Canada: University of Saskatchewan.

Chan, A., MacLean, K. E., & McGrenere, J. (2005). Learning and identifying haptic icons under workload. In *Proceedings of the 1st Joint Eurohaptics Conference and Symposium on Haptic Interfaces for Virtual Environment and Teleoperator Systems* (pp. 432-439). Washington DC: IEEE Computer Society.

Duistermaat, M. (2005). *Tactile land in night operations* (Tech. Rep. No. TNO-DV3 2005 M065). Soesterberg, Netherlands: Netherlands Organisation for Applied Scientific Research

Gallace, A., Tan, H. Z., & Spence, C. (2006). Numerosity judgments for tactile stimuli distributed over the body surface. *Perception*, *35*(2), 247–266. doi:10.1068/p5380

Garzonis, S., Jones, S., Jay, T., & O'Neill, E. (2009). Auditory icon and Earcon mobile service notifications: Intuitiveness, learnability, memorability and preference. In *Proceedings of the SIGCHI Conference on Human Factors in Computing Systems* (pp. 1513-1522). New York, NY: ACM Press.

Golledge, R. G. (1999). Human wayfinding and cognitive maps . In Golledge, R. G. (Ed.), *Wayfinding behavior: Cognitive mapping and other spatial process* (pp. 5–45). Baltimore, MD: Johns Hopkins University Press.

Grabler, F., Agrawala, M., Sumner, R. W., & Pauly, M. (2008). Automatic generation of tourist maps. In *Proceedings of the SIGGRAPH 35th International Conference and Exhibition on Computer Graphics and Interaction Techniques* (pp. 1-11). New York, NY: ACM Press.

Klippel, A., & Winter, S. (2005). Structural salience of landmarks for route directions. In A. G. Cohn & D. M. Mark (Eds.), *Proceedings of the International Conference of Spatial Information Theory* (LNCS 3693, pp. 347-362).

Loomis, J. M., & Lederman, S. J. (1986). Tactual perception. In K. R. Boff, L. Kaufman, & J. P. Thomas (Eds.), *Handbook of perception and human performance volume II: Cognitive process and performance* (pp. 1-41). Hoboken, NJ: John Wiley & Sons.

MacLean, K., & Enriquez, M. (2003). Perceptual design of haptic icons. In *Proceedings of the EuroHaptics Meeting*, Dublin, Ireland (pp. 351-363).

May, A. J., Ross, T., Bayer, S., & Tarkiainen, M. (2003). Pedestrian navigation aids: Information requirements and design implications. *Personal and Ubiquitous Computing*, *7*(6), 331–338. doi:10.1007/s00779-003-0248-5

Michon, P., & Denis, M. (2001). When and why are visual landmarks used in giving directions? In D. R. Montello (Ed.), *Proceedings of the 5th International Conference on Spatial Information Theory: Foundations of Geographic Information Science* (LNCS 2205, pp. 202-305).

Millar, S., & Al-Attar, Z. (2004). External and body-centered frames of reference in spatial memory: Evidence from touch. *Perception & Psychophysics, 66*(1), 51–59. doi:10.3758/BF03194860

Millonig, A., & Schechtner, K. (2005). Developing landmark-based pedestrian navigation systems. In *Proceedings of the 8th International IEEE Conference on Intelligent Transportation Systems* (pp. 197-202). Washington, DC: IEEE Computer Society.

Raubal, M., & Winter, S. (2002). Enriching wayfinding instructions with local landmarks. In M. J. Egenhofer & D. M. Mark (Eds.), *Proceedings of the 2nd International Conference on Geographic Information Science* (LNCS 2478, pp. 243-259).

Ross, D. A., & Blasch, B. B. (2000). Wearable interfaces for orientation and wayfinding. In *Proceedings of the 4th International ACM Conference on Assistive Technologies* (pp. 193-200). New York, NY: ACM Press.

Schiffman, H. R. (1976). *Sensation and perception: An integrated approach.* New York, NY: John Wiley & Sons.

Sorrows, M. E., & Hirtle, S. C. (1999). The nature of landmarks for real and electronic spaces. In C. Freksa & D. Mark (Eds.), *Proceedings of the International Conference on Spatial Information Theory* (LNCS 1661, pp. 37-50).

Srikulwong, M., & O'Neill, E. (2010a). *A direct experimental comparison of back array and waist-belt tactile interfaces for indicating direction.* Paper presented at Workshop on Multimodal Location Based Techniques for Extreme Navigation in conjunction with the 8th International Conference on Pervasive Computing, Helsinki, Finland.

Srikulwong, M., & O'Neill, E. (2010b). A comparison of two wearable tactile interfaces with a complementary display in two orientations. In *Proceedings of the 5th International Workshop on Haptic and Audio Interaction Design* (pp. 139-148).

Tan, H. Z., Gray, R., Young, J. J., & Traylor, R. (2003). A haptic back display for attentional and directional Cueing. *Haptics-e: Electronic Journal of Haptics Research, 3*(1).

Ternes, D., & MacLean, K. E. (2008). Designing large sets of haptic icons with rhythm. In M. Ferre (Ed.), *Proceedings of the 6th International Conference on Haptics: Perception, Devices, and Scenarios* (LNCS 5024, pp. 199-208).

Tsukada, K., & Yasumura, M. (2004). ActiveBelt: Belt-type wearable tactile display for directional navigation. In *Proceedings of the 6th International Conference on Ubiquitous Computing* (LNCS 3205, pp. 384-399).

Van Erp, J. B. F., Van Veen, H. A. H. C., Jansen, C., & Dobbins, T. (2005). Waypoint navigation with a vibrotactile waist belt. *ACM Transactions on Applied Perception, 2*(2), 106–117. doi:10.1145/1060581.1060585

APPENDIX

THE SET OF LANDMARK TYPES USED IN THE USER SURVEY STUDY

Airports, Amusement parks, At the water (ocean and sea), Attractions/Tourist attractions, Bars and Pubs, Bridges, Camping areas, Car rentals, Cash dispensers (ATM), Casinos, Cinemas, Educational institutes, Fairs & Conventions, Ferries, Financial services (Banks), First aids, Golf courses, Government facilities, Hospital healthcares, Hotels, Internet/Wi-Fi, Libraries, Malls and Markets (shopping centre, supermarket), Monuments & Memorials, Mountains, Music & Culture venues, Museums & Galleries, Natural barriers (any object that prevent you from moving forward, e.g. roads.), Parking, Party & Clubbing, Pedestrian lights, Petrol stations, Police, Post office, Public transports (bus/tram/boat stations), Railway stations, Recreation grounds, Religious places (church/cathedral/etc), Restaurants, River, Sports facilities, Stadiums (sports), Taxis, Theatres, Toilets, Travel agencies, Traffic lights, Tourist information, Tunnels, Other landmarks.

Chapter 13
Good Times?!
3 Problems and Design Considerations for Playful HCI

Abdallah El Ali
University of Amsterdam, The Netherlands

Frank Nack
University of Amsterdam, The Netherlands

Lynda Hardman
Centrum voor Wiskunde en Informatica (CWI), The Netherlands

ABSTRACT

Using Location-aware Multimedia Messaging (LMM) systems as a research testbed, this paper presents an analysis of how 'fun or playfulness' can be studied and designed for under mobile and ubiquitous environments. These LMM systems allow users to leave geo-tagged multimedia messages behind at any location. Drawing on previous efforts with LMM systems and an envisioned scenario illustrating how LMM can be used, the authors discuss what playful experiences are and three problems that arise in realizing the scenario: how playful experiences can be inferred (the inference problem), how the experience of capture can be motivated and maintained (the experience-capture maintenance problem), and how playful experiences can be measured (the measurement problem). In response to each of the problems, three design considerations are drawn for playful Human-Computer Interaction: 1) experiences can be approached as information-rich representations or as arising from human-system interaction 2) incentive mechanisms can be mediators of fun and engagement, and 3) measuring experiences requires a balance in testing methodology choice.

DOI: 10.4018/978-1-4666-2068-1.ch013

INTRODUCTION

On a sunny afternoon in mid-July, Nicole and Nick are tourists shopping around Nejmeh Square in downtown Beirut, Lebanon. While Nick insists on seeing the cultural offerings of Saifi Village, a village completely rebuilt as a New Urbanist-style neighborhood after its destruction during the civil war, Nicole has a different notion of what is fun and enjoyable. Familiar with her interests in warm, foreign cities, Nicole's mobile device sets her to experience 'fun' places nearby, suggesting several lively cafés along the Corniche, a seaside walkway with a glittering view of the Mediterranean. Skeptical about the suggestion, she makes a predefined gesture instructing her device to show her different multimedia (photos, songs, videos, text) that reflects people's experiences there. The device presents her with a dizzying nexus of visual and musical perspectives captured by people enjoying themselves, complementing each multimedia message with related past and future events. Leaving Nick, she makes her way toward the Corniche until she reaches a café,

where she sits outdoors, happily absorbing the scorching sun rays. Wondering where Nick went, she decides to capture her current experience. She takes a photo of the clear blue sky and sea (Figure 1), which she annotates with the song by The Cure 'Play for Today' and writes: "That's New Urbanist-style culture too!!" While she awaits her hookah and drink, she scans through other people's experiences at the café she is at, only to realize the place attracts mainly an older crowd, which is no fun at all.

The preceding scenario illustrates ongoing research efforts within the MOCATOUR (Mobile Cultural Access for Tourists, http://mocatour. wordpress.com) project. The aim of the project is to define computational methods that facilitate tourists with contextualized and media-based access to information while they freely explore a city. The provision of contextualized information anytime, anywhere, to the right persons as they go about their daily lives is part of an emerging paradigm dubbed as ubiquitous computing (Weiser, 1991), context-aware computing (Dey, Abowd, & Salber, 2001), pervasive computing

Figure 1. A mockup illustrating the photo Nicole took of the Corniche seaside and the corresponding annotations she added

(Ark & Selker, 1999), embodied interaction (Dourish, 2001), or everyware (Greenfield, 2006). Irrespective of the name given, a central tenet of this paradigm is the promise of populating our everyday lives with context-aware services that make interaction with the world easier, more manageable, and more efficient. This endeavor is made possible through embedding (at times personal and imperceptible) low-cost and low-power sensors and devices into our everyday environment.

A major step in this direction has been the widespread adoption of location-aware technologies such as GPS-enabled mobile devices and automotive GPS. Yet with our cities becoming interfaces for computational experimentation that are intermixed with human activities, we need systems that go beyond location-awareness and towards context-awareness (Dey, 2001). In other words, we need to know more about context (Dey, 2001), its inference from human activity, and how that feeds into our everyday experiences. As Bellotti and Edwards (2001) state, inference and adaptation to human intent in context-aware systems is at best an approximation of the real human and social intentions of people. This raises the need to further explore the kinds of services and usability issues brought forth under real-world usage contexts.

One important shift from computing for the desktop to computing for the world is that systems need no longer be about work-related activities, but also about fun and playful[1] endeavors (Cramer et al., 2010). To realize the system that Nicole in the introductory scenario uses, context-aware systems need to know not only about locations, but about people's lived experiences and their relationship(s) to the location(s) they took place at. To this end, we make use of Location-aware Multimedia Messaging (LMM) systems. Such systems allow people to create multimedia messages (photos, text, video, audio, etc.) that are anchored to a place, which can be received/perceived and interpreted by other people at (approximately) the same place

where the message was made (*cf.,* Nicole's photo portrait in the above scenario made at the café on the Corniche). Given that locations within cities are rich sources of "historically and culturally situated practices and flows" (Williams & Dourish, 2006, p. 43), it is reasonable to assume that LMMs can reflect culturally entrenched aspects of people's experiences at locations.

Given the above scenario, how can a system 'know' what fun or playful experiences are, in general and idiosyncratically as in Nicole's case of not enjoying older crowds? What kind of contextual elements can be automatically acquired (e.g., date, time, place) to infer playful experiences, and are these contextual elements rich enough to disambiguate the meaning of a user's activity, with and beyond interaction with the system (Dourish, 2004)? Should playful experiences be coded as representations to be used as information that the system makes use of (as in Nicole's device), or should fun be understood as an enjoyable open-ended interaction dialogue between a human and machine (Cramer et al., 2010)? If the latter, what kind of mechanisms need to be in place to ensure that not only the information presented is 'about' fun and playful experiences, but the human-machine interaction is itself an enjoyable experience? If fun and enjoyable experiences can indeed be predicted and catered for, how can this be measured?

Below we will try to address the above questions, where the rest of this paper is structured as follows: first, we provide definitions for an experience in general and a playful experience in particular (Section: What is a Playful Experience?). Next, we discuss in detail that inferring playful experiences largely depends on whether context is viewed under a positivist or phenomenological lens (Section: The Inference Problem). Then, we briefly describe past research efforts with using a LMM prototype that allows capturing experiences into different media forms and discuss how the playful experience of capture can be maintained (Section: The Maintenance Problem). Afterwards, we briefly highlight common methodological problems that arise when measuring experiences of people under mobile and ubiquitous environments (Section: The Measurement Problem). In response to each of the mentioned problems, we draw three design considerations for the study and design for playful experiences under mobile and ubiquitous environments (Section: Design Considerations). Finally, we present our conclusions and direction for future work (Section: Conclusions).

WHAT IS A PLAYFUL EXPERIENCE?

We agree with Law et al. (2009) when they state that the high degree of mutual consensus in the current Human-Computer Interaction (HCI) community over the importance of studying and designing for the user experience (UX) is truly intriguing. The trend towards thoroughly investigating the UX is in part a reaction to the traditional HCI usability frameworks that take user cognition and performance as key aspects in the interaction between humans and machines. With the advent of mobile and ubiquitous computing, human-technology interactions, even if they *involve* work settings, need not be *about* work (Cramer et al., 2010; Greenfield, 2006). This computing for everyday 'non-serious' life has shifted the attention of HCI towards user affect and sensations, where the user experience has become a desirable thing to have during the interaction with a system. Yet what exactly is this user experience? As Law et al. (2009, p. 719) write: "...UX is seen as *something* desirable, though what exactly *something* means remains open and debatable." They move on to argue that embracing a unified definition of the user experience can reap valuable scientific and (industrial) design benefits by: a) facilitating scientific discourse within and across disciplines to avoid communicational breakdown b) aiding the operationalization and evaluation of experience-centric applications c) helping understand the term, its importance and scope.

The term 'user experience' is already pregnantly associated with a wide range of fuzzy and dynamic concepts, with attached attributes such as pleasure, joy, pride, etc. (Law et al., 2009; Hassenzahl & Tractinsky, 2006). In a survey conducted to arrive at a unified definition of UX, Law et al. (2009) found that the elements of the UX provided by their participants largely conformed to the ISO definition of UX (International Organization for Standardization, 2010), which states: "A person's perceptions and responses that result from the use or anticipated use of a product, system or service." We find that the ISO definition to be accurate, in part because it provides an abstraction without appeal to specific affective attributes (such as fun or pleasure). However, we find that its accuracy comes at the cost of being overly general in aiding the study and design of context-aware systems. Also, while the definition provides the future aspect of anticipating an experience, it is missing the retrospective aspect of looking at a finished experience.

In attempt to understand experiences, we took the present and past relational temporal properties of experience into account, allowing us to distinguish between prospective experiences (i.e., experiences as they are currently happening) and the retrospective understanding of experiences (i.e., the mental time travel to an experience episode in the past (Tulving, 2002). This is in line with how Hassenzahl and Sandweg (2004) understand an experience, where they make a distinction between instant utility (a moment in product use within a larger experience episode) and remembered utility (a retrospective summary assessment of the product use experience). Not surprisingly, when they asked their participants how they felt towards a product after they used it they found that remembered utility is not necessarily the sum of all measured instant utilities. As will be explained later, the decision to view an experience from within (during its occurrence) or from without (after its elapse) also relates to which epistemological stance (positivist or phenomenological) one adopts in conceptualizing and reasoning about the world[2].

We distinguish between the process of an experience and the memory of an experience, where playful experiences are a subset of both. We define the process of an experience (Nack, 2003) as a sensory and perceptual process that some person undergoes (either through direct participation or observation of events and situations) that results in a change in that person. The high variability and subjective interpretation involved in predicting an experiential process indicates that it is useful to retrospectively capture a given experience; in other words, to consider the memory of an experience. Based on Tulving's definition of an episodic memory (Tulving, 1993), we define an experience memory as the result of an experiential process, which can be manipulated and actively recalled. The memory of an experience consists of one or more actors, a spatiotemporal aspect, a social aspect, a cognitive aspect, and an affective aspect. We use these aspects of an experience memory as a basis for studying experience capture using LMMs. This approach is similar to the one employed by Wigelius and Väätäjä (2009), where they made use of five dimensions of the user experience to study and design for mobile work (e.g., mobile news journalism): social, spatial, temporal, infrastructural, and task context. However, while Wigelius and Väätäjä (2009) separate the characteristics of the user and system from the contextual factors involved, in our understanding of experiences we treat contextual factors as part and parcel of the user's memory of a past episode.

A playful experience, when understood as a process, is characterized by amusement, risk, challenge, flow, tension, and/or negative affect (Csikszentmihalyi, 1990; Nacke et al., 2009). However, we believe that only amusement, which is an affective reaction to a 'playful' activity, is a sufficient condition for playful experiences. While the other attributes (such as tension, risk, flow) can frequently occur in playful experiences, each by themselves, unlike amusement, do not

uniquely give rise to a playful experience. According to the definition of playfulness provided by Cramer et al. (2010, p. 1), playfulness refers to "non-utilitarian (but not necessarily non-useful) aspects of interactions that provide pleasure or amusement." While we do not fully agree with Cramer et al. (2010) that a playful experience is non-utilitarian (as playful experiences serve a practical goal of making one feel better as well as aid child learning and development), we do agree that playfulness is largely based on how an activity is approached, rather than an essential property of the activity itself (Csikszentmihalyi, 1990). While this indicates that playfulness is a mental state brought forth by users to an activity, it does not entail that playful interactions cannot be anticipated for particular user groups and explicitly designed for. In other words, if coupled human-system activities frequently draw users toward playfulness during the interaction process, the designer can reason backwards and identify what it is about the interaction that prompted the playfulness in the first place. As will be shown later, the problem of cleanly delineating the cause of a phenomenon (in this case playfulness) for intelligent inference is subject to what notion of context is adopted.

For fun and playfulness, we believe that the most common elicitors of playful experiences are games (e.g., board games, video games), where most games tend to be challenging, create tension, a sense of flow, induce positive and negative effect, and evoke amusement (Nacke et al., 2009; Poels, Kort, & Ijsselsteijn, 2007). However, something like The World's Deepest Bin[3], a bin that makes an elongated sound to indicate depth when someone throws something in it, only elicits brief amusement. Nevertheless, interacting with the bin qualifies as a playful experience because it elicits amusement. What characteristics of playful experiences (e.g., tension, amusement) are to be elicited in users depends largely on the purpose of the system: is the system designed to carry out tasks that are useful or serious (e.g., a context-aware tourist guide (Cheverst et al., 2000) or context-aware firefighter system (Jiang et al., 2004), or is it meant to entertain (e.g., a location-based game (Benford et al., 2005) or a virtual storyteller (Lim & Aylett, 2009)? While the purpose of the system can aid in helping designers conceptualize the kind of playful experiences desired in interacting with the system, the real problem is how, if at all, a system can infer a playful experience when it happens.

THE INFERENCE PROBLEM: INFERRING PLAYFUL EXPERIENCES

How can a system automatically detect and recognize an experience as playful? What kind of contextual clues are necessary for a system to draw this kind of inference? The answer to these questions we believe lies in revisiting the concept of 'context'. Dourish (2004) argues that the notion of context in ubiquitous computing varies with respect to two distinct schools of thought: positivism and phenomenology.

Positivist vs. Phenomenological Theories

Positivist theories, tracing back to sociologist Auguste Comte (1880), derive from a rich, rational and empirical history that takes the scientific method as the sole arbiter of objectively attainable knowledge. This epistemological stance seeks to reduce complex social phenomena into objective, clearly identifiable descriptions and patterns that are idealized abstractions of the observed social instances and situations that make up such phenomena. Phenomenological theories on the other hand, tracing back to Edmund Husserl (1893-1917), are essentially subjective and qualitative. Objective reality according to the phenomenologists is always channeled through the interpretive lens of human perception and

action; as Dourish (2004, p. 21) writes, "social facts are emergent properties of interactions, not pre-given or absolute, but negotiated, contested, and subject to continual processes of interpretation and reinterpretation."

According to Dourish (2004), the positivist account of context renders context as a representational problem whereas the phenomenological account makes context an interactional one. The representational problem is essentially concerned with how context (such as location, time or date) can be encoded and represented in a system so that the system can intelligently tune its behavior according to what values these precoded contextual factors take in a given situation. The main assumption here is that human activity and context can be cleanly separated. For example, the lighting of a room (a contextual factor) is seen as independent of the series of actions required to make coffee (activity) in the room.

By contrast, the interactional problem is primarily concerned with how and why people, through interacting with one another, can establish and maintain a shared understanding of their actions and the context they occur in? To revisit the coffee example, the phenomenological take on it would be that the lighting of the room and the coffee making within it are inseparable; they are tightly woven into an activity-context coupling that give a unified experience, without which that particular experience could not be said to have happened. For Dourish (2004), this underscores the distinction between viewing context as a set of stable properties that are independent of human actions and viewing context as an emergent set of features that are dynamically generated through common-sense reasoning and culturally entrenched beliefs about the world throughout the course of interaction. In other words, while positivism strives for universals (attained through the method of induction), phenomenology contests that the richness of particulars is irreducible to abstraction.

Playful Representation
or Interaction?

How do the two accounts of context fare into our understanding of playful experiences? In the context of LMM, we make the distinction between playfulness as an information-rich post-hoc representation (*cf.*, experience memory and the positivist claim) and playfulness as interaction (*cf.*, experience process and the phenomenological claim). To illustrate, the kind of playfulness that Nicole's mobile system in the opening scenario affords is retrospective, where the system representation of experiences is composed of a clearly identifiable collection of past, personal and publicized multimedia messages that have been annotated as 'fun'. The very act of conceding that labeling these multimedia messages with an identifiable label such as 'fun' is possible arises from a positivist understanding of the world. Following the sequence of Nicole's activities, the representational vehicle (the media presentation of other people's experiences) which subserves the subsequent experiential process that she undergoes when sitting down at the seaside café (namely, absorbing the sun rays and making a multimedia message of her own) is seemingly no longer within the scope of her interaction with the system (Dourish, 2001). This happens despite that causally, the system representation is what brought her to have the experience at the seaside café in the first place.

Following Nicole's interaction with the system to its interactional finish point, we see that the situation changes when Nicole consults her device while she awaits her hookah: the system's presentation of an older crowd, mistaken about Nicole's notion of fun, has now interfered with and altered her current joyful experience. This unanticipated system response can be seen as a flaw when explicitly designing playful human-mobile interactions, where 'playfulness' is scoped only between the interactional possibilities that

rest between the user and the system. We believe this reflects the deeper issue of whether to treat playfulness as a representational problem independent of the actual activity process involved in playfully perceiving and acting upon it, or on the other extreme, letting the playful process bleed into interaction windows where the interaction is no longer playful. It is this problem of scoping that makes inferring playful experience a hard problem. Since the context-sensitive variables precoded into the system representation do not account for and update dynamically with the unfolding of the human-system interaction process, inferring playfulness becomes entangled between the system representation and the human interaction with this representation, leaving the system with poor inferential precision.

THE MAINTENANCE PROBLEM: MOTIVATING AND MAINTAINING PLAYFUL EXPERIENCE CAPTURE

During past research efforts toward understanding experiences in mobile and ubiquitous environments, we studied using an exploratory approach the experiential and contextual factors surrounding LMM (El Ali et al., 2010). Part of this effort involved field testing a LMM prototype application that allows leaving multimedia messages at locations using three different media types: text, drawing, and photos. The prototype was pilot-tested with 4 subjects where an *in situ* interview method (Consolvo et al., 2007) was used to measure experience capture behavior. By annotating locations, the prototype lets users capture their experiences by allowing them to create a digital *memory* snapshot of their experience (Figure 2a). The generated message remains anchored to the location it was created at for later viewing by anyone who has the application installed on their multimedia-enabled mobile device and is at the same place where the message was created.

Figure 2. Interaction with the LMM prototype

(a) Planning at t_0 (b) Creation at t_1 (c) Viewing at t_n

LMM Prototype

The LMM prototype was installed on the Android Dev Phone 1, a multimedia-enabled mobile device. The interface consists of three functions: Create, Snap, and Explore. In Create, a user can create a free drawing (Figure 2b) using touch-based input or type text using the device's keyboard. Here, the location and orientation of the device is retrieved and the user is presented with a camera-view where she can choose to draw or write something. In choosing either option, a snapshot of the camera view is subsequently used as a background canvas for the user to draw or write on. Once a user is finished, she can save the annotated image. In Snap, a user is taken directly to a camera-view where she can snap a photograph.

After generating a message, a user can view the message by being at the right position and orientation of where the multimedia message was made. In switching to Explore mode, a user is presented with a camera-view, where she is guided to a message by leading her to the creator's original position and orientation. An arrow is drawn on the screen to guide the user towards a message. To indicate the distance between the users's current position and that of the message, the color of the arrow changes within 200m of the message location. Once at the right position, the user can adjust her orientation by looking at a small green indicator arrow shown on the right or left edge of the screen. In doing so, the selected multimedia message appears as an Augmented Reality image overlay on top of the camera-view (Figure 2c). The location-aware aspect of anchoring a message to a location is assumed to provide a deeper contextualization of the message maker's original experience.

Fun, but not Useful

After briefly explaining how the prototype works and how to use it, we let subjects at a café create multimedia messages in all three supported media types: drawings, text, photos. For the drawn expressions, two of the subjects drew a cup of coffee to show that you can get coffee at the cafeteria. The other two made graffiti expressions, where their drawings augmented parts of the environment. For the drawings, we found that drawings were meant only as fun digital augmentations on the physical environment. When asked about his/her drawing message, S1 explained: *"Well that ['Dancer in the Dark' poster] is a poster that I enjoy looking at a lot when I'm drinking, and I always wondered about the frame, so I wanted to draw lines around it, but to do it freely. Doesn't have a purpose but it looks nice."*

For the textual messages, subjects used text for: recommending items (e.g., S4: *"You should try the green tea"*), a means for self-expression (e.g., S1: *"Beer Perspective"* and S2: *"Things are looking up"*), or as a warning to others (e.g., S3: *"Don't confuse gravy with soup"*). For the photo expressions, two of the subjects took a photo of the experimenter, and the other two a photo of the street. All photo messages made were used as a means to contrast the present with the future that others will witness (e.g., friends viewing photos of them with the experimenter at a later time).

When subjects were later asked about their overall experience with the LMM prototype, they all reported that it was fun to doodle over the environment and leave photos to share with public and private networks, but did not find either of them to be useful. On the other hand, they all found that it is useful to share text messages (such as recommendations) with others at a place. Using text for practical purposes is in line with what Persson and Fagerberg (2002) found in evaluating the GeoNotes messaging system and what Burrell and Gay (2002) found for the E-graffiti system. The lack of usefulness in drawing or capturing photos in the LMM prototype hinted that perhaps an incentive mechanism that motivates users to use the application is needed to ensure that the experience of capture using the LMM application is perceived as not only fun, but also useful like the discussion of Greenberg and Buxton (2008) on why designed systems must first be deemed useful, and only then usable). Equipping a system with persuasive techniques to increase personal

and social gain has been explored in social media networks (Cherubini et al., 2010; Singh et al., 2009), where users are provided with a strong incentive to make contributions of a certain type (e.g., high quality media contributions). Likewise, if game-theoretic elements are designed into the interaction process, the playful aspects of using LMM can be maintained beyond amusement reactions, insofar as the LMM contribution behavior of users is reinforced with personal and social rewards.

The Measurement Problem: Measuring Playful Experiences

Finding an appropriate testing methodology to understand playful experiences that can unlock suitable interaction methods in mobile and ubiquitous settings poses a real challenge. This challenge is amplified by the difficulty in probing into the inner subjectivity of the cognitive and emotional lives of people under changing contexts and while on the move. There have been several successful attempts at measuring user's experiences, especially during interaction while immobile. Much work in this respect has focused on interaction with digital (video-)games (Nacke et al., 2009; Bernhaupt et al., 2008; Mandryk et al., 2006).

Subjective and Objective Experience Measures

Broadly, experience measurements can be broken down into subjective and objective measures (Bardzell et al., 2008; Greenberg & Buxton, 2008). Subjective measures typically involve self-reports of a given experience, where methods for obtaining them typically include interviews, surveys, and ethnomethodological techniques in general (Kuniavsky, 2003). Objective measures, by contrast, evaluate *observable* aspects of a person's experience independent of that person's perception. These can range from observations of human posture and gait, button press count and task completion time, to physiological measurements such as Electroencephalogram (EEG) recordings, Galvanic Skin Response (GSR) recordings, Electromyography (EMG) recordings, or eye movement capture using Eye-tracking hardware (Nacke et al., 2009; Bardzell et al., 2008). Such objective metrics however are difficult to generalize to mobile and ubiquitous environments (Kellar et al., 2005), where not only is the user's location subject to change, but also the context at a given location[4].

One methodology that promises to deal with the fuzzy nature of user testing in the wild is the Living Lab methodology (de Leon et al., 2006; Eriksson et al., 2005). El Ali and Nack (2009, p. 23) defined the Living Lab methodology as research grounds for the testing and evaluation of humans interacting with technology in natural settings, to "better inform design decisions sprouted from what real-life users want, so that technology development becomes an intimate three-way dance between designers, developers, and users." Two challenges to this ambitious research agenda raised by El Ali and Nack (2009), the risk of over-measurement and under-measurement, warrant recapitulation here. While these considerations are fairly general, they are stated here to underscore the importance of choosing the right testing methods for measuring experiences in mobile and ubiquitous environments.

Over-Measurement and Under-Measurement

Over-measurement can occur when a user is left to freely use a mobile and/or ubiquitous experience-centric application while on the move. Without informed understanding of what *kind* of data is being collected, extraction of meaning from the continuous flux of data (e.g., interaction history logs) proceeds in an ad hoc manner, and thus risks a loss in interpretation and quality of drawn implications. Consider Nicole's complex behavior in the introductory scenario, where she initially

accepted the seaside walkway recommendation from her device, but retracted the recommendation later in light of new information about the café she is at. Without being explicitly informed about what kinds of media she, or people like her, find enjoyable and fun, it would not be possible for a system to adequately adapt to her needs. This indicates that interaction behavior should be constrained to a small number of measurable units that provide (partial) immunity from the unpredictable nature of unsupervised human-technology interaction. Without minimal supervision exerted on testing conditions during system evaluation and early development, caution should be exercised concerning whether or not the elicited knowledge is trustworthy enough to solicit informed understanding and design of mobile and ubiquitous systems.

At the other end of the spectrum, rigorously controlled laboratory testing can result in under-measurement, where the main problems are: a) testing is confined to the walls of the laboratory. This means that 'natural', mobile behavior is by necessity beyond the scope of the method b) only a handful of experiential variables can be measured. This is due to the complexity and error-proneness of developing multidimensional designs that can properly incorporate several independent variables and tease out the possible effects on the dependent variables of interest. Together, these problems make controlled laboratory testing, by itself, insufficient for measuring playful experiences in mobile and ubiquitous environments.

Given the two highlighted problems, how can a middle-ground be reached for evaluating experiences in unconstrained environments? One immediate response (El Ali & Nack, 2009) is to split the evaluation process into two phases: subjective observation and objective measurement. In the observation phase, the researcher employs outdoor, *subjective* observational methods during the early design stages of application development as a means of reducing the phenomenon dimensionality down to a few *objectively* measurable variables. During the second phase, depending on their nature, these variables can be experimentally teased out under rigorously controlled indoor environments. There are two promising augmentations to the early observation phase, well-suited for dealing with the difficulties in evaluating context-aware applications under mobile and ubiquitous environments: using Urban Pervasive Infrastructure (UPI) methods (Kostakos et al., 2009, 2010) and context-aware Experience Sampling Methods (ESMs) (Consolvo & Walker, 2003; Froehlich et al., 2007).

UPI Methods and ESMs

Without going into excessive detail, the UPI methods (Kostakos et al., 2009) are built on the premise that the city can be viewed as a system, where the variables of interest are the combination of people, space, and technology that together aid in studying and deploying urban pervasive applications[5]. These methods deal with five characteristics of the UPI: mobility (e.g., human distance travelled or visit duration), social structure (e.g., social network analysis metrics such as degree of separation), spatial structure (e.g., space syntax metrics such as integration), temporal rhythms (e.g., time-based distributions of people's activities), and facts and figures (e.g., statistical characteristics such as number of devices detected at a defined area).

Focusing on the above characteristics, Kostakos et al. (2009) have developed methods of observation and analysis that reveal real-world values under these metrics. For example, in their 'augmented gatecount' observation method, gatecounts (using Bluetooth scanners) are used to define the flows of people at several sampled locations within a city. The main point here is that these concepts, metrics and methods can considerably aid in gaining an understanding of a city objectively, which in turn aids in the early design stages of application development. To ground it in context of playful experiences, the understanding

of a city afforded by the UPI methods can identify spatial and social clusters in a city where people meet for entertainment purposes (e.g., the movies or the park), which provides support for narrowing down the objective of playful applications to the right target group or spatial structure.

Other methods that are useful in evaluating and narrowing down the early design space of mobile and ubiquitous application development are Experience Sampling Methods (or ESMs) (Consolvo & Walker, 2003). ESMs work by alerting participants each day to fill out brief questionnaires about their current activities and feelings. Sampling experiences throughout the course of a day make ESMs a great tool to evaluate a given application *in situ*. Moreover, unlike classical self-report techniques, ESMs do not require participants to recall anything and hence reduce cognitive load. Typical studies with ESMs involve a minimum of 30 participants, and are longitudinal. The longitudinal aspect also means the analysis of collected structured data from participant responses is amenable to statistical analyses. Together, these characteristics of ESMs make them not only invaluable tools in uncovering current usage of mobile and ubiquitous applications, but practical methods of investigating human 'technology' needs under different, real-world contexts. An exemplary translation into the opening scenario would be interval-dependent or event-dependent sampling of Nicole's experience of playfulness with her environment and/or with the device. By sampling Nicole's experiences, her device is able to build a predictive user model that probabilistically *knows* what things she finds fun, and can tailor the media presentation accordingly.

Design Considerations

For each of the problems highlighted above (the inference problem, the maintenance problem, and the measurement problem), we provide design considerations that we believe are relevant in the study and design for playful experiences under mobile and ubiquitous environments:

Experience Representation vs. Interaction Experience

As explained earlier (Section: The Inference Problem), a distinction can be made between an experience representation, which is information 'about' an experience, and the experience itself, which is a process emergent from an undertaken activity. This reflects the difference in how one understands context. Under a positivist view, the focus is on capturing experiences while under a phenomenological view the focus is on eliciting experiences through coupled activity-context pairs. For capturing experiences, the aim is to provide an adequate representation of any experience that took place, of which playful experiences are an instance. This requires a computational method for annotating the media-based experience representations with the right kind of information (e.g., affective information about the degree of fun had) for later intelligent retrieval (*cf.*, Nicole's device suggesting fun places nearby given her request of fun things to do).

For eliciting experiences, the aim is to subject users to activities and contexts that would strongly correlate to (if not cause) a desired *type* of experience (e.g., experiencing trust when interacting with a system). The concern here is not about which contextual elements are supported so as to sufficiently re-contextualize the experience of others, but rather about the scoped playful interaction between the user(s) and the system, where the user experience takes place during the interaction process itself. For example, the act of shaking a mobile device to indicate a change in preference for presented location recommendations can itself be a playful experience. In the domain of LMM, one way of enhancing the playful experience would be to provide the right kind of multimodal input and output support (Chittaro, 2009). For example, labeling a media expression (e.g., a photo) by means of textual input (*cf.*, Section: The Maintenance Problem) might be more intrusive and interruptive of a playful experience,

whereas a voice command of 'fun' that achieves the same function can occasion a more seamless interaction experience. In short, researchers and designers alike should be aware of which epistemological stance (positivist or phenomenological) they commit to when studying and designing for experiences in general and playful experiences in particular.

INCENTIVE MECHANISMS AS MEDIATORS OF CONTINUOUS PLAYFULNESS

We discussed earlier (Section: The Maintenance Problem) that our pilot study subjects had reported that their interaction with the LMM prototype for doodling and photo-capture was fun but not useful. This led us to consider that, at least for LMMs, users require an incentive to interact with the system that transcends merely playful interaction. In other words, the fun things such as tension and challenge, risk and unpredictability, positive and negative effect, have to be deliberately embedded in the interaction process. However, the fun aspects should be secondary to the user task of documenting and sharing their experiences as multimedia messages. Simply put, the perceived usefulness of a system should be treated as a first-class citizen.

Notwithstanding the importance of usability issues, this raises an important issue of whether the user should be made aware of the real goal of the performed task (i.e., task transparency), and in what domains does it actually matter to apply such persuasive techniques. For example, implicit ambient light feedback is a useful mechanism to unobtrusively indicate excess electricity consumption during the day. A promising approach for applying incentive mechanisms in the context of LMM is to utilize game-theoretic approaches (Singh et al., 2009) to create competitive game-like environments that persuade users to perform a given task, such as tagging or rating people's generated messages (cf., Facebook's 'Like' button, www.facebook.com). This would not only motivate users to collaboratively rank the generated content, but given the competitive element, would make the experience of doing so fun and engaging.

Balancing Testing Methodologies When Measuring Playfulness

Measuring fun and playfulness is by now a well-known slippery endeavor (Cramer et al., 2010). As discussed earlier (Section: The Measurement Problem), the difficulty arises in deciding to test users in a natural setting, where objective experiential data is hard to acquire. At the other extreme, controlled testing permits objective measurement at the cost of narrowing explanatory scope. While there is no clear prescription for the most effective approach to evaluating experiences, it is likely that a gradual progression from unconstrained to controlled testing in the course of mobile and ubiquitous application design and development is an effective means to measure experience. More concretely, during early design stages, outdoor testing of mobile users can help yield design implications that help narrow down the set of observable phenomena to a few variables, which can then be experimentally teased out in a more controlled environment.

As we have suggested, there are two promising methods to augment understanding, analysis, and narrowing down of the early design space: UPI methods and ESMs. While UPI methods permit objective measurement and analysis of structures (social, spatial, temporal) within the city, ESMs can help shed light into individual human-technology needs under certain places and times. Due to the importance of objective measurement and analysis on the one hand, and the need to systematically understand human subjective responses on the other, we believe that a combination of both methods can strongly aid in both understanding the playground of existing playful interactions, and the subsequent development of future-generation

mobile and ubiquitous tools to enhance these interactions. For example, the duration of a visit at a particular site in a park with a particular social setting (characterized for example by a minimum person co-occurrence frequency count) can be used as a trigger for unobtrusively sampling a person's experience. That person's response includes both the receptivity to the sampling interruption as well as the content of interruption (e.g., what activities he was engaged in at that moment and with how many people). This response in turn can on the one hand provide a useful feedback loop (Kostakos et al., 2009) into the quality and capacity of objectively measuring and inferring people's activities from such measurements, and on the other hand shed light into what kinds of experiences these people undergo at certain locations within a city (such as the park).

CONCLUSION

In looking at what playful experiences are, how they can be inferred, how the experience of capturing them can be motivated and maintained, and how to measure them, we have underscored what we believe to be fundamental problems underlying the scientific study of playful experiences in mobile and ubiquitous environments. Drawing on past research efforts and an envisioned LMM usage scenario, we hope to have drawn attention to the importance of thoroughly examining the different aspects of playful experiences (inference, capture-maintenance, measurement) when designing LMM systems to be used under ubiquitous environments.

As highlighted in the introductory scenario, there are a myriad of cognitive and affective factors intermixed with the system interaction that are difficult to experimentally and computationally disentangle. This in part stems from which epistemological stance (positivist or phenomenological) one chooses to adopt in practicing HCI (Section: The Inference Problem). Intermixing the two views, at least in LMM, makes it difficult for a system to automatically acquire the right kind of experiential information (e.g., media tagged or rated as fun that corresponds to how fun an experience was) and to intelligently retrieve this information in the right situation (*cf.,* Nicole's desire to experience something fun), while at the same time ensuring that interaction with and cognitive processing of this information is itself enjoyable. The latter point, as we mentioned (Section: The Maintenance Problem), can be mediated by explicitly incorporating fun and enjoyable game-like elements in the experience capture process. Lastly, we considered the problems that arise in measuring experiences in general and playful ones in particular (Section: The Measurement Problem), and argued that a gradual progression from controlled to out-in-the-wild testing provides a systematic methodology which can aid in understanding the playground for future experience-centered mobile and ubiquitous systems.

In response to the highlighted problems, we have furnished playful HCI with three design considerations (experience representation is not the same as interaction experience, incentive mechanisms can be mediators of playfulness, and measuring playfulness requires a balance in testing methodology choice) that together serve as useful guidelines for scientifically studying and designing playful experiences in mobile and ubiquitous environments. The need for clear guidelines has been well-articulated by Greenfield (2006, p. 232) when he wrote back in 2006: "Much of the discourse around ubiquitous computing has to date been of the descriptive variety...but however useful such descriptive methodologies are, they're not particularly well suited to discussions of what ought to be (or ought not to be) built." Yet to what extent it is possible to truly design and build mobile and ubiquitous systems that carry out the task of capturing experiences while making the experience of capture itself fun and enjoyable remains an open question.

ACKNOWLEDGMENT

This work is part of the Amsterdam Living Lab project (PID07071), and funded by the Dutch Ministry of Economic affairs and Amsterdam Topstad. The authors thank Amsterdam Innovation Motor (AIM) for their support.

REFERENCES

Ark, W. S., & Selker, T. (1999). A look at human interaction with pervasive computers. *IBM Systems Journal*, *38*(4), 504–507. doi:10.1147/sj.384.0504

Bardzell, J., Bardzell, S., Pace, T., & Karnell, J. (2008). Making user engagement visible: A multimodal strategy for interactive media experience research . In *Proceedings of Extended Abstracts on Human Factors in Computing Systems* (pp. 3663–3668). New York, NY: ACM Press.

Bellotti, V., & Edwards, K. (2001). Intelligibility and accountability: Human considerations in context-aware systems. *Human-Computer Interaction*, *16*(2), 193–212. doi:10.1207/S15327051HCI16234_05

Benford, S., Rowland, D., Flintham, M., Drozd, A., Hull, R., Reid, J., et al. (2005). Life on the edge: Supporting collaboration in location-based experiences. In *Proceedings of the SIGCHI Conference on Human Factors in Computing Systems* (pp. 721-730). New York, NY: ACM Press.

Bernhaupt, R., IJsselsteijn, W., Mueller, F., Tscheligi, M., & Wixon, D. R. (2008). Evaluating user experiences in games . In *Proceedings of Extended Abstracts on Human Factors in Computing Systems* (pp. 3905–3908). New York, NY: ACM Press.

Bulling, A., Roggen, D., & Tröster, G. (2009). Wearable eog goggles: Eye-based interaction in everyday environments. In *Proceedings of the 27ᵗʰ International Conference on Human Factors in Computing Systems* (pp. 3259-3264). New York, NY: ACM Press.

Burrell, J., & Gay, G. K. (2002). E-graffiti: Evaluating real-world use of a context-aware system. *Interacting with Computers, 14*(4), 301–312. doi:10.1016/S0953-5438(02)00010-3

Casson, A. J., Smith, S., Duncan, J. S., & Rodriguez-Villegas, E. (2008). Wearable eeg: What is it, why is it needed and what does it entail? In *Proceedings of the IEEE 30ᵗʰ Annual International Conference on Engineering in Medicine and Biology Society* (pp. 5867-5870). Washington, DC: IEEE Computer Society.

Cherubini, M., Gutierrez, A., de Oliveira, R., & Oliver, N. (2010). Social tagging revamped: Supporting the users' need of self-promotion through persuasive techniques. In *Proceedings of the 28th SIGCHI International Conference on Human Factors in Computing Systems* (pp. 985-994). New York, NY: ACM Press.

Cheverst, K., Davies, N., Mitchell, K., & Friday, A. (2000). Experiences of developing and deploying a context-aware tourist guide: The guide project. In *Proceedings of the 6th Annual International Conference on Mobile Computing and Networking* (pp. 20-31). New York, NY: ACM Press.

Chittaro, L. (2009). Distinctive aspects of mobile interaction and their implications for the design of multimodal interfaces. *Journal on Multimodal User Interfaces, 3*(3), 157–165. doi:10.1007/s12193-010-0036-2

Comte, A. (1880). *A general view of positivism* (2nd ed.). London, UK: Reeves & Turner.

Consolvo, S., Harrison, B. L., Smith, I. E., Chen, M. Y., Everitt, K., & Froehlich, J. (2007). Conducting in situ evaluations for and with ubiquitous computing technologies. *International Journal of Human-Computer Interaction, 22*(1-2), 103–118. doi:10.1207/s15327590ijhc2201-02_6

Consolvo, S., & Walker, M. (2003). Using the experience sampling method to evaluate Ubicomp applications. *IEEE Pervasive Computing / IEEE Computer Society [and] IEEE Communications Society, 2*(2), 24–31. doi:10.1109/MPRV.2003.1203750

Cramer, H., Mentis, H., & Fernaeus, Y. (2010). Serious work on playful experiences: A preliminary set of challenges. In *Proceedings of the 'Fun, Seriously?' Workshop at CSCW*. Savannah, GA.

Csikszentmihalyi, M. (1990). *Flow: The psychology of optimal experience*. New York, NY: Harper Perennial.

de Leon, M. P., Balasubramaniam, S., & Donnelly, W. (2006). Creating a distributed mobile networking testbed environment-through the living labs approach. In *Proceeding on Testbeds and Research Infrastructures for the Development of Networks and Communities* (pp. 134-139).

Dey, A. K. (2001). Understanding and using context. *Personal and Ubiquitous Computing, 5*(1), 4–7. doi:10.1007/s007790170019

Dey, A. K., Abowd, G. D., & Salber, D. (2001). A conceptual framework and a toolkit for supporting the rapid prototyping of context-aware applications. *Human-Computer Interaction, 16*(2-4), 97–166. doi:10.1207/S15327051HCI16234_02

Dourish, P. (2001). *Where the action is: The foundations of embodied interaction*. Cambridge, MA: MIT Press.

Dourish, P. (2004). What we talk about when we talk about context. *Personal and Ubiquitous Computing, 8*(1), 19–30. doi:10.1007/s00779-003-0253-8

El Ali, A., & Nack, F. (2009). Touring in a living lab: Some methodological considerations. In *Proceedings of the Mobile Living Labs Workshop*, Enschede, The Netherlands (pp. 23-26).

El Ali, A., Nack, F., & Hardman, L. (2010). Understanding contextual factors in location-aware multimedia messaging. In *Proceedings of the 12th International Conference on Multimodal Interfaces* (p. 22). New York, NY: ACM Press.

Eriksson, M., Niitamo, V., & Kulkki, S. (2005). *State-of-the-art in utilizing living labs approach to user-centric ICT innovation – a European approach.* Stromsund, Sweden: Luleå University of Technology.

Froehlich, J., Chen, M. Y., Consolvo, S., Harrison, B., & Landay, J. A. (2007). Myexperience: A system for in situ tracing and capturing of user feedback on mobile phones. In *Proceedings of the 5th International Conference on Mobile Systems, Applications and Services* (pp. 57-70). New York, NY: ACM Press.

Greenberg, S., & Buxton, B. (2008). Usability evaluation considered harmful (some of the time). In *Proceeding of the Twenty-Sixth Annual SIGCHI Conference on Human Factors in Computing Systems* (pp. 111-120). New York, NY: ACM Press.

Greenfield, A. (2006). *Everyware: The dawning age of ubiquitous computing.* Berkeley, CA: New Riders Publishing.

Hassenzahl, M., & Sandweg, N. (2004). From mental effort to perceived usability: transforming experiences into summary assessments . In *Proceedings of Extended Abstracts on Human Factors in Computing Systems* (pp. 1283–1286). New York: ACM Press.

Hassenzahl, M., & Tractinsky, N. (2006). User experience - a research agenda. *Behaviour & Information Technology, 25*(2), 91–97. doi:10.1080/01449290500330331

Husserl, E. (1893-1917). *On the phenomenology of the consciousness of internal time.* New York, NY: Springer.

International Organization for Standardization. (2010). *ISO DIS 9241-210: Ergonomics of human system interaction - part 210: Human-centred design for interactive systems.* Retrieved from http://www.iso.org/iso/catalogue_detail. htm?csnumber=52075

Jiang, X., Chen, N. Y., Hong, J. I., Wang, K., Takayama, L., & Landay, J. A. (2004). Siren: Context-aware computing for firefighting. In A. Ferscha & F. Mattern (Eds.), *Proceedings of the International Conference on Pervasive and Ubiquitous Computing* (LNCS 3001, pp. 87-105).

Kellar, M., Reilly, D., Hawkey, K., Rodgers, M., MacKay, B., & Dearman, D. (2005). It's a jungle out there: Practical considerations for evaluation in the city . In *Proceedings of Extended Abstracts on Human Factors in Computing Systems* (pp. 1533–1536). New York, NY: ACM Press.

Kostakos, V., Nicolai, T., Yoneki, E., O'Neill, E., Kenn, H., & Crowcroft, J. (2009). Understanding and measuring the urban pervasive infrastructure. *Personal and Ubiquitous Computing, 13*(5), 355–364. doi:10.1007/s00779-008-0196-1

Kostakos, V., O'Neill, E., Penn, A., Roussos, G., & Papadongonas, D. (2010). Brief encounters: Sensing, modeling and visualizing urban mobility and copresence networks. *ACM Transactions on Computer-Human Interaction, 17*(1), 1–38. doi:10.1145/1721831.1721833

Kuniavsky, M. (2003). *Observing the user experience: A practitioner's guide to user research (Morgan Kauffman series in interactive technologies).* San Francisco, CA: Morgan Kauffman.

Law, E. L. C., Roto, V., Hassenzahl, M., Vermeeren, A. P., & Kort, J. (2009). Understanding, scoping and defining user experience: A survey approach. In *Proceedings of the 27th International Conference on Human Factors in Computing Systems* (pp. 719-728). New York, NY: ACM Press.

Lim, M. Y., & Aylett, R. (2009). An emergent emotion model for an affective mobile guide with attitude. *Applied Artificial Intelligence, 23*(9), 835–854. doi:10.1080/08839510903246518

Mandryk, R., Inkpen, K., & Calvert, T. (2006). Using psychophysiological techniques to measure user experience with entertainment technologies. *Behaviour & Information Technology, 25*(2), 141–158. doi:10.1080/01449290500331156

Nack, F. (2003). Capturing experience - a matter of contextualising events. In *Proceedings of the ACM SIGMM Workshop on Experiential Telepresence* (pp. 53-64). New York, NY: ACM Press.

Nacke, L. E., Drachen, A., Kuikkaniemi, K., Niesenhaus, J., Korhonen, H. J., & van den Hoogen, W. M. (2009). Playability and player experience research. In *Proceedings of Breaking New Ground. Innovation in Games, Play, Practice and Theory.*

Persson, P., & Fagerberg, P. (2002). *Geonotes: A real-use study of a public location-aware community system* (Tech. Rep. No. T2002:27). Gothenburg, Sweden: University of Göteburg.

Poels, K., de Kort, Y., & Ijsselsteijn, W. (2007). It is always a lot of fun!: Exploring dimensions of digital game experience using focus group methodology. In *Proceedings of the Conference on Future Play* (pp. 83-89). New York, NY: ACM Press.

Singh, V. K., Jain, R., & Kankanhalli, M. S. (2009). Motivating contributors in social media networks. In *Proceedings of the First SIGMM Workshop on Social Media* (pp. 11-18). New York: ACM Press.

Tulving, E. (1993). What is episodic memory? *Current Directions in Psychological Science,* 67–70. doi:10.1111/1467-8721.ep10770899

Tulving, E. (2002). Episodic memory: From mind to brain. *Annual Review of Psychology, 53,* 1–25. doi:10.1146/annurev.psych.53.100901.135114

Weiser, M. (1991). The computer for the 21st century. *Scientific American, 265*(3), 66–75. doi:10.1038/scientificamerican0991-94

Wigelius, H., & Väätäjä, H. (2009). Dimensions of context affecting user experience in mobile work. In T. Gross, J. Gulliksen, P. Kotzé, L. Oestreicher, P. Palanque, R. O. Prates et al. (Eds.), *Proceedings of the 12th IFIP TC 13 International Conference on Human-Computer Interaction* (LNCS 5727, pp. 604-617).

Williams, A., & Dourish, P. (2006). Imagining the city: The cultural dimensions of urban computing. *IEEE Pervasive Computing / IEEE Computer Society [and] IEEE Communications Society,* 38–43.

ENDNOTES

[1]   Throughout this paper, we will use the concepts of fun and playfulness interchangeably.

[2]   Throughout this paper, we will use the concepts of fun and playfulness interchangeably. e are here concerned with only these two.

[3]   One of several initiatives taken by Volkswagen to improve people's behavior: http://www.thefuntheory.com/worlds-deepest-bin, last retrieved on 25-08-2010.

[4]   There are exceptions to this: mobile Electrocardiograph (ECG) can measure heart rate while a person is moving, the wearable EOG goggles (Bulling, Roggen, & Tröster, 2009) can measure (saccadic) eye movements in everyday interactions, and Brain-Computer Interfaces (BCIs) such as wearable EEG can measure brain electrical

activity during daily interactions (Casson et al., 2008). While indeed these kinds of tools permit objective measurement, they are not without problems: a) the collected signals are difficult to interpret (especially in noisy environments) and b) these devices are not always feasible for use in user tests.

5   In this context, 'urban pervasive applications' is synonymous with ubiquitous applications deployed in a city.

This work was previously published in the International Journal of Mobile Human Computer Interaction, Volume 3, Issue 3, edited by Joanna Lumsden, pp.50-65, copyright 2011 by IGI Publishing (an imprint of IGI Global).

Chapter 14
A Comparison of Distribution Channels for Large-Scale Deployments of iOS Applications

Donald McMillan
University of Glasgow, UK

Alistair Morrison
University of Glasgow, UK

Matthew Chalmers
University of Glasgow, UK

ABSTRACT

When conducting mass participation trials on Apple iOS devices researchers are forced to make a choice between using the Apple App Store or third party software repositories. In order to inform this choice, this paper describes a sample application that was released via both methods along with comparison of user demographics and engagement. The contents of these repositories are examined and compared, and statistics are presented highlighting the number of times the application was downloaded and the user retention experienced with each. The results are presented and the relative merits of each distribution method discussed to allow researchers to make a more informed choice. Results include that the application distributed via third party repository received ten times more downloads than the App Store application and that users recruited via the repository consistently used the application more.

DOI: 10.4018/978-1-4666-2068-1.ch014

INTRODUCTION

Only recently have we seen mobile phones that are both numerous enough to afford a large trial as well as advanced enough to support downloading and installation of research software. Market research firm IDC (Nagamine, 2010) suggests that, at the end of 2009, 15.4% of the mobile phone market consisted of smartphones, an increase from 12.7% in 2008. So, while still not the predominant type of handset, it can be said that smartphones have been adopted into mainstream use. Running a trial solely with smartphone owners may not be selecting a user-base that is representative of the population at large, it can no longer be seen to be using only the most advanced 'early adopters'.

Evaluation of the use of ubiquitous or mobile computing systems has, as recommended by (Abowd & Mynatt, 2000), moved towards conducting evaluations outside of the laboratory and in the wider world, with all the complexities and challenges that brings. While there have been arguments against the utility and cost-effectiveness of this move (Kjeldskov, Skov, Als, & Høegh, 2004) there have also been arguments presented in support (Rogers et al., 2007). Making use of the market penetration of smartphones and the new App Store style software distribution methods to reduce the cost, in terms of hardware, of recruiting a large group of participants for a trial 'in the wild' while increasing the potential diversity of users is becoming an attractive option for researchers. For those researchers looking to begin to take advantage of these opportunities the range of platforms and distribution methods available to researchers has to be explored, as the cost of re-tooling to develop on a new platform and the purchase of devices on which to develop is not insignificant. The differences in the hardware capabilities and the support given to developers are outlined in Oliver (2009).

However, researchers working in this area in 2007 had less choice. The original iPhone, released in June 2007, was a powerful smartphone that was adopted by the general public at a rate the previous generation of smartphones never hinted at achieving. Within weeks of its launch the development community had produced a method of distributing software directly from the developers to the end users' handsets. In July 2008 Apple launched their App Store and addressed many of the traditional difficulties users experienced in downloading third party applications to their smartphones – all the applications were available in one of two places, compatibility was easy to ascertain and the process was made as painless as possible for the end user to the point where they were able to install new applications without the need for a desktop PC. Faced with these opportunities many research groups made the time and monetary investments necessary to move their development to this platform. The iOS platform, which runs on iPod Touch and iPad devices as well as iPhones, has a larger installed user base of over 90 million units than either Android OS, with 60 milllion, or Blackberry OS, with 50 million (Flurry, 2010).

Oliver (2009) notes that the iOS platform presents a better option for researchers when the device is 'unlocked' from the restrictions placed upon it by Apple. This, however, only took into account the development of the applications and not the difficulties in distributing applications to end-users. Applications developed taking advantage of restricted features are not eligible for distribution via Apple's App Store and therefore must be released via third party software repositories only available to those users who have unlocked their device in this manner. This exclusivity based upon APIs used, among other considerations, means that the decision as to which distribution method to use has to be taken the early in the design and development process. This paper reports on work on practical aspects of research methodology in 'mass participation' trials of ubicomp systems (McMillan, Morrison, Brown, Hall, & Chalmers, 2010). The research goals of such trials include the development of tractable and affordable methods of gathering useful data for evaluation and design, in the context of worldwide software distributions. This paper contributes towards this methodology by focusing upon the two distribution methods available for iOS devices and the affect this choice could have upon a system trial. With the information presented here researchers will be more informed as to the consequences of choosing a distribution channel, and be able to make this choice with more confidence.

RELATED WORK

Due to the practical and technical constraints upon Ubicomp research, large scale deployments are the exception instead of the norm. The distribution methods and processes are rarely described in HCI publications, leaving researchers wishing to conduct complimentary experiments in the dark. Here we survey such large scale deployments and comment upon the information given with regards to how users were recruited.

One of the earliest large-scale deployments of an ubicomp application was *Mogi Mogi*. As reported by Licoppe and Inada (2006), this location-based mobile multiplayer game was released commercially in Japan, and in 2004 had roughly 1000 active players. The distribution was done through a corporate partnership with the mobile phone carrier KDDI in Japan and growth relied upon word of mouth, there was no marketing done to promote the game. In 2008 Nokia Research Centre released Friend View, a "location-enhanced microblogging application and service" (Chin, 2009) via Nokia's Beta Labs. The authors report on statistical analysis of social network patterns based on anonymised log data representing 80 days' use by 7000 users. Again this distribution was done via to the researchers close links with mobile service providers.

In 2006, the trial of Feeding Yoshi was published as "the first detailed study of a long-term location-based game, going beyond quantitative analysis to offer more qualitative data on the user experience" (Bell et al., 2006, p. 418). The research focused on how the study's sixteen participants "interweaved the game into everyday life" and how wireless network infrastructure was experienced as a 'seamful' resource for game design and user interaction, yet was limited by the exotic hardware used and the lack of a simple and user friendly software distribution system. This was revisited by McMillan, Morrison, Brown, Hall and Chalmers (2010), who released Hungry Yoshi, a version of Feeding Yoshi for iOS devices and reported on the running of a trial involving more than 40,000 users.

One of the most long-term deployments of mobile research applications is Cenceme (Campbell, 2008), an application that uses context sensing to automatically update social networking sites with each user's current activity. Initially developed for the Nokia N95 and trialled among 30 locally based participants, the software was then ported to the iPhone and released in July 2008 when the App Store was first launched. The distribution of Cenceme is not detailed beyond noting the need for a re-design of certain features to ensure the application was compliant with App Store rules and regulations. Shapewriter (Zhai et al., 2009) was also released on the App Store at the same time as Cenceme to test a novel form of text input. While the focus of the paper is on the reviews written by end users about the software, the approval delay for submitting to the App Store is mentioned, as is a possible link to a positive blog entry on Time's website and the number of downloads they achieved.

Several research applications have been released on the Android platform, such as AppAware (Girardello & Michahelles, 2010) that allows users to share, via existing social networks or within the application itself, location-tagged information as to which applications they are installing, removing or updating. In doing so users are able to explore applications popular in their current location. AppAware was released for free on the

Android Marketplace, Google's App Store, in February 2010 without advertising or any other form of user base stimulation by the developers. The Android Marketplace gives statistics not only on the number of downloads, but instead for the number of installations, updates and removals. For AppAware these three statistics totalled over 1 million. Henze, Poppinga, and Boll (2010) used a release on the Android Marketplace to conduct a 'controlled' experiment comparing three conditions, exploring how this type of trial and deployment can be used to compliment traditional lab based HCI experiments. This application was released in April 2010 and within 10 weeks had recorded over 5,000 installations, with useable data being returned from 3,934 users. They also note that this application was ran on 40 different devices with varying versions of the mobile operating system – as, for most users, upgrades to the Android operating system are initiated by the mobile carrier at their discretion. Michahelles (2010) outlines a number of applications released on this platform including a barcode-based mobile product discussion application called *My2Cents*, a mobile game to encourage users to scan barcodes and label products called *ProductEmpire* and an application which publishes on Twitter a user's incoming and outgoing calls and SMS messages called *Twiphone*, without details on the distribution method or results. Falaki, Mahajan, Kandula, Lymberopoulos, Govindan, and Estrin (2010) provided 33 Android devices and 222 Windows Mobile devices to participants with unlimited talk time and data for the course of their investigation of smartphone use and its impact upon the network and energy usage. While not distributed in the same manner as the other trials mentioned here, and incurring a significantly higher cost, the scale of this research is comparable.

The Blackberry platform is less utilised as an avenue for research. However it still affords opportunities for such trials via its own software marketplace. Oliver (2010) released a logger to investigate how users interact with and consume energy on their portable devices via the Blackberry distribution system, and achieved 17,300 users providing over a million usage traces.

It is clear that while researchers are increasingly publishing results from applications released for Android devices there is still a large research development community for iOS devices. Oliver (2009) also noted in relation to the iPhone that "Out of the box, iPhone is a substandard research platform; however, unlocking it exposes a rich set of APIs from its Mac OS X foundation."

For researchers choosing to support the iOS platform and following this recommendation they must be aware that the distribution method available to them differs from that available to those who produce software for unmodified devices. To provide researchers with more information to make their choice we now provide information on the distribution options available for iOS, their relative merits and the trial release of an application across both.

DISTRIBUTION METHODS

The large scale deployments mentioned are all influenced by the manner in which they distribute the software to users, which is itself limited by the hardware platform the researchers have chosen to develop on and, in the case of iOS, whether they are developing for jailbroken devices. There are two primary methods for the large-scale distribution of iOS applications; the Apple App Store and the community of third party APT based repositories. Experiences with both will be discussed after an examination of the size and the submission practices of each.

Apple's App Store

The Apple App Store is arguably the best known and most popular mobile software repository in terms of applications available for download and number of applications downloaded, with more than 330,000 applications available for download and a download total topping 10 billion (Apple Inc., 2011).

Each application must go through an opaque review process by Apple in order to be approved for distribution via the store, to pass this review it must be seen not only to comply with the 37 page iPhone Developer Program License Agreement and the 136 page iPhone Human Interface Guidelines documents (iOS Dev Center, 2010) but must also fit within the positioning of the store in the wider market context. The review process itself runs on a sequential failure method, meaning that although an application may break two or more guidelines it will be rejected for one, edited, resubmitted and then rejected for the next. The time between submission and review is not guaranteed, although an estimate of the current load is given on submission. This currently averages at 5.91 business days with a maximum delay of 34 days (App Store Metrics, 2011).

APT Repositories

Only 12 days after the initial release of the iPhone a consumer level method to allow third party software and unrestricted access to the file system was made available online (Ricker, 2007). Called 'unlocking' in (Oliver, 2009) this process is generally referred to as a "Jailbreak", with the process of 'unlocking' being popularly associated with removing mobile carrier restrictions. The security model on iOS is an implementation of the FreeBSD jail mechanism, which is a form of OS level virtualization to compartmentalize the system, both its files and its resources, in such a way that system users can only access their own compartments, or jails. In order to access files and services outside the jail in which user level programs are run bugs in privileged applications or the OS itself are exploited to escalate the privileges of the user. Here we will use the term 'to jailbreak' as it is used by the iOS community not as a description of the exploit used to gain

control, but of the process of taking advantage of one of these exploits to modify the operating system to accept applications from sources other than Apple and installing a repository manager on the device for this purpose. Devices that have had their operating system modified in this manner will be referred to as 'jailbroken'.

Initially homebrew software, a generic term for software developed by a user community for closed platforms, was manually loaded onto the devices, a port of APT, Advanced Packaging Tool, was quickly developed for the iPhone allowing users to manage applications in the same way as on many other *nix based systems. A native GUI, Cydia, was released shortly after, providing much of the functionality of the App Store client, to be released by Apple a number of months later, with combined access to any number of repositories the user cared to subscribe to.

A recent ruling by the copyright office in the U.S.A. (U.S. Copyright Office, Librarian of Congress, 2010) has established the legality of jailbreaking devices in order to run legally obtained software users would otherwise be unable to use, removing any danger of legal action being taken against an end user. The warranty, however, is invalid while a device is in a jailbroken state. The jailbreak process is easily reversible: a device can be restored to its default state with the click of a single button in iTunes. However, as jailbreaking methods become more user–friendly and less the domain of highly technical users, the number of users with jailbroken devices who do not understand the consequences or the procedure to reverse it can be expected to rise. The existence of such users raises questions of the responsibility of researchers releasing software in this manner. Does the act of providing desirable software only to jailbroken devices constitute an encouragement to jailbreak? If that is the case, is it enough for us to inform users of the consequences of, and the procedure to reverse, jailbreaking within an app they will only be able to launch on a jailbroken device?

REPOSITORY CONTENTS AND POTENCY

In order to collate the contents of the 38 most popular repositories the release list of each was downloaded onto the mobile device, copied to the desktop and parsed into a database. The download statistics page for each of the packages in the largest 3 repositories were scraped and parsed. The smaller repositories did not provide public access to download counts, but this still resulted in download statistics for upwards of 80% of packages seen. This, plus information collected from the repository websites, was used in the calculation of statistics for comparison with the Apple App Store.

Applications by Genre

The graphs in Figure 1 and Figure 2 show the distribution of applications and packages available for download by category. Where possible the APT categories have been coded by the author to match those in the App Store in order to allow direct comparison. Due to the nature of the two different distribution methods, the APT repositories' two largest categories have no comparison on the Apple App Store. Ringtones are sold through a different outlet, the iTunes music store, and themes are not available without 3rd party software modifying restricted files.

As can be seen from Figure 1, the vast majority of available downloads center around Themes and Ringtones, neither of which are allowed in the Apple App Store and are only available via this channel. The majority of applications in the rest of the categories can also be seen to fall foul of the rules Apple have set for App Store submission. Modification of the iPhone operating system (e.g. adding folders to pre-iOS4 devices, enabling wifi-only applications to run over the cellular data network), breaking the sandboxing of applications in order to add features and interoperability with other applications (e.g. adding copy & paste sup-port to pre–iOS3 devices), running in the background, duplicating Apple functionality (e.g. 3rd party SMS clients) or using APIs which Apple have deemed private (e.g. directly accessing the WiFi) all cause rejection from the App Store but are the bedrock of this small development eco-system. The number of available downloads not a theme or ringtone is only 2541—less than 15% of the total available.

In comparison, the App Store is dominated by books, games and entertainment applications—with these three of the twenty categories accounting for 46% of the total applications available for download (App Store Metrics, 2011). The difference in scale between the two distribution methods can be seen between Figure 1 and Figure 2, with all but the smallest of categories in the App Store providing a choice of more applications than are available across all the application categories in all the APT repositories combined.

Number of Downloads

The number of downloads per application or per genre is not publicly released by Apple. Most companies keep information about the number of application purchases they have gained private; only the total number of downloads for the App Store as a whole is directly available. Companies like Pinch Media (App Store Secrets, 2009) and Flurry (www.flurry.com) collect aggregate statistics by offering a logging framework to developers for free—giving them details on usage of their own applications in return for the aggregate data that they can leverage in the marketplace. Admob (Mobile Application Analytics, 2010) provide a large proportion of the ads seen in iPhone applications, by some estimations 61% (Duyree, 2009), and also release some aggregate data. Unfortunately most of the data publicly available from these sites is updated infrequently and focuses on paid-for applications, which make up 77% of the applications available for download, but can still be useful in providing insight to the

Figure 1. Distribution of packages across APT repositories on a logarithmic scale

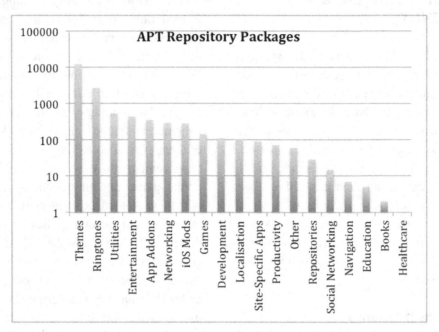

ecosystem. The data for the repository packages is freely available on the three largest community repositories, although only on a package-by-package basis. This information was collated to give information in the same format to give data on 80% of packages across the APT repositories.

The same general shape of trend can be seen for each distribution method with each decile having much less impact than the one before. However, a high number of users downloading an application is no guarantee of a high number of users engaging in the application to an extent to which they can

Figure 2. Distribution of applications in the Apple App Store

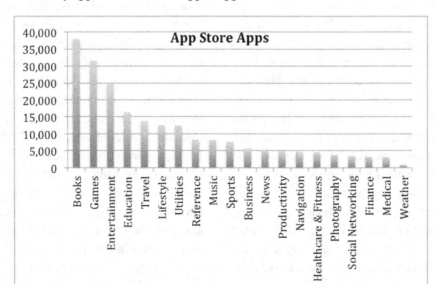

be seen to be a valid or valued trial participant. In order to reach more than 10,000 users an application need only be in the top 50% of applications on the APT repositories as opposed to the top 20% in the much larger set of applications in the Apple App Store. The exposure necessary to achieve this for each distribution method should be taken into account when deciding upon one.

Exposure

A common complaint made about releasing applications on the Apple App Store is being 'lost in the noise' generated by 330,000+ other applications all vying for attention within the 20 categories. A new app, depending on its release time, can have as little as 2 hours and 20 minutes on the first page of the 'New Games' section. As of the end of 2009, updates to an application no longer bump it back to the top of these lists, so paying for featured status or marketing out with the store itself increasingly becomes necessary to achieve a reasonable amount of exposure in a short period of time. The algorithm for computing an application's position on 'most popular' lists in each category is not made public. However, recently, the total number of downloads and the number of recent

downloads seem to be major components in this calculation. Anecdotally, from commercial iOS application developers, each page of applications the user must click through to reach yours on the 'Most Popular' list for your category results in 10x fewer downloads. This can be seen from the distribution in Figure 3 and Figure 4; with the average price of an application on the App Store being $2.89 (as reported by Pinch Media) the vast majority of applications make very little money.

In contrast, the toplists for non-theme related APT packages are easier to appear on, as an application is shown on update as well as on launch, and stays there longer due to the lower number of releases. This increases exposure within the community for new and regularly updated apps.

Another way to achieve exposure in both distribution mechanisms is to collect a number of positively rated reviews from users—although the exact formula used to calculate the overall grading and list position is opaque for both. In our own experience, users have been found to be reticent in producing reviews in comparison to in-application feedback mechanisms. In one of the applications we released which included a direct feedback mechanism, the users made only 2 reviews in the store compared to comments from

Figure 3. App Store Paid applications average downloads per decile on a logarithmic scale

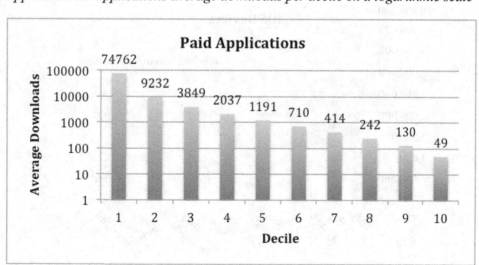

Figure 4. APT Packages average downloads per decile on a logarithmic scale

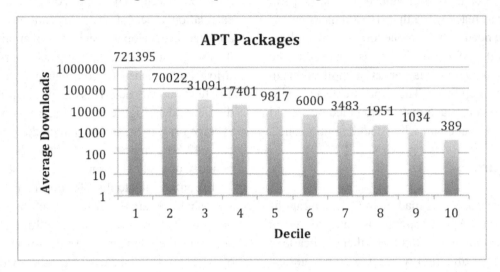

1,224 users in the application—many of which read as one might expect a review on a store to, for example "Everything is good about this app. Very useful."

The ease of exposure must be weighed against the number of users to whom the application is visible. This gives researchers interested in running a trial using one of these methods the option

of lower exposure to around 40 million (Harsh, 2010) devices or higher exposure to 5 million.

The breakdown of the location of jailbroken and non-jailbroken devices can be seen in Figure 5. These numbers come from combining the number of devices seen by Admob and the percentages of jailbroken devices reported by Pinch Media.

Figure 6 shows the average downloads per category of APT package. Unfortunately, no data

Figure 5. Total iOS devices per country including the number of jailbroken devices on a split Y-axis

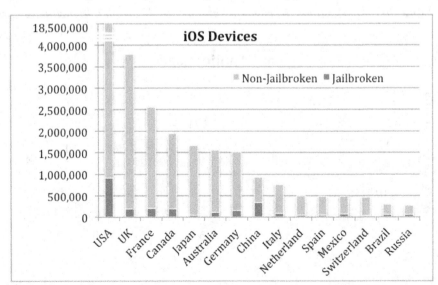

Figure 6. Average total downloads by repository category in thousands

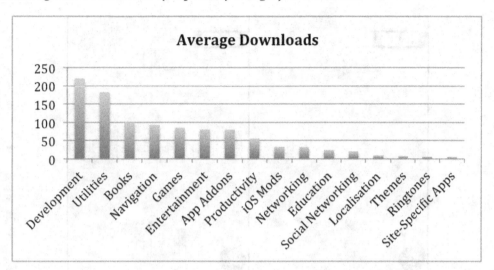

is publicly available to examine the App Store in the same way.

The high number of average downloads in the relatively small category of Development points to the technical literacy of the user base. The large number of utility downloads in comparison to other categories can be explained by the need for utilities to enable the use of themes or application add-ons handled by APT dependency protocols.

TRIAL RELEASE

In order to explore and document the differences in procedure, and results obtained from releasing an application in each method, a memory game application was created. . Rather than the application's design being a demonstration of research concepts in itself, the application design was deliberately very simple, e.g. with no complex use of English so that it could be used worldwide, and a straightforward game design, so that it might easily be taken up and tried out by users. Our aim was to obtain significant numbers of users via each distribution method, and so inform methodology choices for later trials of applications that were more complex research prototypes. Due to the

terms and conditions of both distribution methods stating that any application must be exclusive, the application was 'skinned' to provide two very similar applications. This exclusivity was the source of the question, as any release using one distribution method raises the question of how the results would have been affected by using the other.

When users launch the application they are confronted with a main screen giving them the option of playing a game, looking at the scoreboards and reading the help information provided. The first screen also shows the highest score achieved so far on the device. On selecting the Play option the game board, Figure 7, is shown to the users.

This includes the time they have remaining on the top right, their current score on the top left, the item they must return to its correct place in the centre, and the four locations to which the item can be dragged located at the four points of the compass. In early versions of the game, the timer would start immediately and the overlay of the items each location accepted faded out over 5 seconds. In internal testing this proved to be too difficult for users to understand initially, so the released versions of the game do not start the timer counting down until the user has placed the first item into the correct place. The overlay then

Figure 7. Fruit version (APT) left and Animal version (App Store) right

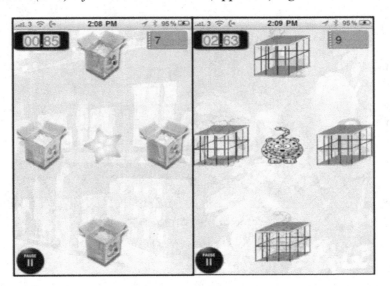

slowly fades out until the user has placed 5 items. If an item is dropped onto the wrong place it is animated back to the centre of the screen. The user is unable to move the item during this animation which, as the timer is continuously counting down, is the penalty for a mistake. The game is over when the timer reaches zero. However for each item placed correctly in under a second an additional second is added to the timer. In order to make the game more challenging the game board changes every time 5 items have been placed in one of two ways: either the items accepted by the 4 locations change to a new random four from the set of twelve, resulting in an overlay of the new items fading away after 3 seconds or the locations are rotated one position clockwise or counter clockwise. Initially, these two conditions were given an equal chance of occurring. However, the internal testing showed that the rotation condition was significantly harder for users to adjust to than the new items condition. This lead to a change in the game whereby the chance of a rotation starts at 20% and increases as the user's score increases to an 80% chance – causing the game to get steadily harder as the user increases in skill.

One version based on animals was submitted to the Apple App Store, and the other based on fruit was submitted to the largest of the APT repositories. Each application was submitted to its respective distribution method on the same day. The App Store version was rejected twice in succession on submission, explaining the slow start to its user numbers. It was first rejected for the artwork of the large and small icons being too dissimilar and then for requesting the user's location without an obvious benefit to the end user.

Both problems were addressed, first with a change in the artwork and then with the addition of a country-based score board which translated the user's GPS location to the country using a reverse geocoding service running on the game server. The application was resubmitted within 24 hours each time. The new scoreboard was also added to the APT version of the game and released on the 7th day of the trial. However, the effect of this on the graph above was small as the application was still experiencing its initial high visibility. These rejections resulted in a 17-day delay before the App Store version was available for download. The application was released on the store late at night on the 17th day of the trial with the large

Figure 8. New users per day for each application

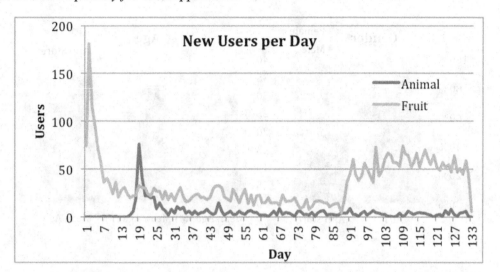

spike on the map being seen on the 18th day. No more publicity was done for either version. As can be seen from Figure 8, the APT version of the game, available only to jailbroken devices, was more popular on all but a single day. In total, the APT version received a 10 times higher number of downloads and continues to gain on average 10 times the users each day.

Jailbreak Effects

The spike shown on the 91st day of the trial on Figure 8 represents a regular cycle seen across all applications released via the APT repositories that have been examined; the release of new iOS versions and the subsequent release of user level jailbreaking applications. As Apple releases each new version of iOS, the community of developers who provide the jailbreaking applications must find new security flaws to exploit in order to alter the operating system to accept homebrew applications. During this lead time a large number of jailbreak users will update to the latest version of the OS ahead of the release of a jailbreak— meaning they no longer have access to the APT repositories. So, when each new OS update is released, the number of jailbroken devices in the

wild reduces for a period of time as a proportion of users who are not using jailbreak software for 'core functionality'—such as unlocking the device from its initial carrier—are likely to update the phone to the latest version of the operating system before an exploit has been identified and released. This causes peaks and troughs in the number of devices that have the ability to download and run software from such repositories, which directly affects user numbers. When a new jailbreak is released the publicity surrounding this drives large numbers of users to enter, or return to, the community at the same time.

Demographics

Each application also asked the user for simple demographic data on the first run and, when they accessed the location-based scoreboard, recorded the country in which they were playing the game.

As can be seen from Figure 9, while the number of users is much greater for the APT version of the game, the gender split is much less balanced than that seen in the App Store. Looking at the age demographics it can be seen that the mean age is consistently higher for users recruited through

Figure 9. Gender spread of users (left) and mean ages by gender(right) for the Fruit and Animal games

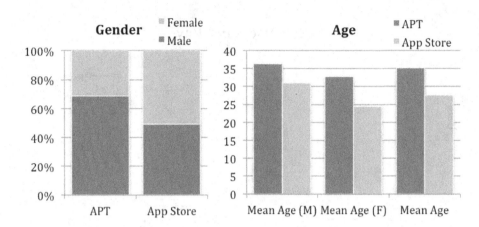

the APT repository – possibly due to the higher technical barrier of entry.

The geographic spread of the users of each application where such information was available was compared to give the charts in Figure 10. 56% of users of the App Store version of the game agreed to share location data compared to 41% of the users of the APT version. As can be seen below, the larger user base of the APT version resulted in a larger spread of countries covered – the number of users in developing countries is higher than may be expected, this could in part

be down to the necessity of users in a country without an official carrier for the device to jailbreak in order to unlock the device from its original, foreign, carrier.

Usage and Engagement

An important consideration when determining how useful any particular set of users will prove over the course of a user trial is the level of engagement they have with the application and with the trial process itself. Any measurement of

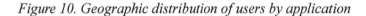

Figure 10. Geographic distribution of users by application

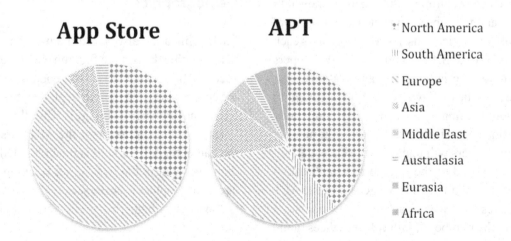

Figure 11. The percentage of users returning to use each application

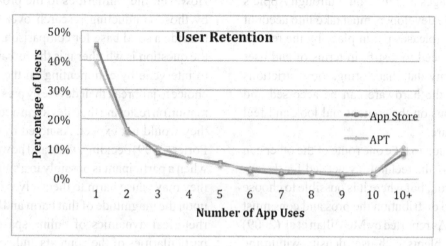

the activity of participants is dependent upon the application they are using, the methods through which they are participating in the user trial and the questions under examination by evaluators.

Figure 11 shows the percentage of total users, along the Y axis, who have used the application more than the number of times shown on the X axis. From this graph it can be seen that 60% of users of the APT version of the game used the application more than once, as opposed to 55% of users of the App Store version returning to the game a second time. The usage tails off with 11% of the users recruited via the APT repositories launching the application more than 10 times compared to 9% of those recruited via Apple's App Store.

The length of each session was also notably different between the two versions of the application, with the users of the App Store version averaging a session length of 62.8 seconds and those playing the APT version of the game s averaging 78.3 seconds per session. The game dynamic being 'beat the clock' means that the more skilful a player becomes the longer their game play will last, suggesting that the population of users with jailbroken devices are more regular game players.

DISCUSSION

There are a number of research groups that initially retooled for Apple's iOS platform on the release of the iPhone in the hope of taking advantage of the powerful hardware, low platform fragmentation and popularity of the devices. This was done at considerable cost in equipment and time. While the advantages of a more open Android platform and the absence of gatekeepers on the Android Marketplace are attractive, the greatly increased fragmentation of the platform (hardware and OS version) and the stores (many carriers provide competing, exclusive distribution channels) must be taken into account. As has been shown there are methods to distribute applications to a wide audience on the iOS platform even if the application falls foul of Apple's App Store policies. However, there are trade-offs that must be made. Applications distributed through the APT repositories are unable to take advantage of Apple services such as their push notification servers, allowing information to be sent to applications without the need for them to run continuously in the background, or the Gamecenter which provides a lightweight social network centred around Apps and their usage as well as achievement badges users can earn by completing in game tasks. In order to partake of these advantages and distribute through Apple's App Store the developers must take into account the delays to releases put in place by the review process, the need to justify in terms of end user experience any data harvesting, the restrictions put on how the hardware can be accessed and the restrictions on the content and look and feel of the software.

The decision to take one route or the other can be dictated by the technology needed for the application to run, but where it is possible to choose either path to distribution the pros and cons must be weighed. As reported by McMillan et al. (2009) there are problems with user density within the jailbreak community both spatially and socially. There the users reported that they were unable to share the application with friends online, or to play collocated with family because potential users not only had to have the correct hardware and be interested in the software genre they had also to have jailbroken their device. As there is, as yet, no way for one user to send a link to another, prompting the install of an application, spreading APT repository packages between peers who have jailbroken their phones is more labour intensive than sharing applications released on the App Store. Users are also unable to share such applications with users who have chosen not to jailbreak their devices, the vast majority in all territories as seen from Figure 5. This becomes a barrier for certain research questions or types of application, those reliant on co-located use (e.g. Bluetooth P2P applications) or on a socially connected user base would be more suited to the App Store.

There are also ethical considerations to be addressed when providing software via APT repositories. As mentioned above, the release of software on these repositories could be seen as encouraging users to jailbreak. As the only contact researchers would have with users would be necessarily after they had jailbroken their devices, addressing this seems to be a classic "catch 22" situation. The ethical responsibilities of researchers conducting large-scale trials are under debate.

However, the similarities to the problems faced by those conducting research over the Internet provide a solid base for comparison. In this case the question is whether it is the researchers place to intervene by commenting on the participants choice to jailbreak their device by presenting information on restoring their device in a context which they would not expect. As noted by McKee and Porter (2009) deciding when and how to intervene when a participant is possibly in a situation where they may cause harm to themselves is dependent upon the magnitude of that harm and "the distinct rhetorical dynamics of online spaces…and the particularities of the contexts and communities" (McKee & Porter, 2009, p. 106). In the case of a jailbroken device I argue that, as the procedure of jailbreaking is much more technically demanding than that of restoring a device the potential harm is low. The harm of intervening where unnecessary and distributing applications within the jailbreak community which encourage members to leave the community would be greater.

CONCLUSION AND FUTURE WORK

This paper gives a comparison of different software distribution methods, increasingly used as powerful tools for research purposes, for Apple iOS devices. There is an initial outline of their characteristics, strengths and weaknesses, and a single application has been used as a comparative example.

The decision as to which distribution method to use for any application must take into account a number of factors.

- If the application needs access to hardware at a level Apple does not approve of, or needs to interact with applications on the device by other developers in order to answer the research questions then the APT repositories are the only option.

- If the application requires Apple's network services, Gamecenter or Push Notifications, to operate then it must be distributed through Apple's App Store
- If the application relies upon colocation of users then, due to the lower density of jailbroken devices, it would be advisable to distribute via Apple's App Store.
- If researchers are looking to explore spread or use across social networks, be they virtual like Facebook or traditional, the density of devices and the difficulty in sharing links to applications on the APT repositories would suggest that in this case Apple's App Store should also be used.
- If the application does not fall into any of the categories above there is no clear choice of distribution method. The researchers must weigh development freedom, faster releases and higher exposure of the APT repositories against the larger, denser, potential user base, Apple's network services and the relative ease with which applications can be shared within social groups seen in the Apple App Store.

The effectiveness of advertising, and the ability to target certain demographics through it, could be significant and is a target for future work in this area. None of the published research in this area has mentioned using advertising to drive recruitment – indeed most explicitly state that they did not advertise – yet researchers regularly advertise locally to recruit participants for trials. As the stores become larger and making a splash becomes more difficult the ability to either advertise for new users or build a relationship with users of one application who can then be brought over to a new research project will become increasingly sought-after.

More investigation into ways to cultivate a relationship with user-participants and their willingness to engage with evaluators is necessary to determine if either method of distribution is more suited to any specific area of research. More users using the application more often does not necessarily translate into more users willing to fill out in-application questionnaires or to be contacted for a more focused form of study, such as an interview.

By exposing the pros and cons of these two distribution methods we hope that more researchers working with the iOS platform will take advantage of the opportunities they present and compliment their local and lab based trials with wider deployments. Not only will this increase the impact of their individual research projects it could pave the way for a greater understanding and appreciation of mobile HCI research with those for whom we strive to innovate.

ACKNOWLEDGMENT

We would like to thank all the members of the Sum Group for their input and the users of all our applications.

REFERENCES

Abowd, G., & Mynatt, E. (2000). Charting past, present, and future research in ubiquitous computing. *Transactions on Computer-Human Interaction, 7*(1).

Anonymous. (2009, February 18). '*AppStore secrets*' - Pinch media blog. Retrieved from http://www.pinchmedia.com/blog/appstore-secrets

App Store Metrics. (2011, January 15). *iPhone development news and information for the community, by the community.* Retrieved from http://148apps.biz/app-store-metrics/

Apple Inc. (2011, January 23). *Apple – iTunes – 10 billion app countdown.* Retrieved from http://www.apple.com/itunes/10-billion-app-countdown/

Bell, M., Chalmers, M., Barkhuus, L., Hall, M., Sherwood, S., Tennent, P., et al. (2006). Interweaving mobile games with everyday life. In *Proceedings of the SIGCHI Conference on Human Factors in Computing Systems* (pp. 417-426).

Campbell, A. T., Eisenman, S. B., Fodor, K., Lane, N. D., Lu, H., Miluzzo, E., et al. (2008). Transforming the social networking experience with sensing presence from mobile phones. In *Proceedings of the 6th ACM Conference on Embedded Network Sensor Systems*, Raleigh, NC (pp. 367-368).

Carter, S., Mankoff, J., & Heer, J. (2007). Momento: Support for situated ubicomp experimentation. In *Proceedings of the SIGCHI Conference on Human Factors in Computing Systems* (pp. 125-134)

Chin, A. (2009). Finding cohesive subgroups and relevant members in the Nokia friend view mobile social network. *Computing in Science & Engineering*, 278–283.

Duyree, T. (2010, May 10). *Beyond AdMob: There's plenty more mobile ad networks to go around.* Retrieved from http://moconews.net/article/419-beyond-admob-theres-plenty-more-mobile-ad-networks-to-go-around/

Falaki, H., Mahajan, R., Kandula, S., Lymberopoulos, D., Govindan, R., & Estrin, D. (2010). Diversity in smartphone usage. In *Proceedings of the 8th International Conference on Mobile Systems, Applications and Services* (pp. 179-194).

Flurry. (2010, October 8). *Mobile application analytics, iPhone analytics, android analytics.* Retrieved from http://www.flurry.com/

Girardello, A., & Michahelles, F. (2010). AppAware: Which mobile applications are hot? In *Proceedings of the 12th International Conference on Human Computer Interaction with Mobile Devices & Services*, Lisbon, Portugal (pp. 431-434).

Harsh, S. (2010, June 30). *May 2010 mobile metrics report.* Retrieved from http://metrics.admob.com/2010/06/may-2010-mobile-metrics-report/

iOS Dev Center. (2010). *Apple developer, developing for iOS.* Retrieved from http://developer.apple.com/devcenter/ios/index.action

Kjeldskov, J., Skov, M., Als, B., & Høegh, R. (2004). Is it worth the hassle? Exploring the added value of evaluating the usability of context-aware mobile systems in the field. In S. Brewster & M. Dunlop (Eds.), *Proceedings of the 6th International Symposium on Mobile Human-Computer Interaction* (LNCS 3160, pp. 529-535).

Licoppe, C., & Inada, Y. (2006). Emergent uses of a multiplayer location-aware mobile game: The interactional consequences of mediated encounters. *Mobilities, 1*(1), 39–61. doi:10.1080/17450100500489221

McKee, H., & Porter, J. (2009). *The ethics of Internet research a rhetorical, case-based process.* New York, NY: Peter Lang.

McMillan, D., Morrison, A., Brown, O., Hall, M., & Chalmers, M. (2010). Further into the wild: Running worldwide trials of mobile systems. In P. Floréen, A. Krüger, & M. Spasojevic (Eds.), *Proceedings of the 8th International Conference on Pervasive Computing* (LNCS 6030, pp. 210-227).

Michahelles, F. (2010). Getting closer to reality by evaluating released apps. In *Proceedings of the Ubicomp Workshop on Research in the Large.*

Nagamine, K. (2010, February 4). *Worldwide converged mobile device market grows 39.0% year over year in fourth quarter, says IDC*. Retrieved from http://www.idc.com/getdoc.jsp?containerId=prUS22196610

Oliver, E. (2009). A survey of platforms for mobile networks research. *ACM SIGMOBILE Mobile Computing and Communications Review, 12*(4), 53–69. doi:10.1145/1508285.1508292

Oliver, E. (2010). The challenges in large-scale smartphone user studies. In *Proceedings of the International Conference on Mobile Systems, Applications and Services.*

Ricker, T. (2007, July 10). *iPhone hackers: We have owned the filesystem*. Retrieved from http://www.engadget.com/2007/07/10/iphone-hackers-we-have-owned-the-filesystem/

Rogers, Y., Connelly, K., Tedesco, L., Hazlewood, W., Kurtz, A., Hall, R., et al. (2007). Why it's worth the hassle: The value of in-situ studies when designing UbiComp. In J. Krumm, G. D. Abowd, A. Seneviratne, & T. Strang (Eds.), *Proceedings of the 9th International Conference on Ubiquitous Computing* (LNCS 4717, pp. 336-353).

U.S. Copyright Office, Librarian of Congress. (2010). *Statement of the librarian of congress relating to section 1201 rulemaking*. Retrieved from http://www.copyright.gov/1201/2010/Librarian-of-Congress-1201-Statement.html

This work was previously published in the International Journal of Mobile Human Computer Interaction, Volume 3, Issue 4, edited by Joanna Lumsden, pp.1-17, copyright 2011 by IGI Publishing (an imprint of IGI Global).

Chapter 15

WorldCupinion:
Experiences with an Android App for Real-Time Opinion Sharing During Soccer World Cup Games

Robert Schleicher
Technical University of Berlin, Germany

Michael Rohs
Technical University of Berlin and Ludwig-Maximilians-Universität München, Germany

Alireza Sahami Shirazi
University of Duisburg-Essen and University of Stuttgart, Germany

Sven Kratz
Technical University of Berlin, Germany

Albrecht Schmidt
University of Duisburg-Essen and University of Stuttgart, Germany

ABSTRACT

Mobile devices are increasingly used in social networking applications and research. So far, there is little work on real-time emotion or opinion sharing in large loosely coupled user communities. One potential area of application is the assessment of widely broadcasted television (TV) shows. The idea of connecting non-collocated TV viewers via telecommunication technologies is referred to as Social TV. Such systems typically include set-top boxes for supporting the collaboration. In this work the authors investigated whether mobile phones can be used as an additional channel for sharing opinions, emotional responses, and TV-related experiences in real-time. To gain insight into this area, an Android app was developed for giving real-time feedback during soccer games and to create ad hoc fan groups. This paper presents results on rating activity during games and discusses experiences with deploying this app over four weeks during soccer World Cup. In doing so, challenges and opportunities faced are highlighted and an outlook on future work in this area is given.

DOI: 10.4018/978-1-4666-2068-1.ch015

INTRODUCTION

Mobile devices are increasingly used for mobile social networking. One explanation for this development is that mobile devices are almost always with their users, have continuous wireless connectivity, and feature increasingly capable user interfaces. They can thus serve as ubiquitous input devices and sensors for user reactions, emotional responses, and opinions around large public events (Diakopolous & Shamma, 2010).

The goal of the work presented here is to investigate mobile social software as a tool for research on opinion sharing in large user communities. We picked the soccer World Cup 2010 as a use case for this research because it is an event with extremely high public attention in many parts of the world and many people have a high emotional involvement to (at least some of) the matches. The matches are also synchronized in time with many simultaneous viewers and thus many potential users. We focus on exchanging spontaneous emotional feedback between users who are part of a virtual fan block.

The particular test application, World Cupinion, is an Android application that lets soccer fans express their opinions about events and moments in soccer matches while watching them. Through this application users can support their favorite teams and share their opinions with other fans. As we expected that users' focus of attention is mainly on the match itself and short bursts of usage occur when interesting events happen, the design focus was on simplicity and quick usage. When not actively used, the app mostly served as an ambient display that conveyed the aggregated opinions of the active users.

This work addresses the following aspects and research questions:

- How to share experiences and opinions effectively in real-time across a large number of mobile devices?

- How to design for awareness of group opinion in a loosely coupled ad-hoc group? How to visualize information related to shared experiences?
- How to distribute and maintain a free Android app for ambient mobile communication?

In the following sections we first discuss the concepts of Social TV, real-time opinion sharing, and the utilization of mobile phones as a research tool. We then give an overview of the design and system architecture of our test application and discuss the distribution and publication channels for the application. After that we present results derived from log files as well as from a subsequent online-questionnaire and report on the experiences we made with the public prototype. We conclude with recommendations for research in the large and with giving ideas for future work.

RELATED WORK

Social TV

Various researches have been exploring the idea of using additional communication channels in parallel with watching TV. "AmigoTV" (Coppens et al., 2004) was an early social TV system that used voice chat communication in combination with broadcast TV. It also provided emoticons and a buddy list with online status. Motorola Labs developed a series of prototypes called "Social TV" system (STV), which allowed users to engage in spontaneous communication with their buddies through text or voice chat while watching TV (Harboe et al., 2008). The system also included an additional display to convey views of the current TV-watching users. Harboe et al. (2008) give a comprehensive overview of social TV systems. Further, various user studies investigated the communication modalities. Geerts (2006) as well as Baillie et al. (2008) compared communication via voice with other modalities. Both studies reported that most users believed that voice chat was more natural and easier to use than text chat. However, Huang et al. (2009) conducted a similar study using the STV system. They found that participants preferred text chat and they often communicated about topics unrelated to the TV content. Geerts and DeGrooff (2009) reported a set of comprehensive sociability heuristics for social TV systems.

Media annotation and sharing while watching TV has been studied, too. Diakopoulos and Shamma (2010) analyzed the sentiments of tweet annotations for a presidential debate to find out their relationship to discussed topics and performance of the opponents in the event. Miyamori et al. (2005) proposed and examined a method for generating views of TV programs based on viewer's opinions collected from live chats on the Web. Affective responses to unstructured video commenting systems were evaluated by Nakamura et al. (2008).

All the mentioned social TV systems require the installation of set-top boxes for supporting collaboration. Since set-top boxes are only available in certain locations, users are restricted to particular environments. To overcome this limitation and attract participants we intended a mobile phone application that would give users the chance to use it for sharing their opinions in any context in which watching the event is possible, even in bars, the stadium, or at public places – a requirement indispensable for the sports domain.

Mobile Phone Sports Apps

Mobile phones are particularly suited to support sports fans that attend such events which frequently take place outdoors: *MySplitTime* (Esbjörnsson et al., 2006) allows users to take pictures of bypassing cars at rallies and obtain additional information about the current ranking of the photographed car. Although not particularly designed for sports events, *coMedia* (Jacucci et al., 2007), an app to

create and share digital memories was also tested at a big rally in Finland. *TrottingPal* (Nilsson, 2004) helps spectators at the trotting track to gather additional information to improve their betting and to coordinate with other visitors who might be dispersed across the area. Information retrieval appears the focus of apps that target in-stadium sports: *eStadium* (Ault et al., 2008) which was later extended to *RISE [Rich Immersive Sports Experience]* provides visitors of football games at Purdue's Ross-Ade stadium with various statistics, replays and other multimedia services (Facwett et al., 2009). *TuVista* is a similar app tested at the Estadio Azteca in Mexico City (Bentley & Groble, 2009) *YinzCam*, a spinoff from the Carnegie Mellon University (http://www.yinzcam.com/about.html) offers related services for various popular college sports including basketball, ice hockey, etc. While the capabilities of these existing apps may vary in detail and may be extended in the meantime, their general intention appears to be to provide a service for mobile phone users that is comparable to the characteristics of professional TV sports broadcasting i.e., detailed background information and multimedia material to guarantee an exclusive viewing experience. We aimed at supporting sports events viewers in another regard, namely sharing their immediate impressions of the game, which may be one of the major reasons to watch such events in a group.

Real-Time Emotion Sharing

Taking a closer look at what information the audience/watchers of sport broadcasting actually wished to share with their friends or fan group, it turned out to be mostly the preliminary evaluation mixed with the personal emotional impact of specific events during the game, much less a "cold" rational assessment of the ongoing maneuvers on the field. This was no surprise, as emotions are known to have a strong social component (Ochsner & Schacter, 2000) and probably even developed to provide a fast and immediate way to communicate the momentary state of an organism to the environment (Ekman, 1999).

The sudden onset and strong expressive component of emotions make them an ideal candidate for mobile communication as it allows the user to somehow extend his/her reach beyond the usual radius of face-to-face communication. On the other hand, these properties also impose a number of requirements for any application: feedback should be quick and if possible "analogous," i.e., nonverbal to avoid the necessity of lengthy formulation to describe a simple and transient affective rush. Emoticons appear to be an appropriate way to communicate these states (Derks et al., 2008). In addition, the provided rating scheme should contain domain-specific labels (e.g., "yellow card") as well as domain-independent features (e.g., "like-dislike") (Pang & Lee, 2008). Relying on such a limited set of means of expression is also referred to as *lightweight communication* (Metcalf et al., 2008) or if it is not restricted to a specific location *ambient mobile communication* (Bentley et al., 2006). The latter authors describe an exemplary prototype, called *Music Presence* that uses a set of domain-independent icons, namely thumbs up/down, and "!" similar to what we offered in our app. The communicative purpose differentiates these approaches from related work subsumed under *experience sampling methods (ESM)* that also utilize mobile phones as digital diaries (Carter & Mankoff, 2005), but mostly let the user create their entries for later analysis by the researcher, not for conveying them to other users. Our app extends the notion of ambient mobile communication by additionally using the mobile phone (and the corresponding app "market") for acquiring people interested in this form of communication.

Mobile Phone Apps as a Research Tool

The approach to recruit participants via mobile phone applications has been deployed by other research labs as well. In the following we will give a short and eclectic overview pointing out the aspects we consider of interest when using apps for research. A detailed description of the mentioned projects can be found in the references as well in the other contributions of this issue.

Oliver (2010) released a Blackberry application to obtain measures such as average usage frequency and duration from more than 17000 users. Main challenges reported were the increase in power consumption caused by the research app and the reliability of time stamps when the data were logged locally on the device. Michahelles (2010) published various social apps, mostly on the Android portal and gives a brief assessment of the varying success in distribution as well as requirements that may be new for research using apps like the need to constantly add new features to maintain the users' interest in the application.

CenceMe, initially developed for Nokia's N95 (Milluzo et al., 2008), later also released for the iPhone (Milluzo et al., 2010), allows users to communicate and share their momentary activity and location with others, including a linkage to Twitter and Facebook. As this app infers high-level status information from the mobile phone sensors, the major concern was which analysis should be done locally and which on the backend server, and how this affects battery lifetime. *CenceMe* is a good example of how extensive the information can be that researchers can derive from mobile app usage. On the other hand, *Hungry* (Windows Mobile) / *Feeding* (iPhone) *Yoshi* (McMillan et al., 2010) shows what innovative data collection paradigms are possible with mobile technology: The basic plot is a location-based game in which users have to collect fruits at various locations. The players can gain additional scores by accomplishing tasks or quests which consist of questions asked by the researchers. The researcher can tailor these questions to specific locations or usage patterns. Since the questions appear to be part of the game, this mitigates the issue of inferring user intentions etc. from log data alone in order not to disrupt the

Figure 1. WorldCupinion screens: Initial screen (left) showing the match list and main screen (right) used during a game

game flow. The authors (McMillan et al., 2010) also discuss the perils of this procedure.

The idea to exploit competition amongst users as a means to increase participation was one of the reasons we chose the World Cup 2010 as a starting point for our research trial. The stimulative nature of this event has obviously also been identified by other researchers: Morrison et al. (2010) offered *World Cup Predictor* for that purpose. Whereas the goal of that app apparently was to forecast the winner of the cup, i.e. the outcome, we focused more on the process of social interaction, namely sharing opinions about ongoing events while watching the game.

APP DESIGN

As mentioned above, the design criteria included simplicity, since the user's focus of attention is on the match itself, and short-term usage, since situations arise quickly an interaction might just involve stating one's opinion about the current event. Moreover one aspect was to visualize the aggregated opinion of a potentially large number of

users and to use the screen as an ambient display. The latter feature gives the user the opportunity to observe how the fan-aggregated opinion evolves even though he or she is not actively interacting, but may react to the updates by rating again.

The app is structured in three screens (Figures 1 and 2). The first screen (Figure 1, left) shows the list of upcoming matches with their starting times and dates in the user's local timezone. The timezone played an important role here, since we intended to deploy the app in the Android Market for worldwide distribution. The game selection could have been automatized, except for parallel games during the first phase of the tournament, but we decided to keep the list in order to allow users to plan their viewing times in advance of the games.

After selecting a game the user would enter the "arena" for that game (Figure 1, right). That screen allows the user to give feedback during the game and to see the aggregated opinions of the fans of the own and the other team. Initially the rating buttons are disabled and the user has to select the team he or she wants to support in order to activate the interface. This design deci-

Figure 2. Geographic distribution of fan opinions

sion means that users have to be fan of a particular team in order to provide input. The input buttons cover most of the display area to be easy and quick to press. Their functionality was either to give soccer-specific assessments of events like yellow/red card, whistle/play on, and in the beginning "offside". The remaining buttons served to express the current mood of the user, i.e., thumbs up/down, invoke a vuvuzela sound to express excitement, and later on a "Yippee" button which replaced the offside icon and was accompanied by the sound of applause. Below each button there is a horizontal bar that indicates the average opinion regarding that input category. For example, if the bar below the "thumbs up" button is half filled that means that 50% of the fans have pressed that button during the last 30s.

A feature that was added one week after publishing the game is to see the aggregated opinions of the fans of the other team as well. The statistics of the fans are shown in green and the statistics of the other team in blue. The blue bar is located behind the green one and the green bar is not fully opaque to see the blue one behind it.

Another change to the main screen (Figure 1, *right*) was the replacement of an "offside" icon with a "yippee!" icon (Figure 1, *right*, center icon). We made this change after we realized that the "offside" icon was rarely used. At the same time, the app lacked a way to express strongly positive emotions, e.g., when the own team scored a goal. Adding the "yippee!" icon provided a way to express that kind of emotional feedback.

The third screen, "world opinion" (Figure 2), shows the geographical distribution of fan opinions of both teams on the map. The underlying idea is that this visualization shows geographical clusters of users having opposing opinions. The map view is based on the standard Google Maps APIs with icon overlays for the feedback that was given at a particular location. Using Google Maps APIs allows interactive panning and zooming of the map. However, we restricted the maximum zoom level for privacy reasons.

SYSTEM ARCHITECTURE

World Cupinion is implemented as a client-server architecture. The World Cupinion mobile application sends two basic request types to the server, update requests, and input requests. Update requests are used to poll the state of the mobile application's user interface, and input requests are used to send user opinions to the server, as soon as an opinion button has been pressed. The map view sends a further request type, to which the server generates a response containing the user inputs of the last 5 minutes.

The server logs all inputs to a SQLite database and maintains statistics of the user opinions received in the last 30s. These 30s statistics are sent to the mobile clients in response to update requests.

We initially used UDP datagrams for communication, as our protocol does not require an active connection. UDP also imposes a lower load on the server, which is beneficial if there are many simultaneous server requests. However, it soon appeared that certain network firewalls and also mobile network providers may block UDP packets that have non-standard destination ports. To remedy this, our mobile application has a fallback mechanism that automatically switches to HTTP requests if UDP communication is unsuccessful. User input events are always sent via HTTP to ensure that they do not get lost.

Supporting HTTP requests has the further advantage of enabling the implementation of platform-independent web interfaces. Although we did not originally plan to use a web interface, we implemented one for evaluation purposes and to fulfill requests from users that did not use the Android platform. It turned out that this was useful, as our statistics show that a substantial proportion of the input originated from the web interface.

A further important issue of mobile phone application is energy consumption (Oliver, 2010; Miluzzo et al., 2008). Over the 90 minutes of a game (plus the 15 minutes break and an optional 30 minutes extension), the application continuously communicates with the server via the mobile phone network or WiFi. There is a tradeoff between the update rate of the interface and energy consumption. In pilot tests we found that one update every 3 seconds is sufficient. A significant contribution to energy consumption comes from continuously using the device as an ambient display for the opinion state. Even if the user is not interacting with the device the community opinion is updated and shown. This is technically implemented with a "wake lock" that prevents the display from switching off completely. Usually, a NexusOne mobile phone that is fully charged at the start of the game has a battery level of about 50% at the end of a game.

DISTRIBUTION AND PUBLICITY

Distribution via the Android Market

From June 4, 2010 onwards (one week before the start of the World Cup), the World Cupinion app could be downloaded for free from the Android Market. An advantage of using the Android Market as a distribution platform over Apple's iTunes AppStore, for instance, is that published applications appear almost instantly for download, and are not subject to a lengthy reviewing process (Miluzzo et al., 2010) with the risk of rejection of the application.

The ability to rapidly push new releases of the application to the Android Market allowed us to publish weekly updates containing bug fixes or new features during the actual soccer World Cup.

Public Relations

It of course does not suffice to simply release a new application into the wild. Potential users need to be informed of the application's existence in order for them to download it.

We used a number of channels to make the application known to potential users. In addition to press releases made by the Deutsche Telekom Laboratories and the TU Berlin, we tried to promote the application in internal events (summer party, weekly lab meeting) and external events (lab open house) of the Deutsche Telekom Laboratories. At these events we distributed flyers advertising the app, containing a QR code linking to the app on the Android Market. We also created a website (www.worldcupinion.com) and actively used social media (Twitter and Facebook) and forum entries to reach as large an audience as possible. Finally, we sent emails to a number of mailing lists in our lectures and posted messages about the app on Android developer forums.

Updates

The Android Market allows to easily publish updates. We took advantage of this feature several times. If a long-term study is conducted it allows to carry out several design iterations while keeping the user base of the app. Besides bug fixes, the changes and updates related to (1) replacing the "offside" icon with the "yippee" icon, (2) adding group-generated sounds, (3) showing the opinion of the opposing fans, (4) sending notifications when a game starts (and restarts after the half-time), and (5) the addition of an in-application questionnaire displayed after the conclusion of the World Cup. These changes were not obvious from the start and reflected insights gained from application usage and user comments. The update mechanism provided a convenient way to do these changes. On the other hand, one has to be careful not to confuse users when features change. If some users do not update the application there might be inconsistencies between deployed application versions, e.g., some users might still have the "offside" icon in the place where users of the updated app have the "yippee" icon. We expected that most users would update their app, and for us having the flexibility to try out different versions was more important than version consistency. Unfortunately, the server protocol did not include the version number, so we could not track the percentage of users connecting to the server with outdated application versions. This is something we will clearly consider in the future when deploying similar apps.

RESULTS

Based on the Android portal at the end of the World Cup, we had registered a total of 1645 downloads and 448 "active" installations (=29% of all downloads). The number of active installations denotes the number of users that still had the app installed on their devices at that point.

The results presented in the following are based on data from two sources: on the one hand the logs of user activities during the matches and on the other from the in-application questionnaire provided with the last update.

Usage Statistics

On average 28.6 (+/- 19.1) users were active during the games, with a maximum of 94 and a minimum of 8 users for a single game. Figure 3 shows the number of participating devices for all of the 64 games of the World Cup. Of course this number was highly dependent on the nationality of the teams playing, but a general decay from the first couple of games (when the app was still "new") could be observed. From match 49 the round of sixteen started, which led to a temporary increase in usage. The most prominent game of the World Cup, the final (game 64) had surprisingly low number participants.

The average participation lasted 681 (+/- 1316.2) seconds during which 17.6 (+/-33.6) actions, i.e., button presses were performed. Based on our database, 71% of inputs were from the Android client and 29% from the Web-based client. Table 1 shows the basic statistics of average number of inputs during a game, and average session length divided by interface type, i.e., the mobile phone client or web interface.

A multivariate analysis of variance (MANOVA) revealed that there was a significant effect of input interface on these parameters: users of the app tended to give less ratings during longer sessions ($F_{2,820}$=6.584, p=0.001). However, univariate comparison revealed that this difference was only significant for #inputs ($F_{1,821}$=9.063, p=0.003), but not for session length ($F_{1,821}$=0.714, p=0.398).

While the means in number of inputs clearly varied for both interface types, the medians were identical, indicating that in the web interface there were some users with very high number of inputs as also shown by the higher maximum (591 total

Figure 3. Number of participating devices during the course of the soccer World Cup 2010

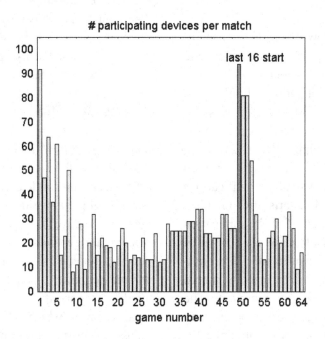

inputs for the web interface vs. 372 for the app). This might be due to excessive clicking using the mouse. To give a better impression of input activity, Figure 4 shows the click distribution for the first 4 games of the round 16 (games 49-52) in all of which more than 50 users participated. For these four games, no clear difference in input patterns between the mobile phone and the web client could be seen, but both occurred during the whole game time. However, the graphs also show that ratings did not happen at a constant rate, but

were closely linked to events in the game. This is exemplified for game 51 (Germany vs England) (http://g.sports.yahoo.com/soccer/world-cup/), and the graphs suggest that users understood the app as intended to communicate moments of high relevance to the other participants. For all games there was also some activity during halftime, which we found to be mostly vuvuzela and thumbs up/ down assessments after examining the log files. This led to the question which buttons were used how often.

Button/Icon Usage

In addition to the amount of activity over time, we were also interested in the usage of the offered icons. Table 2 shows the relative frequency of button clicks per game for games 49 to 52 and across all games. For game 49 to 52 the differences in relative usage frequency where also assessed statistically using a Chi^2 test, which revealed a clear effect of button meaning on usage frequency ($Chi^2(24)=407$, $p<0.001$). Table 2 also indicates to what extent the usage frequency of single but-

Table 1. Usage statistics by interface type, i.e. Android mobile phone app vs. web interface

	Android app		web interface	
	#inputs	session length	#inputs	session length
Mn	15.58	703.79	23.61	615.13
Md	9.00	60.87	9.00	44.86
STD	24.56	1320.22	51.11	1305.39
Min	1.00	.38	2.00	1.44
Max	372.00	6997.30	591.00	6611.56

Figure 4. Number of inputs during the first four games of the round of 16, distinguished by interface type (Android mobile phone app or web interface), over game time (begin and end of halftime indicated by dotte lines). Bars represent absolute numbers. Please notice the adjusted y-axis limits for game 51 (England: Germany), where also the moment of the goals and corresponding rating bursts are labeled

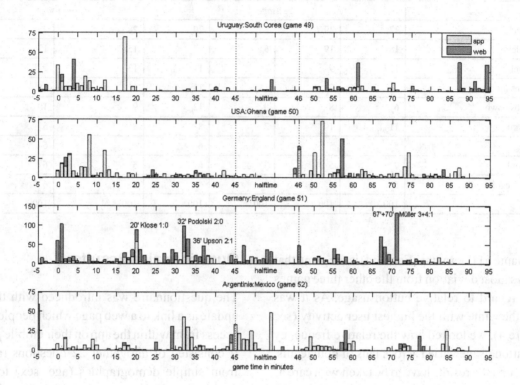

tons varies from game to game and whether this difference is statistically significant (numbers that share the same subscript in a row are not significantly different from each other). While there are slight differences between single games, the vuvuzela was by far the most frequently used button across all games with the highest number in game 51 (Germany: England). The second most frequently used button serves to annotate a typical soccer controversy, namely that the referee should whistle in a specific situation. Yippie!, thumbs up and down were also used regularly with around 9-15% across all games. The remaining buttons that could be used to express boredom with the actual course of the game or whether the referee

should give a yellow/red card or let continue to play refer to events that are rather specific to particular moments in the game and, thus, were used less often.

Values are rounded to whole numbers. Percentages with the same subscript letter are not statistically different from each other in the corresponding Chi² test when comparing row-wise across games, i.e., 27_a% vuvuzela clicks in game 49 is statistically not different from 29_a% in game 50, but both are significantly lower than 36_b% in game 51. $13_{a,b}$% indicates that 13% is neither different from other numbers in the same row with the subscript a nor from ones with the subscript b.

Table 2. Relative frequency of button clicks per game in percent (column-wise), for the first 4 games of the round of 16 and across all games

icon	game nr.				mean games 49-52	mean all games
	49	50	51	52		
	% usage				*% usage*	
vuvuzela	27$_a$	29$_a$	36$_b$	20$_c$	28	23
whistle!	18$_a$	19$_a$	6$_b$	17$_a$	15	19
Yippie!	12$_a$	9$_a$	21$_b$	11$_a$	13	9
thumbs up	9$_a$	9$_a$	15$_b$	13$_{a,b}$	12	15
thumbs down	6$_a$	8$_{a,b}$	8$_a$	12$_b$	9	10
red card	11$_a$	10$_a$	2$_b$	8$_a$	8	5
boring	5$_a$	5$_a$	7$_a$	6$_a$	6	6
play on!	7$_a$	5$_a$	3$_b$	6$_a$	5	6
yellow card	5$_a$	6$_a$	3$_b$	5$_a$	5	6
Total	100	100	100	100	100	100

Game 51 (Germany: England) is the game that shows most deviation from the other three games with regard to relative button usage. As it was also the game with the highest user activity (see Figure 4), we looked how the relative frequency of button usage varied across fans of both teams. Although the results have to be taken with care as both fan groups differed in terms of participating devices (n=16 for England, n=51 Germany, n=6 devices "switched" teams and voted for both), Figure 5 shows that there is a clear effect of team association on clicking activity (Chi^2 (8) =74, p<0.001) which also is reflected in the pairwise comparisons of button selection. While fans of the winning team Germany used the vuvuzela and the Yippie! button significantly more often, the English fans expressed their disappointment with a clearly higher use of the "thumbs down" button and also voted more frequently for the referee to whistle and react with a yellow card.

To avoid misinterpretation about the user intentions and to see whether this idea of opinion sharing was really present, we conducted an additional post-hoc survey which results are described in next section.

Questionnaire Results

The questionnaire was introduced with the last update as a link to a web page which people could access from within the app on their mobile phone. It consisted of altogether 22 questions ranging from simple demographics (age, sex) to open suggestions for improvements. For all evaluative questions, a five-point Likert scale was offered.

In total 46 users (mean age = 20.1 +/- 9.2 years, 6 female users) replied to the questionnaire, of whom 37% followed the World Cup matches frequently and 50% watched occasionally. 55% of the participants considered themselves as knowledgeable fans and 30% as experts (knowledgeable of players' details). 73% of them stated that they normally watched the matches at home and 65% watched with the family or buddies.

World Cupinion Usage

18% of the participants stated that they used the app for most matches, 40% used it regularly, and 42% occasionally or just once. Also, those who considered themselves as knowledgeable or expert

Figure 5. Relative frequency of button clicks during game 51 (England:Germany) split by team. Stars indicate a statistically significant difference between both fan groups (p < 0.05)

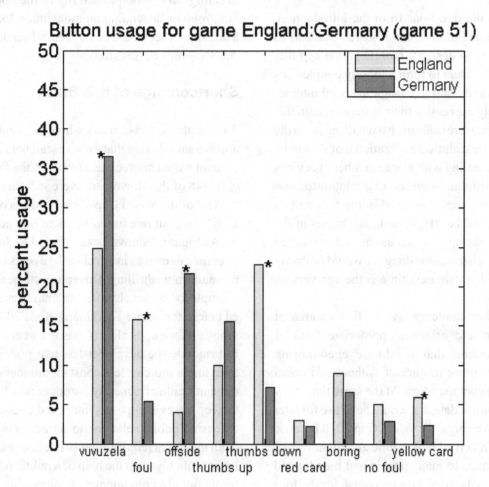

fans used the app more frequently. 60% of those who watched the matches in a group (with family, buddies, or crowd) still used the app regularly or for most matches. Additionally, participants rated the app in general. The average rating was 3.6 out of 5+/- 1.1 (Median=4). Those who used the app more regularly rated the app higher.

Connectedness vs. Fun

Participants were asked to rate if the level of fun and connectedness to other fans changed while using the app. 11% mentioned that the fun aspect did not change at all. 30% believed that the fun aspect increased sometimes and 59% reported to have more fun most of the time or (almost) always. None of the responses indicated that the app reduced the fun of watching. Also, 7 out of 46 participants (15%) did not feel connected to other fans at all. 32% felt (very) strong connection and 53% average/little bit connection. Those who had more fun felt more connected to the other fans (Spearman's ρ = .636, p<0.00).

DISCUSSION

Taking up the questions from the introduction, the overall intention of this research trial was to explore how to design a mobile phone app that would allow users to form a loosely coupled ad-hoc community and to engage in social interaction, namely expressing their opinion, within this group. There were basically two challenges: on the one hand, the technical realization that should be capable of dealing with a large number of devices and provide instant feedback. Our solution to these requirements were described in the first part of this paper, and besides occasional crashes of the map view, the app as well as the server worked stable and reliable. One thing we would probably add to the log data next time is the app version of each user.

The other challenge was to find an area of application and offer an appropriate form of communication that would be encouraging enough for users to interact without additional (financial) compensation. At the same time, they should produce data that would be of use for later analysis. We chose the World Cup 2010 because this event received high public attention, was of high relevance to many people and broadcasted simultaneously. As the rating course for the four games of the round of sixteen showed (Figure 4), the concept of online rating succeeded and already the analysis of rating frequencies alone allowed identifying relevant moments in the game. Thereby, the four expressive icons "Vuzuzela", "Yipie!" and "thumbs up/down" comprise more than half of all clicks across all games (57%, see Table 2), which we take as an indicator that the app was used as intended to share experiences/opinions. Among the soccer-specific icons, the "whistle!" button was used most frequently (19% in Table 2). Fans of opposing teams clearly vary in term of how they evaluate a specific game (Figure 5). This result is not surprising but served as a "sanity check" and was the first step in the direction of identifying group-specific rating pat-

terns. For finer analysis, e.g., evaluative meaning of ratings in reaction to activity of the opposing fan group or depending on momentary location, a larger data set would be desired, which leads to shortcomings of this study.

Shortcomings of this Study

The number of 1645 users who downloaded the app is a sample size that most researchers in HCI domain would appreciate. However, the fact that only 448 of that downloads were still "active" at the end of the World Cup already points out that a high dropout rate has to be taken into account.

As Figure 3 shows, usage tended to decrease over time even if as in our case the events became more and more thrilling. This appears to be another example for the novelty effect this has been reported before for Social TV (Huang et al., 2009) and mobile phone apps alike (Consolve et al., 2008). In principle the developer has two possibilities to ensure a more or less constant number of participants: either frequently introduce new features to keep the existing users interested or engage in extensive public relations to attract new users. Both measures require appropriate resources, the latter probably even the help of a public relations professional as our attempts to attract interest of sports magazines or newspaper editorial boards were not successful and we were informed that they expect a more or less pre-written description that they can publish immediately. As we did not ask it in the subsequent questionnaire, we are not aware how users specifically learnt about our app and could thus not infer which promotion strategy was most successful. We would clearly revise that in a future field trial, and also suggest to other developers to pay attention to this issue right from the beginning on to be able to allocate promotional resources more efficiently in the later stages of a study.

The decision to introduce a web interface and thereby to some extent adulterate the idea of mobile phone app research was a reaction to several requests to also provide the app to other platforms and grounded in the intention to include as many users possible in this first trial. A web interface enabled users outside the Android world to participate. Although rating frequency in general varied between both interfaces, we could not see a clear bias in rating patterns as again exemplified by Figure 4. The fact that the web interface users tended to have shorter sessions (although not statistically significant) was an indicator to us that for opinion sharing while watching TV a mobile phone app might be the more convenient interface as it can be used in almost any viewing situation, e.g. on the couch, in the bar etc. without having to scroll or zoom in a browser.

However, without additional interview or observational data, this assumption cannot definitely be confirmed. In general we had the impression that the data collected via an app that is released into the wild is much more difficult to interpret than data obtained in a controlled laboratory setting. For certain analyses, an incorporation of other (qualitative) data like interviews or observation in addition to the quantitative information obtained from log files appears to be inevitable. Jacucci et al. (2007) may serve as an example how to combine these different methods in a field test. At the same time, their sample size of n=8 also indicates that some kind of pre-selection, which users of an app might be most representative, is required when adopting their approach for research in the large. Our intention was to see whether the recruitment of anonymous participants and data collection of their communication is feasible per se. We will summarize the observations we made and feedback we got beyond the log files and questionnaire data in the next section called "experiences."

Experiences

The first users using the app during an actual game were friends, colleagues and students who were made aware of it by our public relations activity.

Here the focus was almost more some kind of beta testing and providing feedback to the developers than using it for rating. However, the Vuvuzela functionality was quickly used to emphasize important moments or echo the sound coming from the TV, which then led to a more comprehensive usage of the app to comment the ongoing game. Although we were well aware that there were several other "Vuvuzela apps" available in the market, we still thought that this feature might serve as some kind of door opener to attract the user's attention initially. Our observations within our peer group confirmed that assumption. Probably other research projects might also benefit from this approach. However, there is one caveat when adapting popular "recreational apps" or functionalities for research applications: we found that users expect a running software and will give low ratings if an app is not polished or crashes during usage. There were occasional crashes of the map view in our app, which only affected the map view and not the other parts of the application. Some users appeared to be very critical about this, as is documented by one user comment mentioning this particular issue. Moreover, a few handset types had problems for running the Android app. Even though platform fragmentation is a small problem for the Android platform than for other systems, this issue appeared. Of course, providing an industry-strength app as a research prototype is not feasible for most research labs as it requires more development resources than are typically available.

We were surprised to what extent users apparently download and install an application without actually using it for a longer time. It appears that the abundance of available mobile phone applications let them become a disposable article like promotional gifts, an observation that has also be made by McMillan et al. (2010). When research apps are released on the market, they have to compete with these existing products and apparently cannot expect a "research in progress" bonus that people sometimes implicitly grant when testing prototypes etc. in laboratory settings.

In a similar vein, a few weeks after our app was released on the Android market and while the tournament was still going on, we noticed that there was another, unrelated app for soccer fans made available on the iTunes store that featured some of the functionalities we also offered, including the "thumbs up/down" icons and the map view ("world opinion," see Figure 2). Again, these challenges might be pretty new to people who usually conduct laboratory research. Finally, since the app is released to the public, large number of users might use the app which leads to high traffic and performance issues. Users expect that the app works at any moment. Therefore 24/7 maintenance and monitoring are crucial and should already be considered during evaluation.

CONCLUSION

Exploiting spontaneous lightweight communication with a mobile phone during TV shows for annotating these events relies on a critical mass of users. We tried to gain a large number of users for sharing opinions in large user communities in real-time, by picking a popular topic in which people have emotional involvement and simultaneously follow a shared event. The soccer World Cup 2010 provided a good setting for the first trial, because it is quite popular in many parts of the world and extends over four weeks, which allowed us to do several design iterations. We tried to gain as much user attention as possible by publishing the app on the Android Market and announcing it to make it known to potential users. The market update facility allowed us to try out design modifications easily and improve the prototype based on usage statistics and user feedback. The rating course for the first four games of the last 16 showed that the concept of sharing personal opinions about sports events online is feasible. However, conducting an uncontrolled study in the wild has its own

shortcomings. The quality of the data obtained as well as the user experience stands or falls by the number of users involved. We described our experience with the measures we took to attract users in the previous paragraph and will now point out what future work needs to be done.

Outlook

So far, the focus of existing sports apps appears mainly to provide additional statistics and multimedia material to the user as described in "Related Work" Section. If there is any intention to promote a continuous interaction during match time, the lightweight communication we described might be the proper way. The option to comment or vote and present these sentiments to the own as well as the opposing fan group would enrich the so far predominately individual experience of using a sports app with a social aspect. Integration in social media platforms, such as Facebook and Twitter nowadays appears to be standard for applications that aim at social interaction (Michahelles, 2010; Miluzzo et al., 2008), thus, we also intend that for future apps. As Facebook users in particular appear to be willing to share information (Miluzzo et al., 2008) and as such are of special interest for researchers, it would probably be advantageous at some point if that platform could offer a special API to researchers (like *Google scholar* in addition to the standard *Google*) to facilitate integration and to indicate applications free of commercial interest. A connection to Facebook might also answer two other questions: the integration of a chat functionality that was mentioned by some of our users for boring parts of a game. We were not sure whether such a chat extension would distract users too much from the ongoing game (Huang et al., 2009) and interfere with the idea of rating synchronously to the timeline of the TV show. Facebook already provides a chat, and for setting up a contact beyond the game this platform might be more appropriate. The second question is what benefit the app could provide to the user, a crucial aspect for the success of any research app as Michahelles (2010) points out. Morrison et al. (2010) offered prizes of £500 for the winners of their *World Cup Predictor*. So far, we solely trusted in the presumed fun of rating while watching. However, the option to gain some kind of "expert" status with an attached "rating profile" as a consequence of frequent ratings which could then be published somewhere (e.g., Facebook) would probably provide additional motivation. Feelings of competence are among the motives that are currently being discussed as the causes for positive experiences when using interactive products (Hassenzahl et al., 2010). A motivational stimulus like this could be even more important if the idea of a simple shared online rating via the mobile phone is extended to other domains: for example, online rating of sneak previews of movies, where again the reputation of being an expert whose opinion is subsequently publicly available might encourage users to participate. For a stronger focus on research purposes, Diakopoulos and Shamma (2010) analyzed the tweets of a presidential debate. Of course short text messages provide more information than button clicks, but the concept of "thumbs up/down" could in principle also be applied here and if combined with the location information obtained from the mobile phone's GPS sensor enable analysis which statements were of special relevance to viewers in a certain area. Accompanying demographic questions to complete the picture could be presented within the application, following a recommendation of McMillan et al. (2010) "to stay in the app" to increase the response rate.

REFERENCES

Ault, A., Krogmeier, J. V., Dunlop, S. R., & Coyle, E. J. (2008). eStadium: The mobile wireless football experience. In *Proceedings of the Third International Conference on Internet and Web Applications and Services* (pp. 644-649).

Baillie, L., Frohlich, P., & Schatz, R. (2007). Exploring social TV. In *Proceedings of the International Conference on Information Technology Interfaces* (pp. 215-220).

Bentley, F., & Groble, M. (2009). TuVista: Meeting the multimedia needs of mobile sports fans. In *Proceedings of the 17th ACM International Conference on Multimedia* (pp. 471-480).

Bentley, F., Kaushik, P., Narasimhan, N., & Dhiraj, A. (2006). Ambient mobile communications. In *Proceedings of the ACM SIGCHI Conference on Human Factors in Computing Systems* (pp. 1-3).

Carter, C., & Mankoff, J. (2005). When participants do the capturing: The role of media in diary studies. In *Proceedings of the ACM SIGCHI Conference on Human Factors in Computing Systems* (pp. 1-10).

Consolvo, S., Klasnja, P., McDonald, D. W., Avrahami, D., Froehlich, J., LeGrand, L., et al. (2009). Flowers or a robot army? Encouraging awareness & activity with personal, mobile displays. In *Proceedings of the Conference on Ubiquitous Computing* (pp. 54-63).

Coppens, T., Trappeniers, L., & Godon, M. (2004). AmigoTV: Towards a social TV experience. In *Proceedings of European Interactive TV Conference* (pp.1-3).

Derks, D., Bos, A. E., & von Grumbkow, J. (2008). Emoticons in computer-mediated communication: Social motives and social context. *Cyberpsychology & Behavior*, *11*(1), 99–101. doi:10.1089/cpb.2007.9926

Diakopoulos, N. A., & Shamma, D. A. (2010). Characterizing debate performance via aggregated Twitter sentiment. In *Proceedings of the ACM SIGCHI Conference on Human Factors in Computing Systems* (pp. 1195-1198).

Ekman, P. (1999). Basic emotions . In Dalgleish, T., & Power, M. (Eds.), *Handbook of cognition and emotion* (pp. 45–60). Chichester, UK: John Wiley & Sons.

Esbjörnsson, M., Brown, B., Juhlin, O., Normark, D., Östergren, M., & Laurier, E. (2006). Watching the cars go round and round: Designing for active spectating. In *Proceedings of the ACM SIGCHI Conference on Human Factors in Computing Systems* (pp. 1221-1224).

Fawcett, J., Beyer, B., Hum, D., Ault, A., & Krogmeier, J. (2009). Rich immersive sports experience: A hybrid multimedia system for content consumption. In *Proceedings of the 6ᵗʰ IEEE Conference on Consumer Communications and Networking* (pp. 1-5).

Geerts, D. (2006). Comparing voice chat and text chat in a communication tool for interactive television. In *Proceedings of the Nordic Conference on Human-Computer Interaction: Changing Roles* (pp. 461-464).

Geerts, D., & De Grooff, D. (2009). Supporting the social uses of television: Sociability heuristics for social TV. In *Proceedings of the ACM SIGCHI Conference on Human Factors in Computing Systems* (pp. 595-604).

Harboe, G., Massey, N., Metcalf, C., Wheatley, D., & Romano, G. (2008). The uses of social television. *Computers in Entertainment, 6*(1), 1–15. doi:10.1145/1350843.1350851

Harboe, G., Metcalf, C. J., Bentley, F., Tullio, J., Massey, N., & Romano, G. (2008). Ambient social TV: Drawing people into a shared experience. In *Proceedings of the ACM SIGCHI Conference on Human Factors in Computing Systems* (pp. 1-10).

Hassenzahl, M., Diefenbach, S., & Göritz, A. (2010). Needs, affect, and interactive products − Facets of user experience. *Interacting with Computers, 22*(5), 353–362. doi:10.1016/j.intcom.2010.04.002

Huang, E. M., Harboe, G., Tullio, J., Novak, A., Massey, N., Metcalf, C. J., & Romano, G. (2009). Of social television comes home: A field study of communication choices and practices in TV-based text and voice chat. In *Proceedings of the ACM SIGCHI Conference on Human Factors in Computing Systems* (pp. 585-594).

Jacucci, G., Oulasvirta, A., Ilmonen, T., Evans, J., & Salovaara, A. (2007). CoMedia: Mobile group media for active spectatorship. In *Proceedings of the ACM SIGCHI Conference on Human Factors in Computing Systems* (pp. 1273-1282).

Luyten, K., Thys, K., Huypens, S., & Coninx, K. (2006). Social stitching with interactive television . In *Proceedings of Extended Abstracts of Human Factors in Computing Systems* (pp. 1049–1054). Telebuddies. doi:10.1145/1125451.1125651

McMillan, D., Morrison, A., Brown, O., Hall, M., & Chalmers, M. (2010). Further into the wild: Running worldwide trials of mobile systems. In *Proceedings of the Pervasive Computing Conference* (pp. 210-227).

Metcalf, C., Harboe, G., Tullio, J., Massey, N., Romano, G., Huang, E. M., & Bentley, F. (2008). Examining presence and lightweight messaging in a social television experience. *ACM Transactions on Multimedia Computing . Communications and Applications, 4*(4), 1–16.

Michahelles, F. (2010). Getting closer to reality by evaluating released apps? In *Proceedings of the Workshop on Research in the Large*.

Miluzzo, E., Lane, N. D., Fodor, K., Peterson, R., Lu, H., Musolesi, M., et al. (2008). Sensing meets mobile social networks: The design, implementation and evaluation of the CenceMe application. In *Proceedings of the 6ᵗʰ ACM Conference on Embedded Network Sensor Systems* (pp. 337-350).

Miluzzo, E., Lane, N. D., Lu, H., & Campbell, A. T. (2010). Research in the app store era: Experiences from the CenceMe app deployment on the iPhone. In *Proceedings of the Workshop on Research in the Large.*

Miyamori, H., Nakamura, S., & Tanaka, K. (2005). Generation of views of TV content using TV viewers' perspectives expressed in live chats on the web. In *Proceedings of the International Multimedia Conference* (pp. 853-861).

Morrison, A., Reeves, S., McMillan, D., & Chalmers, M. (2010). Experiences of mass participation in Ubicomp research. In *Proceedings of the Workshop on Research in the Large.*

Nakamura, S., Shimizu, M., & Tanaka, K. (2008). Can social annotation support users in evaluating the trustworthiness of video clips? In *Proceedings of the 2nd Workshop on Information Credibility on the Web* (pp. 59-62).

Nilsson, A. (2004). Using IT to make place in space: evaluating mobile technology support for sport spectators. In *Proceedings of the 12th European Conference on Information Systems* (p. 127).

Ochsner, K. N., & Schacter, D. L. (2000). A social cognitive neuroscience approach to emotion and memory . In Borod, J. C. (Ed.), *The neuropsychology of emotion* (pp. 163–193). Oxford, UK: Oxford University Press.

Oehlberg, L., Ducheneaut, N., Thornton, J., Moore, R. J., & Nickell, E. (2006). Social TV: Designing for distributed, sociable television viewing. In *Proceedings of the European Conference on Interactive TV* (pp. 25-26).

Oliver, E. (2010). The challenges in large-scale smartphone user studies. In *Proceedings of the 2nd ACM International Workshop on Hot Topics in Planet-scale Measurement* (p. 5).

Pang, B., & Lee, L. (2008). Opinion mining and sentiment analysis. *Foundations and Trends in Information Retrieval, 2*(1-2), 1–135. doi:10.1561/1500000011

Preece, J., & Shneiderman, B. (2009). The reader-to-leader framework: Motivating technology-mediated social participation. *AIS Transactions on Human-Computer Interaction, 1*(1), 13–32.

Schleicher, R., & Trösterer, S. (2009). The 'joy-of-use'-button: Recording pleasant moments while using a PC. In T. Gross, J. Gulliksen, P. Kotzé, L. Oestreicher, P. Palanque, R. O. Prates, & M. Winckler (Eds.), *Proceedings of the 12th IFIP TC 13 International Conference on Human-Computer Interaction* (LNCS 5727, pp. 630-633).

Weisz, J. D., Kiesler, S., Zhang, H., Ren, Y., Kraut, R. E., & Konstan, J. A. (2007). Watching together: Integrating text chat with video. In *Proceedings of the SIGCHI Conference on Human Factors in Computing Systems* (pp. 877-886).

This work was previously published in the International Journal of Mobile Human Computer Interaction, Volume 3, Issue 4, edited by Joanna Lumsden, pp.18-35, copyright 2011 by IGI Publishing (an imprint of IGI Global).

Chapter 16

SGVis:
Analysis of Data from Mass Participation Ubicomp Trials

Alistair Morrison
University of Glasgow, UK

Matthew Chalmers
University of Glasgow, UK

ABSTRACT

The recent rise in popularity of 'app store' markets on a number of different mobile platforms has provided a means for researchers to run worldwide trials of ubiquitous computing (ubicomp) applications with very large numbers of users. This opportunity raises challenges, however, as more traditional methods of running trials and gathering data for analysis might be infeasible or fail to scale up to a large, globally-spread user base. SGVis is a data analysis tool designed to aid ubicomp researchers in conducting trials in this manner. This paper discusses the difficulties involved in running large scale trials, explaining how these led to recommendations on what data researchers should log, and to design choices made in SGVis. The authors outline several methods of use and why they help with challenges raised by large scale research. A means of categorising users is also described that could aid in data analysis and management of a trial with very large numbers of participants. SGVis has been used in evaluating several mass-participation trials, involving tens of thousands of users, and several use cases are described that demonstrate its utility.

INTRODUCTION

The rise of 'app store' markets is a relatively recent phenomenon. The Apple App Store launched in July 2008 and saw its 10 billionth download in January 2011 (Reuters, 2011). Smartphone usage has also seen a sharp rise in usage in recent years, with market research firm IDC suggesting that 15.4% of the mobile phone market consisted of smartphones at the beginning of 2010 (Llamas, 2010), and several different mobile platforms now offer 'app store' software distribution mechanisms.

DOI: 10.4018/978-1-4666-2068-1.ch016

The combination of this growing potential user base and popular online software repositories provides a relatively simple way to recruit users for worldwide trials of ubiquitous computing (ubicomp) applications and several researchers are beginning to use this 'mass participation' approach (McMillan et al., 2010; Cramer et al., 2010). Distributing ubicomp trial applications in this way provides great opportunities for researchers in terms of potentially reaching a very large number of trial participants. For example, Hungry Yoshi, a game discussed below, has had over 40,000 users in the 12 months since release.

The potential advantages of deploying trial software to a wide audience are numerous. These include getting a larger sample size to provide more certainty to quantitative analyses and, additionally, a global release of software provides the opportunity to reach users from vastly different geographic locations so as to help reduce cultural biases stemming from recruitment of only locally-based participants.

The benefits to be gained from this style of deployment do come with some potential drawbacks; however, as running a trial with a large number of users over a vast geographical area can raise significant challenges. Compared to more traditional local deployments of software, researchers can be further removed from the trial, unable to meet participants and perhaps less able to closely observe the use of the software under examination. Additionally, having such a large user base could lead to an overwhelming amount of data being generated and researchers might not have the resources to study every user in detail.

We have designed several tools to aid researchers in conducting such trials. The previously published SGLog framework (Hall et al., 2009) provides a simple means for developers to instrument their mobile application code and stream log data back to researchers' database. This paper presents SGVis - a complementary desktop analysis tool designed to allow evaluators to study this data. Several modes are available in SGVis, allowing

data gathered from users' devices to be processed in various ways. In one mode an individual's use of trial software can be studied in detail, while in another overall trends for an application can be analysed, such as download patterns over time or the average number of times participants perform certain actions. A 'live' mode allows analysts to watch software use 'as it happens', with a collection of maps and summary statistics showing researchers which of their trial applications are being used anywhere in the world in the last few minutes. The final mode in SGVis allows analysts to categorise the many thousands of users they might have, using multidimensional data analysis, clustering and visualisation techniques. In this way, analysts might be able to more successfully manage categories of participants rather than attempting to study each of the thousands of logged users in detail. SGVis has been used in the evaluation of several mobile applications, handling data sets of tens of thousands of users spread worldwide.

The following section surveys related work in the area of mass participation research applications, and of analysis tools that can be used to study data captured from such applications. A description is provided of the Contextual Software project (UK EPSRC EP/F035586/1), and how SGVis forms part of this research. This is followed by a discussion of the challenges arising from conducting research in this manner and recommendations for the types of data other researchers working in this area might want to capture. Thereafter, the SGVis tools for analysing large scale ubiquitous computing trials are introduced, including a rationale for design and illustrations of their utility with specific use cases, before finally a discussion is presented of the use of SGVis and ongoing work in this area.

RELATED WORK

Researchers in ubiquitous computing have recently begun to take advantage of 'mass participation' methods of distributing software, with several research applications being released via public software repositories.

An early landmark of large-scale deployment of ubicomp applications that pre-dated 'app store' style software repositories was Mogi Mogi. As reported by Licoppe and Inada (2006), this location-based mobile multiplayer game was released commercially in Japan, and in 2004 had roughly 1,000 active players. One of the earliest research systems to be released on Apple's App Store is CenceMe, which uses context sensing to automatically update social networking sites with a user's current activity (Miluzzo, Lane, Lu, & Campbell, 2010). CenceMe was originally developed for the Nokia N95, but re-implemented for the iOS platform in time for the App Store launch in July 2008. Hungry Yoshi, another iOS-based application released in September 2009, is a game that uses detected Wi-Fi access points as a game resource, thus exposing underlying infrastructure through a game mechanic. In McMillan et al. (2010) we show that it is possible to conduct both quantitative and qualitative research with a large, global user base.

A number of research projects have also been released through the Android Market. For example, My2cents allows users to scan barcodes of retail products and discuss them with other users (Karpischek & Michahelles, 2010). WorldCupinion (Rohs, Kratz, Schleicher, Sahami, & Schmidt, 2010), an app for sharing real-time feedback during football matches, was also released on the Android Market and gained 1,645 downloads during the World Cup. Another example, AppAware, is a location-aware application that tracks installs and uninstalls of other software, to inform users of which apps are currently popular among other Android users in the local vicinity. The authors report that AppAware has had over 19,000 unique users, of whom 9,500 were actively using the application at the time the authors wrote the paper (Girardello & Michahelles, 2010).

In many of these studies, reported analysis so far is limited to download numbers and brief reports of usage statistics, often as provided by the software repository. Our own early experience of studying data from mass participation applications was often limited to raw SQL queries, or custom-built tools to handle the data specific to each iOS application. This obviously required considerable time and effort in each case, and the created systems might be tightly coupled to the evaluated system, offering limited potential for re-use.

The goal of SGVis is to create a general analysis toolkit that is reusable across studies of many mobile applications. Commercial organisations such as Flurry offer a similar service, providing developers with a logging framework and displaying online statistics on application usage (Flurry, 2010). However, Flurry's tools only offer aggregate data, and our work with SGLog and SGVis goes further in allowing individual users to be studied in detail and allowing more complex forms of statistical analysis.

Other work in visualising data generated from ubicomp research projects focusses on analysing sensor data. For example, the Cityware system (O'Neill et al., 2006) recorded Bluetooth data and presented a chart of devices, based on aggregate densities and flows of people in particular urban areas. Similar work has looked at coarse-grained city-scale maps of people's density based on concentrations of mobile phone signals sampled from GSM infrastructure (Reades, Calabrese, Sevtsuk, & Ratti, 2007), and scans of Bluetooth data from crowds of people attending sports stadia (Morrison, Bell, & Chalmers, 2009).

Other researchers have explored 'experience sampling' methods, in which a questionnaire appears on-screen when the mobile device detects that it is in a context of interest (Froehlich, Chen, Smith, & Potter, 2006). Carter, Mankoff, and Heer (2007) developed Momento, which supports experience sampling, diary studies, capture of photos and sounds, and messaging from evaluators to participants. It uses SMS and MMS to send data between a participant's mobile device and an evaluator's desktop client. Systems such as Replayer (Morrison, Tennent, & Chalmers, 2006; Morrison, Tennent, Chalmers, & Williamson, 2007) and DRS (Greenhalgh, French, Tennent, & Humble, 2007) have been used for smaller-scale studies, combining data logged on participants' mobile devices with video recorded by trial evaluators in the field. Events detected in system logs could then be used to filter video recorded by roaming evaluators or cameras mounted in fixed locations, thereby easing the process of finding relevant video data from potentially many hours of recorded footage. Although such systems can allow researchers to study users in greater numbers and at larger geographic and time scales than they can directly observe, this approach does not scale up to public software repository-style releases, where analysts are unlikely to gain video footage of participants using the system.

Contextual Software Project

The work presented in this paper is part of the Contextual Software project, which aims to provide tools and practices for developers and evaluators, and so improve the process of creating software that will have sustained contextual fit. The project has several components, such as a logging framework, a mechanism for delivering updates to deployed software and several 'app store'-released applications. These are briefly described here, although the focus of this paper is the SGVis desktop analysis tool.

Several ubicomp applications have been developed for Apple's iOS platform (meaning they can run on the iPhone, iPod Touch and iPad) as part of this project. Two iOS applications in particular are used to provide examples in this paper. Hungry Yoshi, a game that uses Wi-Fi infrastructure as a game resource, was the first application we released through the app-store method. It has had around 200,000 downloads and currently has around 40,000 registered users. The game is described by McMillan et al. (2010), where it was evaluated using a basic set of tools that would later evolve into the SGVis toolkit described here. Our most recent application is World Cup Predictor, a game designed to run alongside the FIFA World Cup, which tried to encourage social interaction with other players through Bluetooth-based data transfers. Evaluation of this trial is still ongoing, based around data analysis in SGVis. Both apps were free to download via an APT-based repository (McMillan, 2010).

SGVis displays data gathered via the complementary SGLog framework (Hall et al., 2009). SGLog provides a simple means to instrument iOS applications with timestamped general or application-specific event logging. An application equipped with SGLog creates local caches of log data on the device while the application is running, and this data is opportunistically uploaded to a server when the device has internet connection. A more traditional ubicomp study might recruit locally-based users, whose devices would log usage data to text files. Devices might be collected at the end of a trial for log data to be retrieved and data analysis to commence. Using an app store distribution method, researchers cannot physically gather the devices, so the SGLog framework captures log data during a trial and regularly uploads these logs to our server over a device's Wi-Fi or cellular data connection. Such a system allows researchers access to more or less live data, eliminating the necessity to wait for devices to be returned before analysis can commence.

The SGLog infrastructure is explained in detail by Hall et al. (2009), along with our mechanism for allowing dynamic updates of deployed applications. Here we focus on the SGVis data analysis tools, designed for use with SGLog.

Large Scale Trials: What Data to Log?

If other researchers wish to run trials in this manner, what types of data should they capture for analysis? A simple answer might be that researchers would like to receive as much data as possible. One of the features characterising research projects that employ app store-style distribution methods is the potential for long-duration trials. If research questions might evolve over the course of such a study, researchers will not be certain in advance of all the data they will require. In such circumstances it would be safer to have logged extra data and not require all of it than to have omitted to log a crucial element. Deciding upon what data to capture might therefore be equivalent to what *can* be logged, dictated by what is technically feasible on the platform.

It was described in the previous section how activity is logged on mobile devices and these logs are uploaded back to the researchers' database. Were this transmission to take place over cellular data networks, participants might incur charges from their network providers, depending on the type of contract they have. If the participant was using the device abroad, this charge could be particularly high. As a workaround to this, the application can detect the current type of connection and the logging software can be configured to only upload data when users are connected to the Internet over WiFi. If it was detected that the application was on a cellular network rather than WiFi connection, the application could be configured to only upload what researchers deem the most 'important' log data. Researchers might view this as a tradeoff, balancing the amount of information they upload with the danger of annoying the users to the extent that they will stop using the application.

Ethics and Privacy Concerns

Of course, considerations on this tradeoff must also be balanced by researchers' ethical responsibilities. An application should make clear any data that is being uploaded and allow users to make a decision on whether the use of the software is worth any potential monetary cost. Further ethical issues extend beyond purely financial considerations. As discussed above, it might be of benefit to researchers to acquire as much as data as possible, but is it appropriate to log vast amounts of data on users, even information that may be irrelevant to the study? We are addressing some of these concerns in ongoing work.

All applications that use SGLog present a privacy statement on first launch, explaining all the data that will be collected. Users must agree to this before they can use the application. Having done so, uploaded data from each user is timestamped and stored on a secured database on a central server. To protect the privacy of participants, this framework uses TLS to encrypt data sent between mobile devices and the server.

As releasing software via an app store means reaching a potentially global audience, it cannot be assumed that all users will be fluent in English. This information is therefore presented in at least four different languages. A contact email address is also supplied for users to opt out of the trial at any time, and it is explained that all data collected from a user will be destroyed on receipt of such a request.

SGLog does not directly address repositories' varied policies as to what one can log and programmers must understand what is acceptable in the repository they use for distribution.

General and Application-Specific Logging

The data that researchers might choose to log can generally be divided into general or application-specific items. Data on location, accelerometer readings or battery levels is generic information that might be useful to record in all applications. It is more difficult to give general advice for application-specific logging, as each specific trial will obviously have a different focus and require different data. In general, we have found that recording screen changes and button clicks provides a useful overview of the usage of trial software.

The particular challenges of running large-scale trials mean that some forms of data might be viewed as being more crucial than others. For example, as researchers do not meet their participants at any point during the trial, it can be important to gain some basic information in order to compile demographic statistics and to understand the composition of the user base. Some forms of demographic information cannot be gathered automatically and would have to be asked of users: age and occupation, for example. Device demographics, such as device model or OS version can be automatically logged. The following sections show examples of these types of demographic spreads.

It is impossible to provide a definitive list of the information necessary to run a large scale trial, just as it is impossible to create an analysis tool that is certain to fully answer all questions researchers will have in every future project. We hope we have provided some general pointers towards the set of information that will be of generic use. In the following sections, we describe the SGVis tool which, used in conjunction with this logged data, aims to tackle some of the difficulties incurred when running trials on such a scale.

Analysis Toolkit Design Choices

As well as the previously noted challenges, large scale trials do offer huge benefits in the ability to gain large amounts of participants, so SGVis has been specifically designed to harness the advantages of large numbers for quantitative analysis, calculating and displaying trends across the large numbers of users.

In the first days following the release of Hungry Yoshi we felt detached from the ongoing trial, and unable to closely follow the game's use. Limited to running awkward SQL queries on the database, there was no simple means of monitoring the application's usage and maintaining 'peace of mind' that everything was running smoothly. We elected to build a 'live' view of the trial to overcome researcher detachment in this way via a set of monitoring and visualisation tools that collect and visualise data in close to real-time. This allows researchers to keep in touch with an ongoing trial, and, as described in an example below, to identify problems as they arise.

Another strategy we adopted early on in the trialling of app store-released software was the use of Flurry (2010) to collect log data and see general patterns of usage. However, it soon became apparent that the overviews provided were insufficient to enable in-depth views of user activity. Only in studying individual users in detail could we select interesting people to interview and to prepare questions for such interviews based on observed behaviour. Therefore, as described in the following sections, SGVis provides analysts with the means to view aggregate data from all users, yet still be able to study an individual participant's activity, and easily move between these views.

Having such a large number of users means it might not be practical to study each individual in detail. We have found that our applications attract tens of thousands of users. Therefore we also describe a mechanism to categorise users based on their recorded usage of the application, via a series of data processing, clustering and visualisation techniques. We suggest ways in which this categorisation might be used in analysis.

SGVIS ANALYSIS TOOLS

Our development work in mobile applications has concentrated on the iOS platform for our large distribution trials, and the SGVis desktop tool is a Cocoa application written in Objective-C. A server backend uses PHP to communicate with SGLog's MySQL database. Hall et al. (2009) describe how the SGLog framework can be added simply to an iOS project, with a single line of code necessary for each event researchers wish to log. SGVis setup is then simply a matter of pointing SGVis towards an SGLog database. The software will automatically configure itself and populate its views with data from all the different trials logged to that database. SGVis has four main modes: viewing quantitative data aggregated by user or app, 'live' mode to monitor current usage, and analysis of derived statistical data. The fol-

lowing sections will describe how each of these aims to facilitate the running and evaluation of large scale ubicomp trials.

Trial Data Viewed by Application

The first feature of SGVis to be described is analysing aggregations of the potentially vast amounts of data generated by a global distribution of an application. Figure 1 shows the SGVis tool, with the 'Apps' option selected from the menu on the left. An analyst first selects an application from the drop-down list at the top, which is automatically populated with all the unique apps that have written to the SGLog database. Several tabs are available, the first being the overall number of users who have used the application each day since release. Separate trends are shown for the number of users who played for the first time that

Figure 1. SGVis shows several summary statistics on app usage. Here, the number of users running a trial application per day is plotted, with separate series for new and returning users

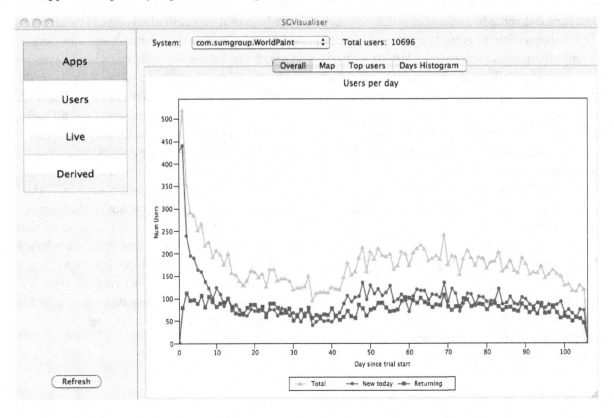

day (the green series), returning users who had used the application on a previous day (blue), and the total of these numbers combined (yellow).

The graph shown in the figure is typical of the general trend we have observed for application releases. The peak of usage usually occurs on the first days of release. We have questioned participants on how they typically discover new applications, both via telephone interviews and through in-app questionnaires, and many have reported that they regularly browse the app repositories' 'Newest' sections, where recently released applications are listed. This would help explain the early peak, with the number of new users gradually declining as an application falls further down the Newest list. The number of users per day often reaches a plateau—after around two weeks in Figure 1. A rise in the trend can be expected again when new versions of the application are released. Different repositories have varying policies on this, with some putting a release of a new version of an application at the top of the Newest list again. It is clear from the graph that there is always a steady stream of new users using the application, as represented by the green line. After the initial peak has plateaued, there is an average of around 100 new users every day. As the number of returning users (blue line) shows no significant increase, it can be inferred that there is a high degree of churn in the user base, with new users trying the application every day but the majority not returning regularly for several consecutive days or weeks.

Another section in the 'Apps' mode displays a map that presents an aggregated view of all the locations at which the selected application has been used. Maps are viewed via a live connection to Google Maps, with SGVis-supplied markers. In the left image in Figure 2, one unique marker is shown for every user who has used the application in the British Isles, placed at the location at which that user was most recently seen. Data can be filtered to show, for example, only those users exhibiting a specified minimum amount of usage. After a data

set reaches a certain size, it becomes prudent to cluster the data before rendering it. This step aids in both keeping the Google Maps rendering time acceptable and creating a readable map, where it is possible to see user numbers. Figure 2 shows an example of this, plotting users of Hungry Yoshi in Europe. Zooming into a map region expands a

Figure 2. Map showing the geographical distribution of users of Hungry Yoshi, a mobile game released through an app store. Location data is clustered for large numbers of users to decrease rendering time and to aid readability

A

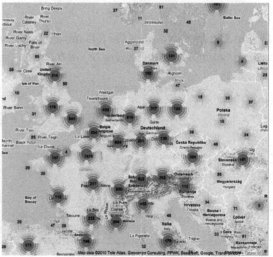

B

cluster to see the data represented as individual markers again. These maps use the open source MarkerClusterer utility (MarkerClusterer, 2011).

Viewing the geographic spread of an application's users in this way provides an insight into the countries in which an application is popular. Researchers might like to act on this information in a number of ways. For example, iOS allows localisation of user interfaces. Developers can supply translations of all strings used in the user interface of an application, and these will replace the default English values if a user has selected that language on their device. If an application is seen to be receiving a lot of usage in a particular country, a researcher might consider it worthwhile to translate the application into that country's native language in order to maximise usage. Indeed, it might be considered to be a researcher's responsibility to the participants to translate terms and conditions into the languages of countries where the application is receiving a lot of use, in order that the users can make informed consent about taking part in a trial.

Demographic information is available in another tabbed section. This includes device type and operating system version, which is automatically collected for every user, and age and gender of any participants who have chosen to submit that

information via an in-app questionnaire. Figure 3 shows some of this information for Zoo Escape - a simple puzzle game application. The chart on the left shows the total number of each type of iOS device on which the application was used. The total number of different days on which a participant used the game has also been counted, and the chart on the right shows the average number of days for each device type. From these two charts it can be seen that the application has been used by more iPod Touch users than iPhone users, but that on average the iPhone users returned to the game more often. iPad users play the game for the fewest number of days on average.

An application's users can also be sorted and displayed in SGVis. The 'top' 50 users can be displayed, using such measures as those making the greatest number of launches or accumulating the greatest total minutes' use. These values can be measured since release date or queried by a specific time period.

Such a view of the user base is often useful in selecting users to contact for interviews, or in selecting users to whom researchers can push specific questions via an in-app questionnaire. We showed in (McMillan et al., 2010) that qualitative data can still be captured from users in a mass participation trial, but that challenges exist in selecting the par-

Figure 3. A breakdown of the iOS device types on which the Zoo Escape game was played (left) and the average number of days users of each device type returned to the game (right)

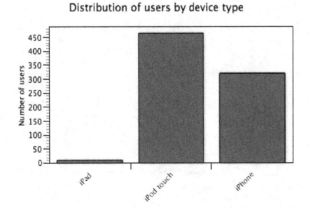
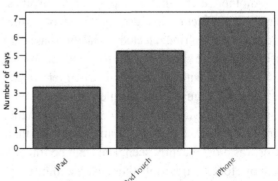

ticular participants to interact with in this manner from the large numbers available. A particular challenge is in contacting users at appropriate times in their trajectory of usage of an application. In a more traditional study, it might be common to do interviews shortly after a trial's conclusion, while the application will still be fresh in the mind of participants. In a mass participation deployment, however, there is not a clear definition of what 'after' a trial means. Depending on whether the application has a server-side component, which may have maintenance costs, an application might continue to work indefinitely. In that sense, a trial doesn't have a fixed end point. And during this time players will continuously start and stop using the application, so the overall highest points scorers in a game might have ceased playing or uninstalled the application months or even years earlier. In order to select those participants for whose use is still fresh in the mind, SGVis can be used to see who is still using the application, perhaps by looking at the most active users in the last week, and selecting potential interviewees from this subset.

Lists such as this, and maps showing markers for each user, can also be used to link to the second main mode in SGvis, which allows for analysis of individual users in detail. An analyst might begin by looking at an aggregate summary of an application's data, then select an interesting-looking user to smoothly drill-down to see visualisation of this single user in detail, as described in the following section.

Trial Data Viewed By User

Where SGVis is more powerful than systems such as Flurry is in allowing detailed analysis of individual users. SGVis includes a number of tools for displaying information for a single user's activity, which can be useful in identifying those that are unusual or interesting cases, or to prepare for further qualitative analysis, such as interviews.

Figure 4 shows an illustration of SGVis operating in this mode. An analyst can select a user to drill-down into from aggregate views, as described in the previous section, or simply enter a unique device ID (UDID). The analyst is then presented with information this user has declared, such as age and gender, and a table summarising of all the SGLogged applications that the user under scrutiny has run. This table contains information such as last launch time and the number of days on which the user has run the application.

Clicking a row in this table displays a visualisation in the lower part of the screen that shows details of the selected participant's usage of the selected application. The chart in the centre of the figure shows the times of day at which the user was running the application and the length of time it was running each day. Different days are shown along the x-axis, and the hours of the day are shown on the y-axis. A blue vertical bar is drawn to represent each period of usage. Underneath this, another graph shows the minutes of use per day, using the same x-axis as the chart above. The data in the figure is from a user of the Hungry Yoshi mobile game. It is notable that this user is a keen player of the game, using the application for almost 7 hours one day.

Whereas the tools described in the previous section aided evaluators in selecting appropriate users for interview, we suggest that these user-specific tools can aid evaluators in preparing for the interview. As an example of this, in studying usage of Hungry Yoshi, we were interested in the ways in which users fit playing the game into their everyday routines. Through use of these tools we noticed that one player had been using the application in short bursts during the evenings. On telephoning her to enquire about this behaviour, she revealed that she worked as a poker dealer in a casino, and used the game as a way of filling in time during her enforced breaks.

Another usage of these tools is in seeing when a participant has stopped using the application. On

Figure 4. Individual users can also be studied in detail. The table at the top shows all the logged apps this user has run. The lower charts show the times of the day the selected app was run (the 24 hours of the day on the y-axis, shaded blue when in use) and below, the length of time the app was running each day

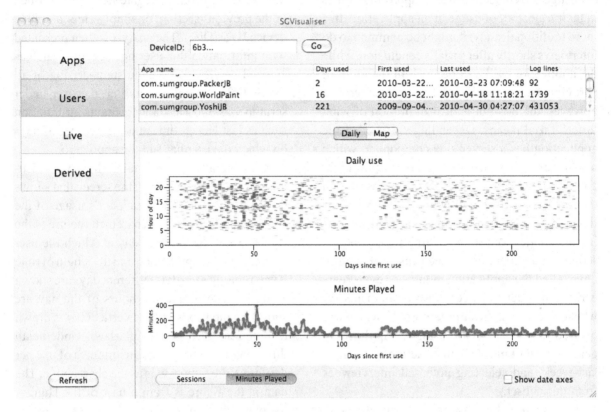

one occasion we noticed that one of our most keen users of a game had abruptly stopped playing. We attempted to telephone him to enquire why and although he did not have time to speak to us, he emailed us to say "I was in a rush on my break at work, where btw I'm banned for 30 days from bringing my iPhone into the building because I was caught playing Hungry Yoshi."

User data can also be viewed by location. Whereas the maps shown in the previous section showed aggregated usage of an application and rendered one marker for each user, Figure 5 shows an example of mapping a single user's data, where every location at which that user has run the application is displayed. If the user has run the app while moving or travelling, the map will show the route. This has allowed us to identify for example users who play the game while

commuting, compared to those who only play at home and we have been able to target specific questionnaires towards these participants to learn more on this observed behaviour.

In general, the tools for analysing a single user's activity are also useful for contextualising interesting behaviour. For example, when analysing our World Cup Predictor application, we were particularly interested in usage of the 'head-to-head' mode, which transferred data over ad hoc peer-to-peer Bluetooth connections. Occurrences of this activity were quite rare, only being performed by 45 of the 10,806 registered users. On each occasion the head-to-head feature was used, an analyst could study the context in terms of location and time of day, and also see if this contrasted with the player's general pattern of usage. For example, it could be seen whether the player

Figure 5. A participant has used the application while travelling, and SGVis displays the route taken

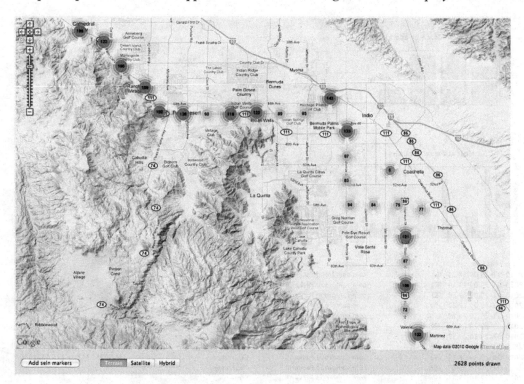

had travelled to a friend's house to perform a Bluetooth transfer.

Similarly, our applications all have a section for reporting bugs to developers. SGVis reports the user's device type and operating system version and it can be invaluable in diagnosing a bug to view a user's context when the problem was discovered, as well as a history of usage.

Live Trial Data

The previous sections described analysis of data accumulated over the duration of a trial. SGVis can also visualise 'live' data, which is a regularly updating view summarising recent activity in SGLogging applications. Figure 6 has an example of this. A slider on the bottom of the application allows the analyst to set the period of time to view, ranging from activity recorded in the last 1 minute up to the last 24 hours.

Two graphs in the top left of the screen show respectively the number of users to have used each application in the specified time period and, below, which users have been active, charting the number of logged actions. A map in the top right shows the locations where people have recently been playing. Again, this is zoomable to see areas of interest in more detail. Finally, a table at the bottom of the screen shows data in a rawer format as it comes in. Data is uploaded to the database continually by people using the apps worldwide, and SGVis refreshes its view every 10 seconds.

All the sections in the live mode are linked, to afford brushing and linking between views (Becker & Cleveland, 1987). This allows users to make selections in one view which will then be reflected in the others. For example, an analyst can select an app by clicking on a bar in the bar chart, and this will filter all other views to show only the data recorded from within that app. Similarly, the analyst can draw a box to select a particular area

Figure 6. SGVis has a constantly updated view of recent activity

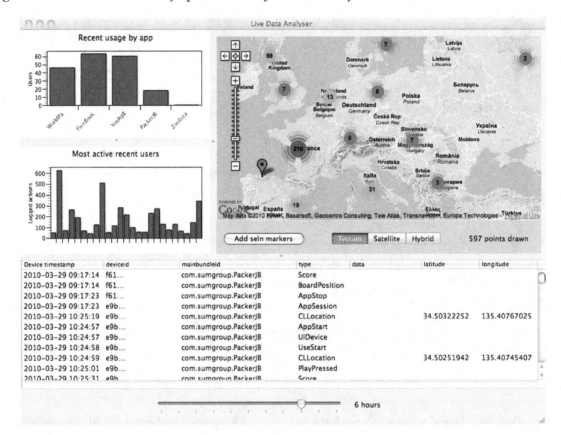

of the map and this will filter the other views to show only activity recorded in the selected region.

By using the SGLog and SGVis toolkit, analysts are no longer required to wait until the conclusion of a trial before retrieving log data, which means they can react to events they see in the log data and possibly change aspects of the trial as it is ongoing. This could include updating the software based on observed usage, or perhaps altering the logging code to fine-tune the specific data being captured.

There are also uses in monitoring a trial to ensure that everything is running smoothly. For example, when a new version of an application has been released, it can be important to ensure that everything is working as intended for all users. As an example, using the live tool we once observed that the Hungry Yoshi application was showing

an uncharacteristically small amount of activity. On further investigation, by moving back to look at data from the whole user base, it appeared that users with one particular combination of model of iPhone and operating system version were unable to use the application following a recent update. Once the problem had been correctly diagnosed and the bug removed, a further update was released and we could use the live tool again to observe usage return to normal.

Although such monitoring of a trial is useful, it is probably impractical for a research project to have someone dedicated to looking at this data all the time. Future work will look at ways to automatically detect changes in patterns of activity and push out alerts to analysts.

Derived Data and User Categorisation

The final menu option in SGVis allows for more in-depth statistical analysis of data gathered from mass participation studies. A previous section described tools for studying individual users in detail. However, when there might be tens or hundreds of thousands of people using an application, with hundreds joining every day, it becomes impractical to study every user in detail. Tools included in SGVis aid analysts in this regard by clustering participants into different groups based on characteristics of usage. Thereafter, different groups of users could be handled differently, for example sent questionnaires with only those questions pertinent to their specific use of the application.

To illustrate this functionality, a use case is described showing results from analysis of Hungry Yoshi. The SGLog database is processed to create a number of statistical measurements for each participant. These include, for example, the cumulative time spent playing the game, time between first and last uses, type and form of response to feedback question, cumulative distance travelled when actively playing the game, and the size of the geographical area in which users play the game.

This data is used to create a high-dimensional vector for each user. An analyst can plot each of these dimensions x vs. y to see trends or correlations, but practical experience has suggested it might be more useful to consider the full high dimensional vector as a whole, and look for ways of being able to analyse the vectors all together.

One way to achieve this is through the use of 'spring model' dimensional reduction algorithms (Eades, 1984; Chalmers, 1996; Morrison, Ross, & Chalmers, 2003) to map the high dimensional vectors to a 2D layout, which we can then search for patterns and relationships across all aspects of the data. Such algorithms take as input the set of high-dimensional vectors representing the trial users, and create a 2D layout such that the high-dimensional pairwise relationships between vectors are well represented in 2D space. This is achieved through simulating a system of mechanical springs connecting objects, with each spring's ideal rest distance proportional to the high-dimensional dissimilarity between the pair of users. By iterating a process of calculating and applying the forces exerted upon the objects by each spring, objects are pulled and pushed towards a state of equilibrium that should be a good representation of the relationships between each pair of users.

Figure 7 shows an example of this in SGVis. The full data is shown in a table and a scatterplot is on the right. Here an analyst can choose to plot any pair of dimensions as x against y or, as shown in the figure, perform a dimensional reduction routine to derive a 2D layout. Again, the views are linked, so that making a selection in one highlights the corresponding subset of objects in the other.

To perform this type of analysis in greater detail, data can be exported for analysis in the Hybrid Information Visualisation Environment (HIVE) (Ross & Chalmers, 2004), a Java-based multi-purpose visualisation tool that has been extended to handle SGVis data files. Figure 8 shows an example of a HIVE analysis, where various clustering and layout techniques have been used to create a basic categorisation of users. 16 logged features from each user are processed in a spring model to generate a 2D layout. The analyst has then used Voronoi-based clustering (Okabe, Boots, Sugihara, & Chiu, 2000; Ross, Morrison, & Chalmers, 2004) to group coherent subsets of this layout, resulting in four distinct categories, as shown in the scatterplots in the top right of the figure.

Having assigned each user to one category, the high-dimensional centroids of each of the created clusters can be computed, which show the analyst the trends shown by the 'average' user in each group. Cluster centroids were studied by creating a parallel coordinate plot (Inselberg & Dimsdale, 1990), where the dimensions under

Figure 7. A spring model layout of data derived from the Hungry Yoshi game. Views are linked, so that selecting an area of the layout on the right highlights the corresponding rows of the table on the left

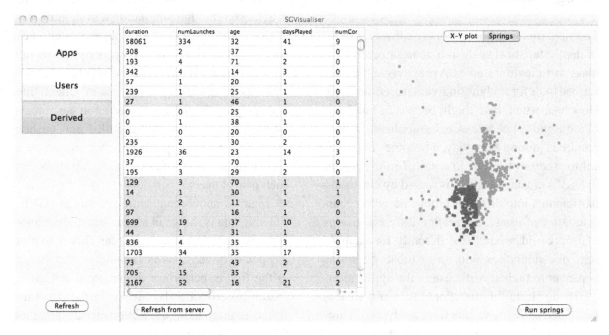

Figure 8. Each dot in this spring model layout (top middle) represents 16 logged features from one of 13,000 users of the Hungry Yoshi game. A Voronoi diagram was created from this layout, and then an interactive clustering algorithm (based on thresholding Voronoi cell sizes and spacing) was used to create four major categories of styles of play

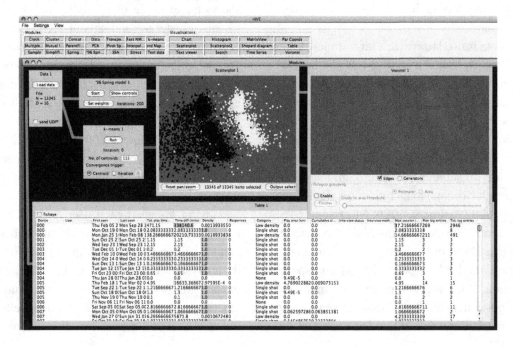

Figure 9. The centroids of the four clusters identified in Figure 8 have been calculated, and displayed here in a parallel coordinates plot. Studying the characteristics of the four groups leads us to label these categories 'top players', 'commuters', 'static players' and 'beginners'

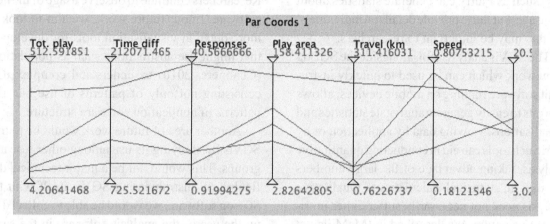

scrutiny are stacked as vertical axes and each of the four clusters is represented by a polyline intersecting these axes at the appropriate value. This is shown in Figure 9.

In this example, one group is clearly distinct from the others, rating highly in terms of physical activity as well as temporal activity. It would appear that these are the game's most enthusiastic users, who spend the most time playing, and who also play over the widest area and travel the furthest while playing. This group also has the high response rate to questionnaires. Another group (characterised as 'commuters') are distinguishable from the others by having a higher mean speed, due to greater amounts of travel while playing. A third group ('static players') has spent the smallest proportion of time moving while playing, and have the smallest play area, and a fourth group ('beginners') comprises those who have played the game for a briefer period and who have cumulated less distance playing and offered fewer responses to questionnaires.

By creating such categories, it is hoped that steps can be taken to help cope with an unmanageable volume of users. In a more traditional trial with a smaller user base, a researcher would possibly want to study each user in detail. This becomes less practical when user numbers are in the tens of thousands, and an analysis tool that can summarise the major different ways in which participants have used the trial system could greatly reduce this burden. Of course, researchers may then wish to study individual participants from each of the identified categories, to see usage in greater detail.

Categories can also be used as a basis for further evaluation, with, for example, different questionnaires being shown to each of the four groups. Once categories have been identified and questionnaires written, this could be an automated process, with a process on the server analysing users' behaviour and distributing the appropriate questionnaire at preset times. In the future our aim is to go further still, experimenting with deploying different software updates to users based on their categorisation. Other ongoing work is on coupling the SGVis toolkit more tightly to HIVE.

DISCUSSION

As ubicomp trials scale up in size, with a vast global user base replacing the more traditional local deployment of software, analysts can be challenged by the large amount of data generated by tens or hundreds of thousands of users. Online tools such as Flurry can generate statistics about app usage, but fail to provide details on individuals' use that may be important to researchers.

The SGVis analysis tool, alongside the SGLog framework which can be used to quickly instrument software running on mobile devices, allows analysts to easily generate aggregate statistics and visualisations, viewing data by application or by user. Such tools can aid in conducting quantitative analysis, taking advantage of the large numbers for use in statistical calculations. Qualitative analysis need not necessarily suffer either in trials of this nature. We described in (McMillan et al., 2010) how researchers running a large scale trial still performed interviews arranged in social networking applications and conducted via VoIP services, with the sort of data visualisations provided by SGVis valuable in allowing analysts to select interviewees and brief themselves on a participant's usage of the application before calling him or her. A 'live' mode allows for near real-time monitoring of application use across the globe. Ongoing work in this area is looking to augment applications with a messaging feature, so that analysts could push a query out to a user in real-time, which would pop-up on a device running a trial application. In this way, researchers could ask users about a specific event they had just observed, while the context of the event was still fresh in the user's mind.

It is hoped that tools of these forms, used together, can support the software development lifecycle. We suggest a scenario where, in monitoring live data, researchers witness an interesting event. Logging code is altered to capture this event in more detail, and the updated logging code deployed to participants. Collected data is processed through dimensional reduction and clustering techniques to identify a category of users behaving in the observed manner, some of whom are contacted for interview. New software updates are created to support the observed behaviour and deployed to the members of the appropriate categories. Researchers continue to observe usage of the new software. In our future work, we aim to look in more detail at visualisation tools for developers that might use abstractions such as populations (Chalmers, 2010) to understand complex data consisting not only of patterns of use but also patterns of application software structure.

Another area of future work would be to trial SGVis by studying its use among other research groups. This would not be a major technical difficulty; by instrumenting SGVis itself with the SGLog software, we would be able to collect data on the use of the analysis software in the same way we do with the mobile software. By collecting this type of data we could then conceivably use SGVis in the analysis of itself. Although we have found it very easy to get large amounts of users for mobile applications, researchers running large-scale trials of iOS-based research software are obviously a far smaller target population. Therefore we encourage anyone interested in using SGLog and SGVis to help in the analysis of iOS-based research software to contact us.

In summary, it is hoped that through the use of tools such as SGVis, researchers will be able to manage the vast amounts of data that can be generated by app store-style deployment. In particular, we aim to combine both quantitative and qualitative forms of evaluation, to support timely response to ongoing trial activity as well as retrospective analysis of temporal patterns, and directly assist new software development as well as evaluation work. In such ways, we aim to obtain the full benefits offered by mass-scale research.

REFERENCES

Becker, R. A., & Cleveland, W. S. (1987). Brushing scatterplots. *Technometrics*, *29*, 127–142. doi:10.2307/1269768

Carter, S., Mankoff, J., & Heer, J. (2007). Momento: Support for situated Ubicomp experimentation. In *Proceedings of the SIGCHI Conference on Human Factors in Computing Systems* (pp. 125-134). New York, NY: ACM Press.

Chalmers, M. (1996). A linear iteration time layout algorithm for visualising high-dimensional data. In *Proceedings of the IEEE 7th Conference on Vizualisation* (pp. 127-132). Washington, DC: IEEE Computer Society.

Chalmers, M. (2010). A population approach to ubicomp system design. In *Proceedings of the ACM-BCS Visions of Computer Science Conference* (p. 1).

Cramer, H., Rost, M., Belloni, N., Chincholle, D., & Bentley, F. (2010). Research in the large: Using app stores, markets and other wide distribution channels in UbiComp research. In *Proceedings of the 12th ACM International Conference Adjunct Papers on Ubiquitous Computing* (p. 511-514). New York, NY: ACM Press.

Eades, P. (1984). A heuristic for graph drawing. *Congressus Numerantium, 42*, 149–160.

Flurry. (2010). *Flurry homepage*. Retrieved from http://www.flurry.com

Froehlich, J., Chen, M. Y., Smith, I. E., & Potter, F. (2006). Voting with your feet: An investigative study of the relationship between place visit behavior and preference. In *Proceedings of the 8th International Conference on Ubiquitous Computing* (pp. 333-350).

Girardello, A., & Michahelles, F. (2010). AppAware: Which mobile applications are hot? In *Proceedings of the 12th International Conference on Human Computer Interaction with Mobile Devices and Services* (pp. 431-434). New York, NY: ACM Press.

Greenhalgh, C., French, A., Tennent, P., & Humble, J. (2007, October). *From ReplayTool to digital replay system*. Paper presented at the 3rd International Conference on e-Social Science, Ann Arbor, MI.

Hall, M., Bell, M., Morrison, A., Reeves, S., Sherwood, S., & Chalmers, M. (2009). Adapting Ubicomp software and its evaluation. In *Proceedings of the 1st ACM SIGCHI Symposium on Engineering Interactive Computing Systems* (pp.143-148). New York, NY: ACM Press.

Inselberg, A., & Dimsdale, B. (1990). Parallel coordinates: A tool for visualizing multi-dimensional geometry. In *Proceedings of the 1st Conference on Visualization* (pp. 361-378). Washington, DC: IEEE Computer Society.

Karpischek, S., & Michahelles, F. (2010, November-December). *my2cents - Digitizing consumer opinions and comments about retail products*. Paper presented at the Internet of Things Conference, Tokyo, Japan.

Licoppe, C., & Inada, Y. (2006). Emergent uses of a multiplayer location-aware mobile game: the interactional consequences of mediated encounters. *Mobilities, 1*(1), 39–61. doi:10.1080/17450100500489221

Llamas, R. (2010). *Worldwide converged mobile device market grows 39.0% year over year in fourth quarter*. Retrieved from http://www.idc.com/getdoc.jsp?containerId=prUS22196610

MarkerClusterer. (2011). *Gmaps utility library*. Retrieved from http://gmaps-utility-library-dev.googlecode.com/svn/tags/markerclusterer/

McMillan, D. (2010, September). *iPhone software distribution for mass participation*. Paper presented at the Research in the Large Workshop, Ubicomp, Copenhagen, Denmark.

McMillan, D., Morrison, A., Brown, O., Hall, M., & Chalmers, M. (2010). Further into the wild: Worldwide trials of mobile systems. In *Proceedings of the 8th International Conference on Pervasive Computing* (pp. 210-217).

Miluzzo, E., Lane, N., Lu, H., & Campbell, A. (2010, September). *Research in the app store era: Experiences from the CenceMe app deployment on the iPhone*. Paper presented at the Research in the Large Workshop, Ubicomp, Copenhagen, Denmark.

Morrison, A., Bell, M., & Chalmers, M. (2009). Visualisation of spectator activity at stadium events. In *Proceedings of the 13th International Conference on Information Visualisation* (pp. 219-226). Washington, DC: IEEE Computer Society.

Morrison, A., Reeves, S., McMillan, D., & Chalmers, M. (2010, September). *Experiences of mass participation in Ubicomp research*. Paper presented at the Research in the Large Workshop, Ubicomp, Copenhagen, Denmark.

Morrison, A., Ross, G., & Chalmers, M. (2003). Fast multidimensional scaling through sampling, springs and interpolation. *Information Visualization, 2*(1), 68–77. doi:10.1057/palgrave. ivs.9500040

Morrison, A., Tennent, P., & Chalmers, M. (2006). Coordinated visualisation of video and system log data. In *Proceedings of the 4th International Conference on Coordinated & Multiple Views in Exploratory Visualization* (pp. 91-102). Washington, DC: IEEE Computer Society.

Morrison, A., Tennent, P., Chalmers, M., & Williamson, J. (2007). Using location, bearing and motion data to filter video and system logs. In *Proceedings of the 5th International Conference on Pervasive Computing* (pp. 109-126).

O'Neill, E., Kostakos, V., Kindberg, T., Schiek, A. F. G., Penn, A., Fraser, D. S., & Jones, T. (2006). Instrumenting the city: Developing methods for observing and understanding the digital cityscape. In *Proceedings of the 8th International Conference on Ubiquitous Computing* (pp. 315-332).

Okabe, A., Boots, B., Sugihara, K., & Chiu, S. N. (2000). *Spatial tessellations: Concepts and applications of Voronoi diagrams* (2nd ed.). Chichester, UK: John Wiley & Sons.

Reades, J., Calabrese, F., Sevtsuk, A., & Ratti, C. (2007). Cellular census: Explorations in urban data collection. *Pervasive Computing, 6*(3), 30–38. doi:10.1109/MPRV.2007.53

Reuters. (2011). *Apple's app store hits 10 billion downloads*. Retrieved from http://www.reuters.com/article/idUSTRE70L0Y220110122

Rohs, M., Kratz, S., Schleicher, R., Sahami, A., & Schmidt, A. (2010, September). *WorldCupinion: Experiences with an Android app for real-time opinion sharing during World Cup soccer games*. Paper presented at the Research in the Large Workshop, Ubicomp, Copenhagen, Denmark.

Ross, G., & Chalmers, M. (2003). A visual workspace for constructing hybrid MDS algorithms and coordinating multiple views. *Information Visualization, 2*(4), 247–257. doi:10.1057/palgrave.ivs.9500056

Ross, G., Morrison, A., & Chalmers, M. (2004). Coordinating components for visualisation and algorithmic profiling. In *Proceedings of the 2nd International Conference on Coordinated and Multiple Views in Exploratory Visualization* (pp. 3-14). Washington, DC: IEEE Computer Society.

This work was previously published in the International Journal of Mobile Human Computer Interaction, Volume 3, Issue 4, edited by Joanna Lumsden, pp.36-54, copyright 2011 by IGI Publishing (an imprint of IGI Global).

Chapter 17
Experimenting Through Mobile 'Apps' and 'App Stores'

Paul Coulton
Lancaster University, UK

Will Bamford
Lancaster University, UK

ABSTRACT

Utilizing App Stores as part of an 'in-the-large' methodology requires researchers to have a good understanding of the effects the platform has in the overall experimental process if they are to utilize it effectively. This paper presents an empirical study of effects of the operation an App Store has on an App lifecycle through the design, implementation and distribution of three games on the WidSets platform which arguably pioneered many of the features now seen as conventional for an App Store. Although these games achieved in excess of 1.5 million users it was evident through their App lifecycle that very large numbers of downloads are required to attract even a small number of active users and suggests such Apps need to be developed using more commercial practices than would be necessary for traditional lab testing. Further, the evidence shows that 'value added' features such as chat increase not only the popularity of an App but also increase the likelihood of continued use and provide a means of direct interaction with users.

INTRODUCTION

Adopting the methodology of 'in the large' advocated by a number of researchers in the field of ubiquitous computing (McMillan et al., 2010; Rodgers et al., 2007) is particularly attractive to

researchers creating applications, now simply referred to as 'Apps', on commercial mobile phones as it offers the potential of accessing a large audience of users from a broad demographic compared to lab based studies. However, if adopting this methodology it is important that

DOI: 10.4018/978-1-4666-2068-1.ch017

Figure 1. Top: WidSets dashboard with minimized widgets. Bottom: maximized widget views

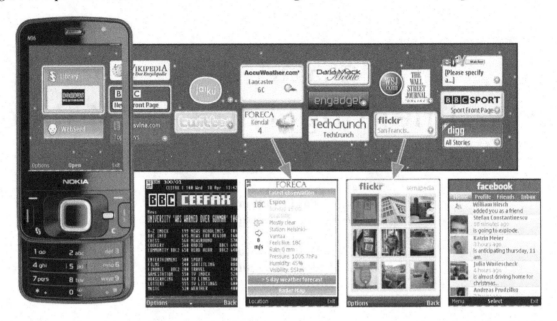

researchers are fully aware of the effects that the distribution platform or 'App Store' has on the users' consumption of applications.

Initially 'in the large' distribution was limited, due to restricted availability through operator portals or aggregators using systems that had their roots in the ring-tone and wallpaper markets built around Short Message Service (SMS) payments and Wireless Application Portal (WAP) push (Garner et al., 2006). As these channels generally required a commercial working relationship with the companies operating portals they were effectively closed to mobile researchers and many smaller mobile development companies. This often led to many researchers having to make applications available through their own websites and hoping for a 'viral' uptake to occur. Things improved somewhat with the emergence aggregator sites such as GetJar™ (initially targeted at Java Micro Edition (J2ME) applications but now one of the few stores offering Apps for the majority of platforms) which allowed free Apps to be shared easily by researchers and small developers. The main difficulty for researchers with such services in the early days was the collation of data relating to application as it was often marred by others users of the site 'high-jacking' an application rising in popularity by re-posting it on the store in an attempt to drive traffic to some particular web service (Chehimi et al., 2008).

When the Nokia WidSets™ platform (Figure 1) appeared in 2006 it was the first adopting what would now be considered as 'the App Store approach' made famous by Apple™ for its iPhone™. These characteristics being the simple search, installation, and rating of applications for the users and the ability for developers to easily push new versions of the applications to those users.

One of the strong features of this service was the IT was largely device agnostic running in J2ME meaning that it was available across a wide range of handsets not just those which would be considered as 'Smart phones'. To put this in perspective in August 2010 Oracle suggested that 2.1 billion of the approximate 4 billion phones in the world are JME enabled. The consumer base of a particular App Store or device is an important factor for researchers to consider when considering its use for in-the-large research. For example, whilst some have suggested that Apples iPhone

Figure 2. iPhone ownership worldwide (AdMob, 2010)

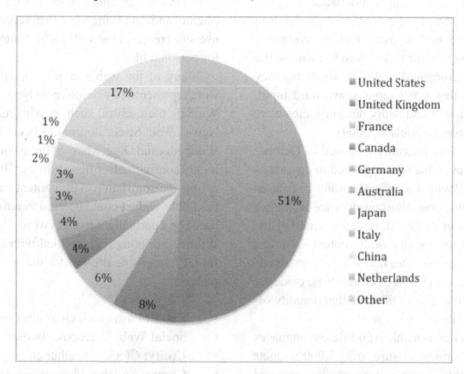

and its App store enables access to a wide audience (Morrison et al. 2010, Miluzzo et al. 2010) the AdMobs May 2010 report on mobile metrics highlight that iPhones represent less the 2% or the world phone market. Further from the breakdown of percentage of these phones within a particular country, shown in Figure 2, it is clear that it is a device primarily used affluent developed countries. This distribution will clearly have an effect on the adoption of particular applications and services.

In the following sections we present an empirical analysis of the WidSets platform operation through the creation and distribution of three games that achieved in excess of 1.5 million users.

WIDSETS PLATFORM

WidSets was created through the Nokia Emerging Business Unit in 2005 as a run-time technology and service developed for the distribution and

execution of widgets across multiple mobile platforms. The service was device agnostic – it was designed to run on any handset supporting Java ME MIDP 2.x. The platform was basically an engine on which applications, referred to as Widgets, could run. Widgets are small applications designed specifically to perform a single function in a user-friendly manner thus sharing the same ethos as Don Norman's Information Appliances (Norman, 2009) and echoing Apples tag line on its commercials "There's an App for That".

The initial goal of WidSets was to enable mobile users to cope with the huge amount of information, and data "feeds", being generated on the web – they identified that there were too many identities, contexts, social networks, interfaces and information sources to handle using conventional mobile phone browsers of the time (Tero-Markus et al., 2007). Additionally, the characteristics of the mobile phone often made the experience of accessing sites designed for the desktop PC a very poor experience. Indeed the situation does not seem to have changed dramatically in the intervening period – a recent study by Nielson (2009) found mobile users had on average a 59% success rate for tasks given to them on the mobile web compared to 80% for desktop – they highlighted that small screens, awkward input, download delays and badly designed sites were the key problems (Nielson, 2009).

WidSets was officially launched in October 2006. However, the service ceased to operate as of June 2009 with well over 10 million registered users. This was possibly a consequence of Nokia's push toward using Ovi Store as the central distribution platform for all mobile content – a notice on the website indicates that WidSets "evolved" into Ovi Store, but there is currently no evidence of any real merging of the core functionality offered by WidSets.

WidSets was available in 56 different languages and at the time of closure, over 5,000 widgets were available. These widgets can be separated into two categories:

- **User:** Those generated through the online "WidSets Studio" (which allowed users to create a skinned "wrapper" widget based on any RSS/Atom feed) or through the mobile client itself. Both methods allowed widgets to be created in minutes and then shared and downloaded by other users in the WidSets community.
- **Developer:** Those created by WidSets themselves or the WidSets developer community. These widgets involved the development of assets to define the look, feel and behaviour of the widget. Whilst they required greater technical expertise and development time, and hence were more difficult to create, they allowed for far greater customisation and functionality.

The former category accounted for around 90% of widgets generated (Tero-Markus et al., 2007). Widgets created through the WidSets Studio

were not configurable beyond defining a "skin" (theme) and data source, but they served to allow users to create widgets for niche content sites or for personal blogs.

Many of the web's large content providers were represented by bespoke widgets created by WidSets themselves, such as widgets for BBC News, Wall Street Journal, Flickr, Engadget, Wikipedia and Digg. Many of these widgets were highly configurable; often allowing filtering and/ or other customisation of the content returned e.g. the Flickr widget could be used to return images based on a particular set of keywords. In addition, WidSets scripting API coupled with the powerful UI framework allowed for more feature-rich widgets:

- **Email:** Push email client for Gmail etc.
- **Social Web:** Facebook, Twitter, Jaiku etc.
- **Utility:** Clocks, Weather etc.
- **Games:** Sudoku, Memory Game etc.

Users were able to personalise (add new, sort, configure) their collection of widgets that were shown on their dashboard either online (by manipulating a virtual representation of the dashboard), or through the mobile client itself. Any changes made client-sidewere automatically synchronized with the mobile client in a very similar manner to iPhone App updates.

The mobile client provided a "Library" shown in Figure 3 that acted as a portal through which other widgets could be discovered (through text search, navigating top rated, most popular, newest, etc.), installed to the dashboard, and rated.

BOMBUS

Bombus was a WidSets game conceived when the platform was relatively new. Only a handful of games were available at the time and these were all puzzle-style games. As part of the research goal was to ascertain how the platform could be

Figure 3. WidSets Library widget flow

 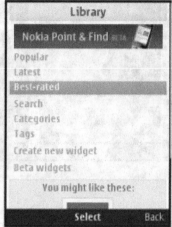 

used to support a wide-range of content such as fast action games. Bombus was designed to be a side scrolling 'arcade' style game.

Having decided on the type of game, the next step was to choose the design criteria. The performance characteristics of mobile phones varies between handsets, therefore this is often the primary consideration when creating games for mobile (Coulton et al., 2005). This was particularly the case for WidSets widgets because it was not possible to create multiple 'builds' for the various makes and models of mobile phones available (Coulton et al., 2005). Device fragmentation is major consideration for researchers considering the use of an App Store and one of the major benefits of the iPhone is currently fragmentation levels are extremely low as Apple only updates its device once a year at most.

Whilst confident that it was possible to create a game that would run on all WidSets handsets, it was determined that the cross device compatibility should extend to the gameplay itself to ensure it was available to the widest possible audience. Providing a consistent experience to all players is particularly important when players can post their scores to ascertain their global standing; all players should feel that with they have any equal chance of winning (Salen & Zimmerman, 2003).

The three main factors that determine player equability in mobile games are: Screen resolution; Input mechanism; and Frame rate.

In order to adapt the game to different screen resolutions we created a simple fixed-point vector game engine that allowed for the 2-D affine transformation (rotation, scaling, translation) of arbitrary points. This enabled game objects to scale appropriately to fit a range of display dimensions. Additionally, the use of vector graphics helped ensure the engine maintained a high frame rate and allowed the arbitrary rotation to the game scene to become part of the game play as illustrated in Figure 4.

The vast majority of mobiles phones still have an interface predominantly designed for navigating menus and numeric input; few have made any accommodation to the needs of game players. This limitation led to the emergence of the one-button mobile game concept (Nokia, 2006) still evident on even the latest touch screen devices whereby the whole action is controlled via through a single button. The challenge in these games does not come from mastering the interface, as in many console games (Wigdor & Balakrishnan, 2003) but rather from precise timing, fast reactions, and/or the rhythm of actions. Achieving a good rhythm or completing a series of difficult actions pre-

Figure 4. Bombus Screenshots

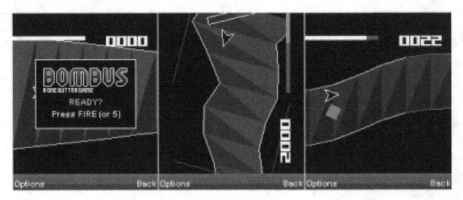

cisely on time is often a key element in a good one-button game experience.

In Bombus, players take control of a spaceship that travels through a scrolling tunnel. Players must press and hold fire to apply an upward thrust to the ship thus avoiding obstacles and the tunnel walls (Bamford, 2007). There are two different types of tunnelled section: in the first the walls of the tunnel are procedurally generated; in the second the tunnel is constructed of multiple compound sine waves (this section moves faster than the first). Various other aspects of the game are generated randomly within certain parameters (game object movement speeds, sizes, positions, background colours etc.). Generating the game environment procedurally ensured that the game size was minimal (WidSets recommended that all application resources including compiled code, images etc. should be 30KB or less) and allowed the game to evolve differently depending on the players' decisions thus increasing re-playability. Although application size is less an issue many App stores place a limit for Over The Air (OTA) delivery. The game ends when the ship's 'life force' reaches zero (colliding with obstacles and the walls of the tunnel depletes life) and it is therefore similar to games such as Tetris, or Bejewelled in that there is no specific win state and players are continually engaged through trying to better their previous score (Juul, 2009). When the game is over, users are given an option to submit their score to the server in order to compare their performance with others.

BOOM!

Whilst Bombus was extremely popular in terms of the game performance and the overall number of downloads, it was anticipated that the relatively static nature of the game would result in the dashboard "lifespan" being shorter than for widgets with more dynamic forms of content that may be considered for some research projects. Therefore another game was considered this time exploiting more of the key advantages of the platform such as the ability to dynamically update content and the possibility of incorporating user generated content (UGC). This game became "BOOM!" which was designed to explore support for the dynamic game content related to the Web 2.0 paradigms of "perpetual beta" and UGC.

For BOOM! a lightweight scrolling platform engine was created in which the player controls the eponymously titled "Boom", character in the form of a bomb with a lit fuse. The goal was to reach water, thus dousing the fuse, before Boom detonated i.e. within a certain time limit. Players used the left and right D-pad keys (or '4' and '6') to move the character and fire/OK/5 to jump. Points were earned for completing the level in as quick a time as possible, collecting items (coins,

Figure 5. Boom! Screenshot

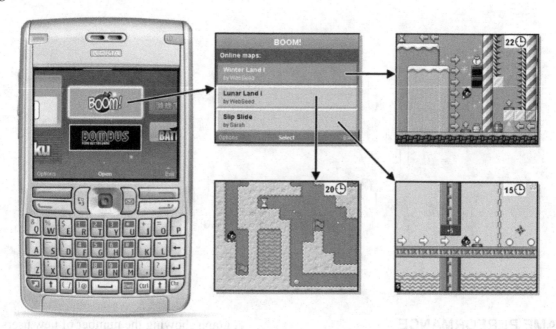

jewels, presents etc.) along the way – "combo" bonuses could be earned for collecting items in quick succession (Bamford, 2008).

To allow dynamic content to be delivered, the list of levels (maps) in the game was updated whenever the widget was maximized. This was hoped to extend the life of the game (with users returning to play new levels as they were released) and also allowed seasonal or themed content to be released. A technique now maximized by iPhones games such as Angry Birds™ and DoodleJump™ For example, the "Winter Land" map was released in time for Christmas 2007 and a San Francisco level was created to coincide with the Game Developers Conference 2008 as shown in Figure 5.

The game itself uses the concept of scrolling tile-based graphics which was a stalwart technique of early video games (Salen & Zimmerman, 2003) as it allows large game worlds to be created out of a relatively small set of graphical components. In a tile-based game the playing area consists of a uniform grid of small rectangular, square, hexagonal, or any other tessellatable graphic images, referred to as tiles. Each tile has certain physical

or player attribute based properties e.g. empty, solid on top, solid all sides, points, extra time, water, wind, jump etc. The complete set of tiles available for use in a playing area is called a 'tile set' and the tiles are laid out according to a specified 'tile map'. New play areas could be created either by changing the tile map, the tile set or both. The game extended this concept by also allowing game attributes such as the gravitational constant and time limit to be altered on a per-level basis. The original idea concept was users would be provide with an level editor to create their own maps but unfortunately the demise of the platform put paid to this particular feature.

Given that new tiles and maps must be downloaded over the mobile network, an important consideration was to keep the overall file size quite low, both to reduce potential cost to the user and allow for the lower bandwidth connections offered by the 2G which may be encountered in countries who are still in the early phases of deploying their mobile infrastructure. Access to an App Store does not dictate the quality of the underlying mobile infrastructure.

Figure 6. Final user numbers for Bombus and Boom! as per 01/06/2009

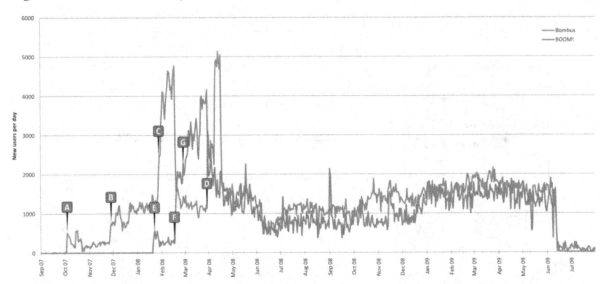

GAME PERFORMANCE ON PLATFORM

In total, Bombus had just over 820,000 users and BOOM! just over 630,000 users by the time the service closed as shown in Figure 6.

A graph showing the number of new users to both Bombus and BOOM! per day over each game's lifecycle is shown in Figure 7. It is possible to correlate some of the patterns evident in the graph with certain events (labeled A to K) on the WidSets platform which affected their lifecycles:

Figure 7. App lifecycles for Bombus and Boom!

1. Bombus was released into the public library in early October 2007. This also coincided with it winning the WidSets developer award. There is an initial spike in new users per day (nupd) due to the competition win and because the widget appeared both in the "Latest" widget section on the web and in the client library. A few weeks after launch there is another small jump in new users as the widget is also promoted as a recommended "Widget from users" on the website. By the end of October the widget no longer features on any lists and the nupd drops back down to a low level.

Figure 8. Promoted widget on the client ("You might like these")

Figure 9. WidSets pre-install widgets grid

2. The WidSets client is officially declared out of beta and Bombus was subsequently one of the widgets promoted through the client libray as shown in Figure 8. Additionally it reaches the 5 "Top Rated" widgets list. As one of the most prominent means of promoting a widget this results in a significant (over 5 times) increase in nupd.

3. Bombus is provided as an option for including on new users' dashboard as a pre-install (Figure 9) – this has a dramatic effect on nupd. After one month BOOM! replaces [F] Bombus as a pre-install and subsequently nupd decreases.

4. Bombus trades places with BOOM! as a pre-install resulting in another spike in nupd. After a period of a few weeks it is removed from the pre-install grid.

5. BOOM! launches and is shown on the "Newest" widget list and "Widgets from users list" (as with Bombus).

6. BOOM! wins developer competition and subsequently trades places with Bombus as a pre-install option. It is also promoted through the client library and from an article on the WidSets homepage itself.

7. BOOM! reaches 5 "Top Rated" widgets list.

After promotion for both widgets had finished (May, 2008), and they had reached the 5 Top Rated widgets list, the number of nupd showed a steady increase until the demise of WidSets early June 2009. Although difficult to quantify, this gradual increase in nupd is thought to be due to the fact that Nokia began to pre-install WidSets on some of their S40 range of handsets, but it could also point to an increase in popularity of the WidSets service as data contracts became more affordable and WiFi-enabled handsets were becoming more mainstream

In terms of the community ratings for the two widgets, Bombus received approximately 20,000 positive votes whilst BOOM! received approximately 16,700. By taking into account the number of users for each game this corresponds to 2.5% of users voting positively for Bombus and 2.6% for BOOM!. By comparison, the two most popular rated widgets "Push email" and "Wikipedia" received 0.7% and 0.2% respectively. Many other game widgets on WidSets also received a very high quantity of positive ratings in comparison with other categories of widget.

Table 1 shows the percentage of users who submitted a single score to either Bombus or BOOM!. Whilst 23% of users submitted a score to Bombus, only 5% of users submitted a score to BOOM! This reflects the fact that, unlike in Bombus, scores could only be submitted to BOOM! after completing the level. This also perhaps suggests that whilst many users felt positively about both games, only a small percentage invested significant time in playing the games or learning the mechanics to be able to complete a particular level..

FOUR IN A ROW

After we had completed the development of both Bombus and BOOM! WidSets released a beta version of the "Channel API". This API allowed widgets to create a connection between multiple users through channels (communication hubs) thus enabling real-time communication between clients for features such as chat between users and multiplayer gaming. Having witnessed the demand for casual games through WidSets the Channel API presented an interesting opportunity as a potential research tool to enable greater interaction with the community of users. Note these features are very similar to those introduced by Apple for its Game Centre which was launched in June 2010 as part of iOS4.

As the fundamental game mechanics of Four-In-A-Row are well understood, they will not be discussed here. However there was one fundamental addition to the gameplay – users were forced to make their turn within a given time period (chosen by the challenging player) from "Slow" [15s], "Medium" [10s] and "Fast" [5s]).

In terms of operation users are first presented with the game lobby as shown in the first two screenshots in Figure 10. The lobby shows all of the other users in the channel who are not already engaged in a game. From the lobby users can either wait for a challenge by another user or instigate a challenge directly. Once a challenge has been accepted the widget creates a game room shown on the right in Figure 10 in which the players compete. During the game, each player can send pre-defined or custom messages (often playful taunts) to the other player.

Table 1. Bombus and BOOM! score statistics

Game	Total Number of Users	Number of users who submitted scores	Percentage of users who submitted scores	Total scores submitted
Bombus	820,000	185,000 (65% anon)	23%	578,000
BOOM!	626,000	32,000 (53% anon)	5%	41,000

Figure 10. Left: Game Lobby, Right: Game Operation

PLAYER BEHAVIOUR

Four-In-A-Row was the first multi-player game released on WidSets. During the first four weeks of launch, the widget obtained in excess of 10,000 users and was achieving a positive aggregate rating of around 9%. . However, in an update made to the WidSets API, the beta version of the Channel API did not function correctly. Therefore, the community of users was lost until two months later when the problem was rectified in a client update. Subsequently, the observations presented here are taken from 100+ games played by our research group against external players during the initial period (Coulton et al., 2008).

One interesting aspect of the user base was that there were a high proportion of Asian origin names as illustrated in Figure 10. This was also very apparent in the social community features introduced into the client by WidSets in version 3 such as the widget recommendations and forum. This perhaps suggests that many users from countries where broadband and PC penetration levels were low compared to mobile ownership were using their mobile phone to access content through WidSets (Coulton et al., 2008).

The in-game chat between users also displayed common patterns. As well as the preset taunts such as "Hurry up" and "You lose" we began to see that players were interested in establishing first the location, then the gender of their opponent. This was particularly evident with players who declared themselves to be from an Asian country. The limitations imposed by the chat facility meant we could not fully explore the player's intent, but such actions would seem to align with numerous reports over the last few years indicating that young people in predominantly Muslim countries flirt via their mobile phones using Bluetooth (Abu-Nasr 2005) and this game would seem to have offered similar possibilities.

Indeed, we found that the usernames themselves began to follow certain naming conventions; users began to adopt names that summarized their key characteristics (location/gender/date of birth etc.) and perhaps likely attract interest from others (Coulton et al., 2008). Figure 11 shows two screens from the lobby that were quite representative of activity as a whole. Many users chose their username in order to encode key characteristics e.g., "The Saint [f]" (encoding gender), "Moh3n_1987" (encoding year of birth / age), "Iran2008" (encoding location).

EFFECTS OF BUSINESS MODELS

Initially WidSets attempted to monetize their service by allowing content/service providers to promote or pre-install widgets through the client.

Figure 11. Typical chat messages during game play (Coulton et al., 2008)

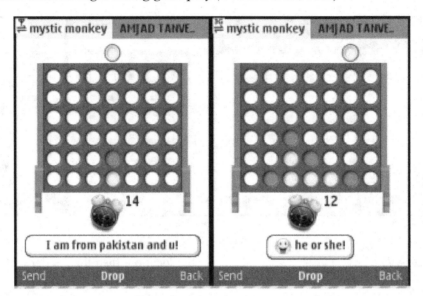

The main proposition in this case was to use the widgets themselves as advertisements (WidSets, 2008). Content/service providers were given a number of means of paying for the service e.g. pay-per-pick, pay for pre-install, pay for premium promotion etc.

In March 2008 we were invited by WidSets to participate in their efforts to further monetize the WidSets platform through a shared-revenue advertising model. Although not initially a research question considered as part of this project it was decided that the revenue generation model has a direct effect on the operation of the platform and therefore it would be beneficial to partake in this trial as this is ultimately the driving force for all such App Stores that may be used by researchers. To avoid ethical concerns over income generation from an academic project all the money generated was subsequently donated to charity.

As WidSets were able to access information about users that would be useful in the delivery contextual advertising e.g. location, language, phone model, and their preferences by examining their installed widget base (WidSets, 2008), they proposed that targeted advertising would become their primary business model (Kumlin, 2007).

The revenue was split as follows: the advertising agency (Nokia Interactive) receive 30% of the cost of sales and the remaining revenue would be split between Nokia WidSets and the developer/partner (35% each).

In April 2008 WidSets began to show advertising in both Bombus and BOOM! The WidSets client was able to display the advertisements either on a widget splash screen, or as a banner in the client as seen in Figure 12.

Figure 12 shows the advert as displayed in BOOM! Note that clicking/selecting the advert could trigger a number of actions depending on the advertising campaign. Upon starting the game, the banner was removed from view thus not obstructing the game play.

During the period from April to September 2008, the advertisements displayed were mainly "dummy" ads used to promote internal Nokia products and services, or charities, such as the World Wildlife Fund, for which no income was generated. Therefore only the advertisements shown after this date through to May 2009 when WidSets ceased to operate are considered. Over this period the combined in-game advertisements for Bombus and BOOM! received 1,367,514 im-

Figure 12. WidSets in-game advertisement showing possible actions

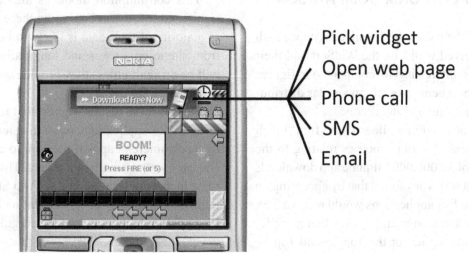

pressions, with 34,202 click-throughs. This results in a click-through rate (CTR) of approximately 2.5% on average. By extrapolating these figures, the total revenue generated for a year would have been around €1,000.

One interesting observation was that over the first month, the CTR was 6.9%, whereas in subsequent months this figure quickly came down to around 2%. This suggests that initially users were either curious about such adverts or clicked through accidentally. In general, however, this figure is high when considering that online CTRs have fallen to around 0.2-0.3% and perhaps hints at the motivations of companies such as Google in investing heavily in mobile. This pattern reflects other anecdotal evidence from mobile ad networks on the iPhone whereby CTRs were an order of magnitude higher initially than after a period of a few months. According to AdMob reports 0.5-1% CTR for mobile advertisements through platforms such as the iPhone is now typical.

There are two perspectives to consider when evaluating the success of the advertising trial: that of the developer/partner and that of WidSets themselves.

- **Developer/Partner:** Considering that both Bombus and BOOM! were amongst the most popular widgets on WidSets, the income through advertising would seem quite low and probably not enough to attract commercial developers or service/content providers.
- **WidSets:** It is hard to determine if advertising revenue was generating sufficient income for WidSets themselves as obviously our games only formed part of the large portfolio of widgets. However, given the high cost of operation that WidSets must have presented along with the fact that WidSets is no longer running, it is probable that advertising on WidSets was not a commercial success. However, the value of WidSets to Nokia could be considered as more than the income generated through advertising as the platform helped to promote the Nokia brand and may have attracted people to buy Nokia handsets.

CROSS PLATFORM COMPARISON

In terms of the pattern of downloads (nupd) both games received whilst on the WidSets and their usage, it is interesting to compare our figures against subsequent research on similar distribution platforms such as the iPhone.

In February 2009, mobile analytics firm "Pinch Media" released various metrics relating to the download of 30,000,000+ iPhone app downloads (Yardley, 2009). They found that by appearing on the Top 100 list, applications would receive 2.3x more downloads on average and often an order of magnitude higher for the Top 25 and Top 10 list. This reflects the experience of of Bombus and BOOM! when they reached the Top 5 rated. This highlights that while platforms such as iPhone and WidSets make it possible to easily search/find any application within the library the important aspect in gaining new users is through editorialisation by content managers through their recommendations/ suggestions and in reaching the Top 'n' lists for popularity and ratings.

Significantly, the research from Pinch Media (Yardley, 2009) also shows that most mobile applications have a relatively short "shelf life" with on average less than 25% of users returning to the app one day after download, dropping to around 5% after 30 days. The category to which an application belongs seems to have a strong effect on return rate – applications that are typically more dynamic e.g. sports (results, league tables etc.) and entertainment apps fared better than games, utility, and lifestyle applications. This seems to closely follow our findings from analysis of the scores submitted for Bombus and BOOM! with users generally posting only one or two scores on the day of download and subsequently never submitting again. This suggests that the range and depth of content on platforms such as WidSets is as important as personalization. In other words users want to experience a wide range of content, with only a relatively few applications becoming part of their daily activity.

This consumption model is also extremely significant for researchers using these platforms as majority of their data is likely to be obtained from single time users and longitudinal studies will be difficult unless they provide users with a good reason for doing so.

Further these results suggest that researchers must create applications at similar levels quality as commercial apps if they are to attract and engage users at significant levels. This is going to be increasingly important as App Stores seek improve the quality of its content as illustrated by Apple in its recently published guidelines to developers (Apple, 2010):

If your App looks like it was cobbled together in a few days, or you're trying to get your first practice App into the store to impress your friends, please brace yourself for rejection. We have lots of serious developers who don't want their quality Apps to be surrounded by amateur hour.

The high positive ratings received by both games suggested a demand for casual games. The lower number of scores submitted for BOOM!, a more involved game suggests many users were looking for a very accessible, casual experience – one button game/ familiar puzzle style games. However, the ratio of positive ratings per user was highest for Four-In-A-Row, thus suggests that users found real value in being able to socially interact and compete with others. Therefore utilizing facilities such a chat within and app through services such as Game Centre could be highly beneficial to researchers and provide a useful method of obtaining the in-situ information from the users that is so often difficult to obtain in these situations (Church & Cherubini, 2010).

CONCLUSION

This paper has presented an analysis of using Apps on Apps stores as an in-the-large research methodology through the longitudinal study of three games on the WidSets platform which attracted in excess of 1.5 million users. Despite the fact that the WidSets platform was discontinued (arguably prematurely) it clearly utilized many of the features we now attribute to successful App Stores such as simple search, installation, and rating of applications for users and the ability for developers to easily update applications and push these to their already installed user base. The main differentiator of WidSets from the majority of current App Stores is that it specifically targeted low and mid range phones as well as Smartphone's. This may have ultimately contributed to its demise as this would understandably limited the capabilities that could be developed and such users are often perceived as less likely to download content. However, the performance of applications on the platform in terms of factors effecting ratings, downloads etc was entirely consistent with what has been observed subsequently on the Apple store. The consequence of concentration on the Smartphone market though for researchers means that App stores are currently skewed to predominantly affluent western markets.

In examining the Apple iPhone platform we found many of the same characteristics of WidSets implemented at the developer API and operating system level. We would argue that the coupling of native provision of Mobile 2.0 enabling technologies with comprehensive third-party libraries that enable the integration of rich social functionality, has accounted for much of the iPhone's success with users and developers alike. For example, the iPhone used a similar democratic rating model to WidSets that results in content being ranked by certain user determined and defined criteria. When WidSets introduced this in 2007, this represented something of a departure from the operator portal model that was prevalent at the time, whereby the visibility of an application was wholly dependent on the developer's relationship with the network operators. Through analysing user downloads, we found that this was a key factor in determining

the success of an application, as whilst it was possible to easily find any widget within the WidSets library (either on the web or through the client), the important aspect in gaining new users was through both editorialisation by WidSets content managers (through recommendations/suggestions) and more importantly in reaching the "top lists" for popularity and ratings. Whilst others have correctly pointed out that downloads alone do not give a true picture of the relationship with users (Morrison, 2010) may be key to obtaining significant number of users if longitudinal study of usage is required. Additionally success is relative in the first few months of Bombus it was exciting to be obtaining a few hundred new users per day but in reality we were still below the 'noise floor, of the platform where applications may be difficult to find and as the platform gets more popular this floor get higher as we later observed with a 3D version of Bombus (Mobile Radicals, 2008) which only achieved tens rather than hundreds of thousands of users.

The research also demonstrates that the so-called value added features such as chat can, not only increase the popularity and usage of applications, but may also facilitate emergent forms of behaviour and provide easier access to users and as such deserves further study.

These values added features could also address the issue that majority of data is likely going be obtained from single time users and longitudinal studies will be difficult unless they provide users with a good reason for doing so.

In terms of the quality of the apps developed for in-the-large research it needs to be significantly higher than would be traditionally developed for lab based research. Indeed App stores are already reacting to commercial developers concerns that too much amateur content is making it difficult to find new professional content. Indeed widespread disillusionment amongst developers could lead to an App-ocalypse whereby insufficient content appears on the stores. Therefore if researchers don't match the App stores expectations it will prove increasingly difficult to get research Apps accepted unless they match the required quality. It will thus become increasingly important that researchers need to reflect requirement for greater quality when considering timescales and budgets for in-the-large projects.

Overall experimenting through Apps and App Stores offers novel and exciting prospects for researchers but they must fully understand the nature of these stores and their users including any commercial considerations as it will ultimately influence their results and this longitudinal research project has revealed much about not only about the usage of Apps but also that the evolution of the App Store also affects the App lifecycle.

ACKNOWLEDGMENT

The authors would like to express their thanks to Nokia for the provision of software and hardware to used in the implementation of this project. Further we are particularly grateful to the team at WidSets for their wholehearted support in providing us the opportunity to experiment on their platform.

REFERENCES

Abu-Nasr, D. (2005, August). *Bluetooth takes a bite out of Saudi anti-flirting rules.* Retrieved from http://findarticles.com/p/articles/mi_qn4188/is_20050812/ai_n14886549

AdMob Mobile Metrics. (2010). *Metric highlights.* Retrieved from http://metrics.admob.com/wp-content/uploads/2010/06/May-2010-AdMob-Mobile-Metrics-Highlights.pdf

Apple. (2010). *App store review guidelines.* Retrieved from https://developer.apple.com/appstore/resources/approval/guidelines.html

Bamford, W. (2007). *Bombus: One button mobile game.* Retrieved from http://www.youtube.com/watch?v=aVQIUtLzZAg

Bamford, W. (2008). *Boom! A mobile phone game for WidSets.* Retrieved from http://www.youtube.com/watch?v=E5V3L_sVtF8

Chehimi, F., Coulton, P., & Edwards, R. (2008, July 10-13). 3D motion control of connected augmented virtuality on mobile phones. In *Proceedings of the International Symposium on Ubiquitous Virtual Reality* (pp. 67-70).

Church, K., & Cherubini, M. (2010, September 26). *Evaluating mobile user experience in-the-wild: Prototypes, playgrounds and contextual experience sampling.* Paper presented at the Workshop Research in the Large: Using App Stores, Markets, and other Wide Distribution Channels in UbiComp Research, Copenhagen, Denmark.

Coulton, P., Copic Pucihar, K., & Bamford, W. (2008, November). Mobile social gaming. In *Proceedings of the International Workshop on Social Interaction and Mundane Technologies* Cambridge, UK (pp. 20-21).

Coulton, P., Rashid, O., Edwards, R., & Thompson, R. (2005). Creating entertainment applications for cellular phones. *Computers in Entertainment, 3*(3), 3. doi:10.1145/1077246.1077254

Garner, P., Coulton, P., & Edwards, R. (2006, June 29-July 1). XEPS – Enabling card-based payment for mobile terminals. In *Proceedings of the Tenth IEEE International Symposium on Consumer Electronics*, St. Petersburg, Russia (pp. 375-380).

Juul, J. (2009). *A casual revolution: Reinventing video games and their players.* Cambridge, MA: MIT Press.

Kumlin, K. (2007). *Mobilize your campaign.* Nokia Emerging Business Unit, WidSets.

McMillan, D., Morrison, A., Brown, O., Hall, M., & Chalmers, M. (2010, April 23). Further into the wild: Running worldwide trials of mobile systems. In P. Floréen, A. Krüger, & M. Spasojevic (Eds.), *Proceedings of the 8th International Conference on Pervasive Computing* (LNCS 6030, pp. 210-227).

Miluzzo, E., Lane, N. D., Lu, H., & Campbell, A. T. (2010, September 26). *Research in the app store era: Experiences from the CenceMe app deployment on the iPhone.* Paper presented at the Workshop Research in the Large: Using App Stores, Markets, and other Wide Distribution Channels in UbiComp Research, Copenhagen, Denmark.

Mobile Radicals. (2008). *3D Bombus: WidSets game.* Retrieved from http://www.youtube.com/watch?v=RSi-LndShE8

Morrison, A., Reeves, S., McMillan, D., & Chalmers, M. (2010, September 26). *Experiences of mass participation in Ubicomp research.* Paper presented at the Workshop Research in the Large: Using App Stores, Markets, and other wide distribution channels in UbiComp Research, Copenhagen, Denmark.

Nielson, J. (2009). *Mobile usability: Mobile usability test.* Retrieved from http://www.useit.com/alertbox/mobile-usability.html

Nokia. (2006). *Turn limitation into strength: Design one-button games version 1.0.* Retrieved from http://www.forum.nokia.com/info/sw.nokia.com/id/8dff4326-3979-4149-96c0-5fa95a14a3cb/Turn_Limitation_into_Strength_Design_One-Button_Games_v1_0_en.pdf.html

Norman, D. (2009). *The invisible computer: Why good products can fail, the personal computer is so complex, and information appliances are the solution.* Cambridge, MA: MIT Press.

Rogers, Y., Connelly, K., Tedesco, L., Hazlewood, W., Kurtz, A., Hall, B., et al. (2007). Why it's worth the hassle: The value of in-situ studies when designing UbiComp. In J. Krumm, G. D. Abowd, A. Seneviratne, & T. Strang (Eds.), *Proceedings of the 9th International Conference on Ubiquitous Computing* (LNCS 4717, pp. 336-353).

Salen, K., & Zimmerman, E. (2003). *Rules of play: Game design fundamentals.* Cambridge, MA: MIT Press.

Tero-Markus, S., & Lumivuori, M. (2007). *WidSets platform for mobile Web 2.0 users and developers.* Retrieved from http://fruct.org/sem2/WidSets%20Platform%20for%20Mobile%20Web%202.0%20Users%20and%20Developers.pdf

WidSets. (2008). *WidSets business model.* Retrieved from http://research.nokia.com/publication/11214

Wigdor, D., & Balakrishnan, R. (2003, November 2-5). TiltText: Using tilt for text input to mobile phones. In *Proceedings of the 16th Annual ACM Symposium on User Interface Software and Technology,* Vancouver, BC, Canada (pp. 81-90).

Yardley, G. (2009). *App store secrets: What we've learned from 300,000,000 downloads.* Retrieved from http://www.slideshare.net/360conferences/pinch-media-app-store-secrets

This work was previously published in the International Journal of Mobile Human Computer Interaction, Volume 3, Issue 4, edited by Joanna Lumsden, pp.55-70, copyright 2011 by IGI Publishing (an imprint of IGI Global).

Chapter 18

My App is an Experiment:
Experience from User Studies in Mobile App Stores

Niels Henze
University of Oldenburg, Germany

Benjamin Poppinga
OFFIS - Institute for Information Technology, Germany

Martin Pielot
OFFIS - Institute for Information Technology, Germany

Torben Schinke
Worldiety GbR, Germany

Susanne Boll
University of Oldenburg, Germany

ABSTRACT

Experiments are a cornerstone of HCI research. Mobile distribution channels such as Apple's App Store and Google's Android Market have created the opportunity to bring experiments to the end user. Hardly any experience exists on how to conduct such experiments successfully. This article reports on five experiments that were conducted by publishing Apps in the Android Market. The Apps are freely available and have been installed more than 30,000 times. The outcomes of the experiments range from failure to valuable insights. Based on these outcomes, the authors identified factors that account for the success of experiments using mobile application stores. When generalizing findings it must be considered that smartphone users are a non-representative sample of the world's population. Most participants can be obtained by informing users about the study when the App had been started for the first time. Because Apps are often used for a short time only, data should be collected as early as possible. To collect valuable qualitative feedback other channels than user comments and email have to be used. Finally, the interpretation of collected data has to consider unpredicted usage patterns to provide valid conclusions.

DOI: 10.4018/978-1-4666-2068-1.ch018

INTRODUCTION

Mobile application stores such as Apple's App Store and Google's Android Market revolutionized the distribution of applications for mobile devices. This distribution channel lowered the gateway hurdle dramatically and opened the market for small companies and engaged hobbyists. Mobile application stores -- for the first time -- enable virtually any developer to easily reach hundred thousands of mobile users. Recently researchers discovered this opportunity and began to publish prototypes via mobile application stores.

It has been argued that the "easy access to such a potentially wide audience could radically alter the nature of many UbiComp trials" (Morrison et al., 2010). In the tradition of UbiComp research most attempts to distribute prototypes via mobile application stores focus on the evaluation of prototypes (e.g. Zhai et al., 2009; Girardello, 2010; Michahelles, 2010; Gilbertson et al., 2008). Proof-of-concept prototypes are developed and the large number of users is used to demonstrate the successfulness of the respective application. Feedback is mainly gathered to understand the nature of the respective prototype.

In the tradition of psychology and social sciences Human Factors and Human-Computer Interaction research in contrast focus on understanding the human. Commonly, controlled experiments, quasi-experiments and observations are used to derive general findings. As in psychology, prototypes are often just the apparatus to investigate a research question. The psychologist Danziger describes an apparatus as a tool for "exposing experimental subjects to controlled and precisely known forms of stimulation" and "for recording and measuring responses" (Danziger & Ballantyne, 1997). In previous work we showed that Apps distributed to thousands of users can successfully be used as an apparatus for controlled experiments (Henze & Boll, 2010; Henze et al., 2010).

In this paper we report our findings from five studies we conducted by publishing Apps in the Android Market. The paper first presents these Apps, the research questions they address, and the outcomes. In the subsequent sections we then discuss our general findings and conclusions on the participants, the quantitative and qualitative data, and ethical aspects. We conclude with aspects that should be considered when conducting experiments in mobile application stores.

Case Studies

In order to investigate different mobile HCI topics we conducted five studies, which actually use an App as apparatus and were published via mobile application stores. All Apps have been implemented for the Android platform and are therefore available for a range of users and devices. Table 1 gives an overview of the studies that are described in the following.

SINLA: Off-Screen Visualizations for Augmented Reality

In Augmented Reality the visualization of nearby points of interest (POIs) is commonly done by displaying a small mini-map to provide an overview as the user moves around. However the 3D augmented environment and the 2D mini-map have different reference systems. Therefore, interpreting the mini-map and align it with the augmented environment demands special mental effort. A

Table 1. Overview about the five conducted studied

Name	Installs	Samples	Time	Type
SINLA	~1,737	8	8.5 months	quasi-experiment
Pocket-Navigator	9,149	670	6 months	quasi-experiment
MapExplorer	6,372	4,197	6 months	experiment
Poke the Rabbit	5,708	5,103	5.5 months	experiment
Tap It	7,811	6,907	2 months	observation

number of techniques have been developed for digital maps to visualize off-screen objects that are currently beyond the screen (Zellweger et al., 2003; Baudisch & Rosenholtz, 2003; Burigat et al., 2006). We adapted an existing arrow-based technique for visualizing off-screen handheld Augmented Reality. In a lab study we compared this technique with a state-of-the-art mini-map (Schinke et al., 2010). Based on our findings we included three off-screen visualizations, a mini-map, 3d arrows and a combination of a mini-map and 3d arrows in our prototype (shown in Figure 1) resulting in three conditions. The aim of the study was to validate whether users have age or gender specific preferences for a visualization technique.

First of all the prototype is a simple handheld Augmented Reality application that displays nearby POIs as blue balls located at the appropriate virtual position on a live camera viewfinder. To make the prototype (shown in Figure 2) useful for real users we created a function to search for nearby POIs. The user can filter the results by searching for keywords and selecting a maximum search radius. As soon as POIs are in range the user becomes supported by displaying the off-screen graphics.

Every user is considered as a potential participant, thus all relevant data is recorded locally without any remark. After five minutes of total usage a message appears that the user can take part in a study and as a reminder a new button is shown in the main menu. If the user took part the button disappears but after another five minutes of use the button reappears and the user can take part again.

Using SINLA we want to collect long-term usage data based upon a quasi-experiment as the study design. We continuously measure the time, each visualization is used. To avoid a systematic influence the initial visualization is chosen randomly on the first start up but the user can freely select another one. If the user decides to take part, a short questionnaire is shown to select a gender and the age from six categories. This data and the measured times for each visualization, in addition to a unique device hash and the current date, are submitted to our server.

The application was first published without any logging functionality containing earlier drafts of our off-screen visualizations in the Android Market on August 30, 2009. It has been updated on November 14, 2009 to provide the described features for the study and was accessible until September 29, 2010. Despite the fact that the application was available in the market for over one year, only 2,853 users installed one of the two versions and 1,737 of them were able to participate using the updated prototype. We collected only 8 samples, which made it impossible to derive any conclusions. Also no users took part more than once. Revising our study design it is likely that the attempts to get the user's attention for participating in the study are not obtrusively enough. The prototype probably also does not motivate long term usage. Another conclusion, taking the Android Market comments into account, is that our prototype does not work well on different android devices and platforms.

PocketNavigator: Conveying Geographic Cues with Tactile Feedback

More and more often, we can find pedestrian navigation systems integrated in modern smartphones. Inspired by the established car navigation systems they are able to show a user's location on a map and highlight a route to a destination. Some of these applications provide turn-by-turn instructions through text, visualisations or speech output. However, in some situations visual or audio feedback is not appreciated by a pedestrian. Tactile feedback as navigation aid has been proposed and studied by several groups (e.g. van Erp et al., 2005; Pielot & Boll, 2010). However, existing studies mostly focus on artificial settings and tasks. The question that remains unanswered is if

Figure 1. SINLA screenshots of the two visualization techniques arrows and mini-map. The third condition is the combination of arrows and mini-map

Figure 2. SINLA's menu, preferences, and questionnaire

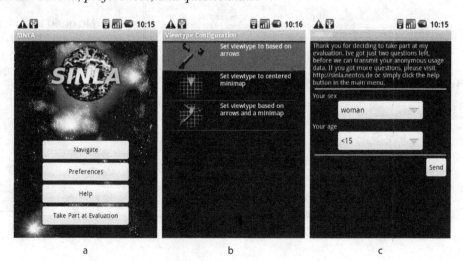

tactile navigation feedback works in non-artificial everyday situations.

At a first glance, the PocketNavigator (Figure 3) is an ordinary pedestrian navigation application. A scrollable map, like e.g. available in Google Maps, is shown and the user's location is displayed. Furthermore, a waypoint-based shortest route can be calculated for any destination. The route is displayed as an overlay on the map, but is also shown as a visual arrow pointing towards the next waypoint, using the device's integrated compass. In addition, the PocketNavigator provides tactile patterns, conceptually showing towards the next waypoint (Pielot et al., 2010). The aim of the study was to analyse how the tactile feedback will be used in the wild and how it affects the navigation performance (e.g., navigation errors, disorientations). For the PocketNavigator we tried different techniques to ask for permission to log data. In early versions of the application an opt-in checkbox in the about/tutorial view is used. In later versions the logging is also advertised through re-appearing popups, asking for participation in the study.

For the application we are interested in long-term everyday use results. Thus, no artificial data collection task has been integrated, but the participants are expected to use the application when they actually need navigation assistance. The tactile feedback serves as independent variable with the conditions on and off. The PocketNavigator uses a quasi-experiment as study design, as every participant is free to turn the tactile feedback on and off as desired. Important settings and configurations are logged into a file every second and are transmitted to a server every 120 seconds. In detail we log: touch interactions, speed, device orientation, compass angle, loudness, light level, and the current configuration and state of the application. We do not log any personal or private data (e.g., phonebook, SMS, user location). Some of the logged parameters are based on frequently

Figure 3. Screenshots of the PocketNavigator's main view, about screen, and information about the tactile feedback

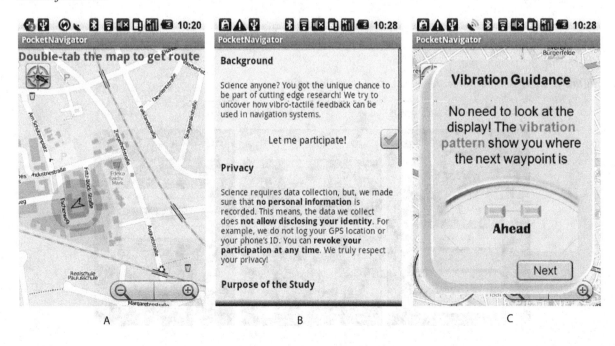

occurring data (e.g. acceleration), which we do not log as raw data. Of particular importance are the logged navigation errors, disorientations, and the device posture, which we identified as an important measure for our results. Each of these high-level values is based on a combination of basic values.

The application is available in the Android Market since April 15, 2010 and has been installed 9,149 times. 670 participants agreed to participate in the user study. However, most of them did not navigate in the foreseen way or not at all. Most of the users tried and tested the application once. If the application has been started multiple times, reoccurring use-cases were e.g. driving in a vehicle or watching the map. 19 participants fulfilled the criteria of seriously navigating as pedestrian at least twice. On average these participants used 2.47 routes. They spent 390.79 seconds per route on average. In 93.90% the participants followed the route presented by the navigation system. The tactile feedback has been used by 8 users

on 9 different routes. Having collected such a small amount of data only shows that conducting experiments via mobile application stores does not automatically yield in a large data set. Here, we believe, the reasons were that people do not navigate as often as they e.g. play games and that our method of asking for the users' consent was very conservative.

MapExplorer: Comparing Off-Screen Visualizations with a Tutorial

The aim of this study was to compare different off-screen visualization techniques for digital maps e.g. Halos (Baudisch & Rosenholtz, 2003) and arrows (Burigat et al., 2006). Previous work conducted studies with static maps and did not consider tasks where users can dynamically interact with the map by panning it (Baudisch & Rosenholtz, 2003; Burigat et al., 2006). Furthermore, the conclusions are based on studies conducted in a lab with few participants that share similar

Figure 4. Screenshots of the three visualization techniques Halo, stretched arrows, and scaled arrows used by the MapExplorer

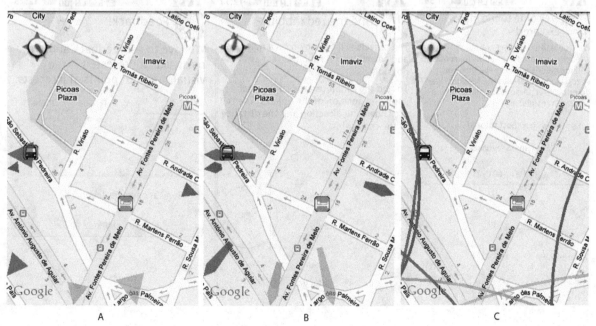

A   B   C

backgrounds. To compare the three previously studied off-screen visualizations shown in Figure 4 we implemented a location based App.

To collect usage data as early as possible and to be able to compare the visualizations we decided for a tutorial which mimics a well defined task similar to the tasks in lab experiments. Using one defined task should improve the repeatability and reduce the effect of other influences. The tutorial appears when the application starts for the first time. After an introduction users should execute a simple find-and-select task using each visualization. While executing the task a map containing 10 randomly distributed POIs is shown. The tutorial's task is to select the red rabbit. The map can be explored by panning it with the finger. A POI is selected by tapping on it. After completing the tutorial the application offers the standard functionalities of a location-based application. Users can search for nearby POIs and access details about them. When starting the MapExplorer for the first time the user learns that the

App logs data (Figure 5). The user can opt-out by deselecting a preselected checkbox.

The App's tutorial is designed as an experiment with repeated measures. The off-screen visualization technique is the independent variable resulting in three conditions. The order of the conditions is randomized to reduce sequence effects. While the user executes the tutorial the task completion time, the number of map shifts, and the number of errors are logged. After finishing the tutorial the time spend with each off-screen visualization is measured and we also measure if the user interacts with the application or not. Furthermore, users can fill the feedback shown in Figure 5. In addition, we collected the user's time zone and the selected locale (e.g. en_US or de_DE).

We published the application in the Android Market on April 1, 2010. Till September 29, 2010 the App was installed 7,664 times and we collected data from 4,197 users. Analysing the data we found that users need significantly more time and map-shifts with Halos (p<.001) to complete

Figure 5. Screenshots of the MapExplorer's introduction, tutorial instruction, and the feedback form

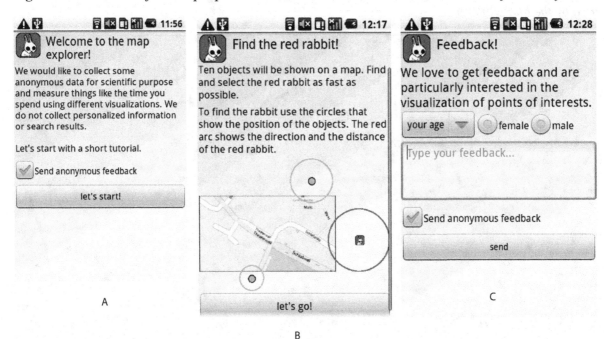

the tutorial (further details can be found in Henze & Boll, 2010). However, investigating the data in more detail shows that a number of users needed much more time to complete the tutorial than one would expect (e.g. longest time spend using Halos was 100 seconds). Reconsidering our design it might be assumed that instead of measuring the pure task completion time the results are affected by the "interestingness" of the visualizations. From informal tests we can report that some users explore the map much longer using Halos than using the other visualizations. Furthermore, our results are limited because users had no previous training and most users performed the tasks only once with each visualization.

Poke the Rabbit: Evaluating Off-Screen Visualizations with a Game

Based on the ambiguous results from the tutorial-based approach we decided to repeat the evalu-

ation of off-screen visualization techniques with a different apparatus. The main shortcoming of the previous experiment is that it is not clear if users' tried to accomplish the task in an efficient and effective way. Therefore, we designed a game using the same three visualization techniques (see Figure 6) as conditions. Compared to a tutorial a game has the advantage that it is natural to confront players with variations of the same task.

Before starting the game a short introduction explains how to play. The game starts with a stage of three levels each containing 30 objects, represented by "cute" rabbit icons. The objects are randomly distributed on a plane (much larger than the actual screen size) that can be paned much like a digital map. Each level uses a different off-screen visualization. The task of the player is to "poke" as many objects as possible by tapping them with the finger in a certain time frame. Once an object is poked it fades to gray and a new object appears. If a player finishes the three lev-

Figure 6. In-game screenshots of the three visualization techniques Halos, stretched arrows, and scaled arrows from Poke the Rabbit

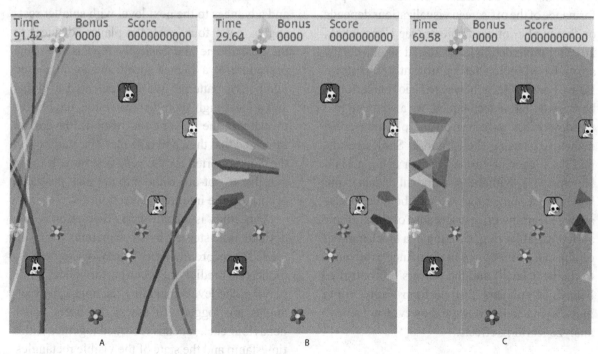

els he or she goes to the next stage where 20 objects are used and afterwards to a stage with 10 objects. The visualizations are randomized within a stage to reduce sequence effects. After finishing three stages the game starts from the beginning with more time to complete a level but also with more objects needed to successfully finish a level. All players are directly considered as participants. The user is never informed that the App is the apparatus of a study and that usage data is transmitted to our server.

Poke the Rabbit uses the same study design as the MapExplorer's tutorial. The study is an experiment with repeated measures and task repetition. The off-screen visualization technique is the independent variable resulting in three conditions. The order of the conditions is randomized to reduce sequence effects. We recorded the number of poked rabbits for each level played. In addition, we collected the user's time zone and the selected locale.

We published the game in the Android Market on April 14, 2010. Till September 29, 2010 the game was installed 6,098 times and we collected data from 5,103 devices. We found that the performance of the off-screen visualizations depends on the number of used objects (see Henze et al., 2010 for more details). For 20 and 30 objects the arrow-based approaches significantly outperform Halos. For 10 objects, however, Halos outperforms both arrow-based techniques. We also found that the device has an effect on the players performance (e.g. using the Motorola Sholes results in 13% higher performance than using a HTC Hero $p<10^{-9}$). With the vast amount or data in our hands we assumed that we will be able to also analyze learning effects. Because of the multiple varied variables (e.g. duration of a level, number of objects, and required performance to advance to the next level) and the players' uncontrolled behaviour (players that perform badly might quit playing soon) we were, however, not able to analyze learning effects.

Tap It: Assessing Users' Touch Performance

Following Poke the Rabbit, we conducted another study with the aim to investigate the touch behaviour of smartphone users. With Tap It we want to assess the touch performance for different target sizes similar to the work by Park et al. (2008). By collecting a huge amount of data we aim not only at determining the error rate and reaction time for different screen locations and target sizes. More, we want to derive a model for predicting the users' performance that takes the touch history into account.

After starting the game, the player has to touch appearing white rectangles before they are fully visible (Figure 7). Different patterns of rectangles appear from a single rectangle, over a number of connected or randomly distributed rectangles, to the whole screen filled with rectangles. If the user touches a rectangle it disappears and points are added to the user's score depending on his/her speed. As soon as a pattern is completed the next pattern appears. After completing a number of patterns the player is rewarded with a "badge" and advances to the next level with smaller rectangles. After four levels the player advances to the next theme and basically repeats the same procedure at a higher speed and an increased number of patterns. We implemented a global and a local high score list as well as the badges to increase the players' motivation. Players are informed that they will take part in a study when the App is started for the very first time. It is not possible to opt-out without to not play the game or turn off the internet connection.

This study is a controlled observation with a defined task steered by an apparatus. It is intended to record the users' behaviour to derive a model that predicts the touch performance. While playing, the levels including the appearing rectangles are logged. To assess the users' touch behaviour each touch is recorded together with a timestamp and the state of the visible rectangles.

Figure 7. Screenshots of Tap It. The images on the left and in the centre show in-game screenshots with two and four appearing rectangles and the right image shows the global high score list

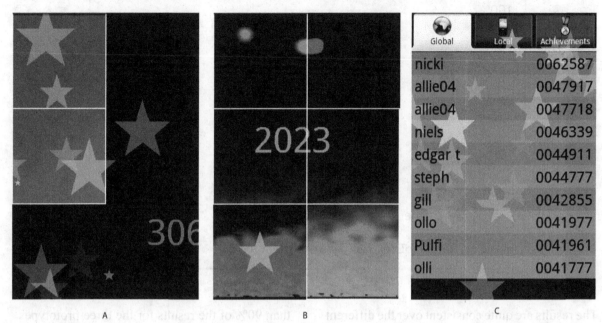

A   B   C

We also collected the user's time zone and the selected locale.

Tap It has been published to the Android Market on July 31, 2010. Till September 29, 2010 the game was installed 8,495 times and we collected data from 6,907 devices resulting in 7,284,263 touch contacts. Not surprisingly we found significant differences between players that use different devices. We also found that the position of the preceding rectangle affects the touched position for the following rectangle. The average touch distribution is skewed towards the previous target and the position of the preceding rectangle affects the error rate as well as the users' speed.

Distribution of Users

To derive general conclusions from user studies that are globally applicable it can be argued that the sample of persons must reflect the whole population. A number of studies investigated the importance of the diversity of participants.

Evers and Day, for example, analyzed the role of culture in interface acceptance (Evers & Day, 1997). They showed that fundamental differences exist between cultures regarding the user interface beyond obvious factors, such as, language and characters. Another example is Simon who showed that gender and culture has an important effect on the perception of web sites (Simon, 2000). Young reviewed the literature that addresses the integration of culture in the design process and points out that "there is room for improvement" (Young, 2008). Our initial expectation (Henze & Boll, 2010; Henze et al., 2010) was that by deploying Apps via a mobile application store we will get access to a worldwide audience (and we are not the only ones Korn, 2010; Morrison & Chalmers, 2010). This would enable to derive conclusions that are not only applicable to a particular country or culture but to the global population.

Analyzing our data we found that the participants are less diverse than we expected. Figure 8 shows which locale the prototypes' users use.

Figure 8. Percentage of users with the respective locales for three of the prototypes (n=15739)

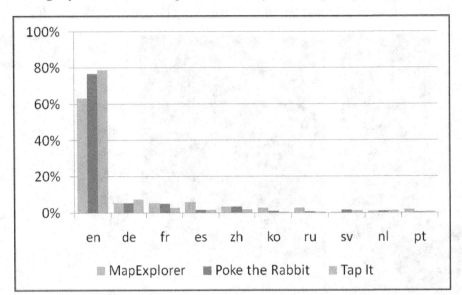

The results are quite consistent over the different prototypes (we only collected the user's locale for three of the prototypes). With 63%-79% English is by far the most common language followed by German (5.42%-7.45%), French (2.72%-5.29%), and Spanish (1.62%-6.04%). The most common non western languages are Chinese (1.93%-3.44%) and Korean (0.32%-2.84%). In general western languages accounted for more than 90% of the results for the three prototypes. This is consistent with the time-zone that users use (Figure 9). 85.89% of all users have an American or European time zone.

Market research shows that 66% of the Android users are in the United States (AdMob, 2010b). Contrary to our data the report from May 2010 says that 13% of the Android users are in China (compared to 1.93%-3.44% for our prototypes).

Figure 9. Fraction of users from different continents based on the respective time zone (n=15608)

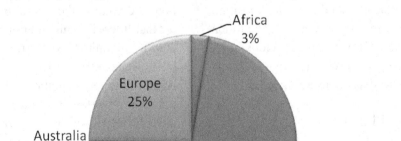

There are a number of potential reasons for this divergence (e.g. our Apps address a particular audience). We, however, assume that the main reason is that we did not translate any of the Apps into Chinese. On the other hand, we internationalized the MapExplorer to German without a noticeable effect. Further looking at market research we see that in January 2010 Android users were mostly male (73%) while other platforms had an almost equal gender split (iPhone: 57% male and webOS: 58% male) (AdMob, 2010b). Market research from Nielsen (Kellogg, 2010) analyzed Smartphone users in the United States and it can be seen that this sample does not even reflect the US population at all. In particular, the average Android users as well as smartphone users in general have a considerably higher income than the average US citizen.

Looking at market research but also by looking at our data, it can be concluded that general claims about participants' diversity are simply misleading. Smartphone users, and in particular Android users, are obviously far from being a perfect sample of the global population. As users have to actively install the respective apparatus themselves this further shifts the sample towards more tech-savvy people. General conclusions about the global population cannot be derived. However, depending on the aims of a study the market population might be more relevant than the overall global population.

Collected Data

When using Apps as an apparatus it is necessary to collect data about the users' performance and behaviour. Compared to lab studies, researchers cannot directly influence the users' behaviour. It cannot be ensured that users perform the tasks as often as desired and in the intended way. Furthermore, in contrast to most lab studies no personalized data is collected (at least this is the case for the Apps discussed in this paper). It might

therefore not even be necessary to ask users to fill an informed consent form an ethical point of view.

Informing Users

On the one hand, researchers want to collect as much data as possible, but on the other hand following the principle of data economy only the necessary data should be collected. Ethical consideration and possibly even legal requirements might also impair the options to inform users. There are no clear guidelines about the way to inform users about the fact that data is collect or if it is necessary to ask if the App is allowed to collect data. Therefore, different techniques have been used by the Apps used for the five studies. Figure 10 shows the ratio of installation (according to the Android Market) to the number of samples received on our server. The fraction of users that contributed data differs dramatically for the five prototypes. While the Apps differ in many aspects the most important factor is certainly the way to inform and ask the user about collecting data.

SINLA uses the most conservative and complicated approach compared to the other Apps. If the App is used for more than five minutes the user is informed that s/he can take part in a study and a new entry is added to the main menu. After selecting the new menu entry, filling an additional form, and finally pressing the send button, data is transmitted to our server. With this approach only 0.46% of the installations resulted in a log file. During the iterative development of the PocketNavigator we tried different techniques. In early versions an opt-in checkbox hidden in an about view was used. In later versions an additional re-occurring popup asks for participation resulting in an average return rate of 7.32%. The MapExplorer asks the user for permission to log data on the App's very first screen using a preselected checkbox (Figure 5). This approach resulted in a return rate of 54.76%. Tap It also informs the users when the App is started for the first time using a popup. There is, however, no option to

Figure 10. Percentage of installations that lead to a data sample

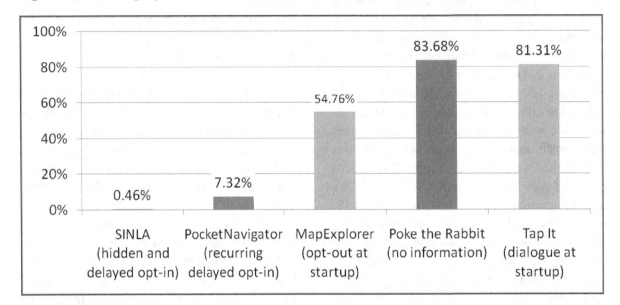

opt-out but quitting the game. This results in a return rate of 81.31%. With 83.68% Poke the Rabbit has the highest return rate. The game never asks or even informs the player. The fraction from the total installation for Poke the Rabbit is thus likely the upper boundary of what amount of data samples can be expected. The missing 16.32% likely consist of persons that only installed the App without ever playing a single level and persons that never played with an active internet connection.

Even though the five Apps differ and probably attract different user groups we assume that the most important difference regarding the number of collected data samples is how the user is informed and asked about data logging and the true purpose of the App. The results of the PocketNavigator and especially SINLA show that hiding the option to opt-in the study dramatically reduces the amount of collected data. Presenting the option to opt-out when the App is started as we did it for the MapExplorer can be a good compromise that reduces the number of data samples only by around 30%. Comparing the results of Poke the Rabbit and Tap It shows that telling users that they

are going to be part of a study without the option to opt-out only marginally reduces the amount of results. Thus, a simple popup at the Apps start is an opportunity to act ethically without losing too much data. Future work should, however, validate our results by comparing different ways to inform the user with prototypes that randomly choose one of the alternatives when starting the App. Thereby, the potential effect of the App on the number of samples could be cancelled out.

Amount of Collected Data

The amount of people that contributed to the results is only one factor that must be considered if looking at the amount of collected data. The second factor is the amount of data collected from the individual users. Depending on the study, data samples from persons that use the App for a short time (e.g. SINLA or the PocketNavigator) can be meaningless. Figure 11 shows the fraction of users that played a certain number of levels for Poke the Rabbit and Tap It. For the MapExplorer the Figure shows the time users spend with the App in seconds/20. In all three studies most users or players are engaged in the App for only a short time. 50% of the Tap It players played seven levels or less and 46.02% played only one or two levels of Poke the Rabbit. However, on 115 devices (1.67%) more than 100 levels of Tap It where played and on nine devices (0.18%) someone played more than 100 levels of Poke the Rabbit. 40.39% of the persons that used the MapExplorer used it for less than 60 seconds and only nine (0.21%) used it for more than one hour.

For the three analyzed studies the majority of the users only tested the Apps for a short time. This prevents from analysing which off-screen visualization users prefer in the long-run using the data collected by the MapExplorer. The two games (and the MapExplorer's tutorial) on the other hand are designed in a way that even short term users contribute meaningful results. Especially with Tap It even the very first played level provides relevant data. Thus, for studies that do not need long-term user involvement it is important to collect results as early as possible and design the App accordingly. Thereby, it is feasible

to collect the results before users even notice that they do not like or do not want this particular App. If long-term involvement is needed for a study it is necessary to increase the total number of users (e.g. by addressing multiple platforms) and/or increasing the number of long-term users (by increasing the quality and usefulness of the App). Both involve putting major effort in the development of an App and it might be considered to use conventional studies instead.

Qualitative Feedback

In their textbook Cooper, Reiman and Cronin claim that understanding the user "cannot be achieved by digging through the piles of numbers that come from quantitative study" (Cooper et al., 2007). While quantitative data collected in experiments is used to identify that a cause results in an effect, qualitative data can help to understand why the cause results in the effect. When conducting experiments in the lab it is common

Figure 11. Amount of feedback received from users of the respective Apps. We normalized the amount of feedback for comparison purpose. For Poke the Rabbit and Tap It the x-axis represents the number of played levels and for the MapExplorer the x-axis is the time in seconds divided by 20

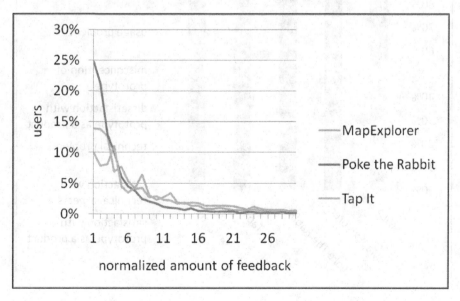

to collect qualitative data either during or after the experiment.

In order to collect qualitative data we used four different approaches to receive feedback. The used feedback channels are comments from the Android Market, a feedback from inside the MapExplorer, providing email contact information in the market description and submission forms on the PocketNavigator's website. The key findings are summarized in the following two subsections.

Comments from the Market

The Android Market allows users to write comments for installed Apps by filling a form provided in the Android Market. Users can also rate installed Apps on a 5-point scale. We collected comments and ratings for the five Apps and clustered them into the following seven categories shown in Figure 12: *nonsense, usage problem, misconception of prototype, dissatisfaction with prototype as a product, technical problem, satisfaction with*

technical aspects and *satisfaction with prototype as a product*. If a comment contains ambiguous statements we decided upon the rating to which category the comment belongs.

In total only 0.4% to 0.8% of the users who ever installed an application also rated it and from these users only 16%-51% left also a comment. In general, most users rated the Apps as real products. Therefore, they report technical problems but provided little insight for the addressed research. It is noticeable that many users (36%-66%) that commented on an App reported technical incompatibilities with their Android phone or with a particular Android version. This shows how difficult it is to implement an App that runs on a variety of devices and different versions of a platform (e.g. four widespread Android versions). No comment mentioned privacy issues or worried about data protection. There is also no comment that provides insight in the addressed research questions. Only one comment reveals that the user is aware that Tap It is used for a study. A likely

Figure 12. Distribution of clustered comments among the prototypes

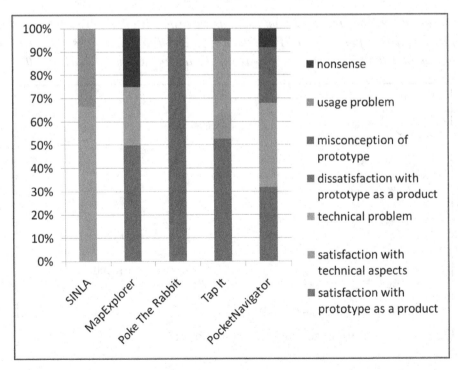

reason that we did not collect useful comments for our research is that information about the addressed questions is not prominently displayed inside the Apps. A lack of interest in the scientific background could be another reason. It can, however, also be argued that comments from the Market are not the adequate tool to collect qualitative feedback.

Feedback from Other Channels

A constraint to publish applications in the Android Market is to supply an email address as a contact option for users. Thus, user of all Apps can send comments and requests via mail. However, we received only emails for the PocketNavigator and besides SPAM all mails are feature requests.

Another concept for collecting feedback was used by the MapExplorer that contains a simple feedback form. In total 67 comments were collected, however, the results are as useful as the Android Market comments without any valuable feedback. Nevertheless, we were surprised what people entered into the feedback form. The clustered results are shown in Figure 13. Besides a

high amount of nonsense comments, 25% percent submitted a name or their address although they have never been asked for it.

Instead of using a form inside the application, the PocketNavigator provides a form on the projects website. People are encouraged to visit the site by a description and a button inside the App that opens the phone's browser. In total 22 comments were logged but half of them were advertisements and the others were multiple submissions of redundant comments that are identical to comments from the Android Market.

All in all, four different feedback channels were used to collect qualitative feedback from users. However, none of these channels provided useful experiment-specific feedback. Most of the feedback contains either real garbage which does not fit at all to the App or assessed the prototypes as real products.

It can be concluded that qualitative feedback does not come for free and none of the obvious options provides valuable information per se. A viable approach described by McMillan et al. is rewarding users for providing feedback using in-game badges or bonus (McMillan et al., 2010).

Figure 13. Distribution of clustered comments of the MapExplorer's feedback form

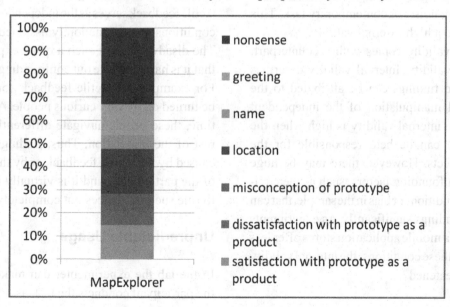

They also describe that it is feasible to get in direct contact with participants, e.g. using Facebook and providing vouchers for participation in phone interviews. If, however, in-deep qualitative feedback is needed it should be considered if lab studies can not only provide richer feedback but are also more cost efficient.

Validity

When experiments are conducted in the lab the number of participants is often limited and the tasks are artificial. This can make it difficult to generalize the findings to other populations or situations other than the artificial tasks. Since applications evaluated in mobile application stores reach a large number of users that use the App in "natural" settings, findings in the study can be generalized more confidently. How well findings can be generalized is expressed by the term external validity. The external validity is threatened, when findings have only been observed in very controlled settings as in a lab study. The mobile HCI community, for example, usually conducts studies in the lab even though a mobile or natural context would influence the outcome (Kjeldskov & Graham, 2003). Using applications in "natural" settings means highly varying conditions, such as time, location, noise, or contemporary task. This contributes to a high external validity.

External validity comes with a counterpart, the internal validity. Internal validity describes how well the findings can be attributed to the experimental manipulation of the independent variable. The internal validity is high when the manipulation can be held responsible for the observed effects. However, there may be huge wealth of confounding factors, such as the environmental conditions or bias in the sample, that can distort the findings significantly. As experiments conducted via mobile application stores offer less control over these conditions, the internal validity is highly threatened.

For the conducted studies we found that the design of the experiment and the unpredictability of the usage are the two biggest threats to the internal validity.

Experiment vs. Quasi-Experiment

In a true experiment, conditions are randomized. This means, that the participants are assigned to a condition randomly rather than choosing one. For an application, this means that it would switch between the conditions automatically. This, however, requires defying the participants' control over which condition is active. In some cases, such as Poke the Rabbit and MapExplorer, where the condition changes as part of the game or the tutorial, this will not bother the user. If the application is however meant for productive use, such as SINLA and the PocketNavigator, it is difficult to force the user into a certain mode, and would surely cause many users to uninstall the application.

Consequently, some experiments conducted via mobile application stores cannot be designed as true experiments, but have to be quasi-experiments. In the case of the PocketNavigator, for example, people are allowed to turn the tactile feedback on and off whenever they want. As the tactile feedback serves as the independent variable, conditions are not randomly assigned anymore. The disadvantage of such a quasi-experiment is that it is harder to rule out confounding variables. For example, the tactile feedback could mostly be turned on by very curious people. At the same time, these people navigate differently than the rest of the population. Thus, findings could be caused by the tactile feedback or by the curiosity of the participants, and it is virtually impossible to rule such influences out completely.

Unpredictable Usage

In the lab the experimenter can make sure that the participant conducts the task as scheduled. If conducting experiments in an application store, the experimenter has no control over how the task is conducted.

When the experimenter designs the experiment, s/he often has a certain usage pattern in mind that is being tested. With respect to the games Poke the Rabbit and Tap It, the designer anticipates that the user plays the game. For the PocketNavigator, we anticipated that it would be used as route guidance by pedestrians.

However, the unsupervised use of the apps offers many opportunities for unforeseen usage. For example, user may get interrupted while playing the game and they may not shut the game down before putting the mobile device aside. The application still assumes the experiment is running. If such events are not detected, the experimenter might infer that some users had severe difficulties in performing the given tasks.

In the MapExplorer example, we observed an increase in the usage time in one condition. However, this does not necessarily mean that the participants had difficulties using it, but they also could have been fascinated by the visualization and tested it out.

In the case of the PocketNavigator we envisioned the PocketNavigator to be a pedestrian route guidance system and wanted to study its use as such. However, from the log data we learned that the majority of the people never compute a route, so it is not used as a route guidance system at all. Other data shows travel speeds beyond 50 km/h, which is the speed that cars usually travel outside of the city. So the application was not used by a pedestrian but in the car. Our approach was to discount all data where the user had not computed a route and was not moving with walking speed. Nevertheless, it is impossible to be sure that no unforeseen usage is biasing the data.

This bias is a huge threat to the internal validity. In general, an experimenter that chooses to conduct experiments in an application store has to analyze the data very carefully for such unforeseen usage and process it accordingly. All conclusions drawn from the findings should clearly highlight this limitation.

Ethical Considerations

As for every conducted user study, legal regulations and ethics need to be considered. However, there are several ways how ethics can be approached, e.g. through official regulations, law or personal experience. While official regulations are not yet explicitly covering studies through mobile phones in the large scale, they look quite defensive and do not leave much room for interpretation.

To create a comprehensive data protection system throughout Europe, the Organisation of Economic Cooperation and Development (OECD) developed the "Guidelines on the Protection of Privacy and Transborder Flows of Personal Data" with the aim to protect the individuals' privacy (Organisation for Economic Co-operation and Development, 2002). Seven principles evolved and are incorporated into an EU directive (Directive 95/46/EC). Each of these principles addresses the collection of personally identifiable information. A subject should be given notice that data is being collected. The collected data should only be used for a specific purpose stated to the subject. Each subject has to give consent for the data collection. The collected data should be kept secure. The participants should exactly know, who is collecting the data (disclosure). Furthermore, the subjects should be able to access their own data and make corrections to inaccurate data. Finally each subject should have a method available to hold the data collector accountable for these principles.

Some information that is stored or is available on a mobile phone is definitely considered as personally identifiable information. A prominent example is the phone number or the phone's international mobile equipment identity (IMEI). Both are unique and can probably be tracked to a personal identity. This data should obviously not be accessed or logged in any way without asking the user for permission. A user's location is something which can be traced to particular buildings, which then might be identified as work place or home. The addresses can be looked up in address books and identify a user as a specific person. Therefore, none of our applications log any exact location information.

Most of the information we observe is either already non-person specific (e.g. the touch behaviour) or the information is abstracted to a level that an identification of a specific person is not possible (e.g. loudness instead of complete audio stream). With respect to the users, some of our applications explicitly consider some of the principles from the EU directive and notice the participants about the logging or even request a consent before the observation starts. A good trade-off between a satisfying return rate and a reasonable consent from the participant is to inform the user about the logging, while he agrees implicitly by continuing using the application. As no information that allows the identification of subjects is collected for the studies described in this paper we assume that the Apps comply with EU regulations.

CONCLUSION AND FUTURE WORK

In this paper, we reported from five experiments we conducted by publishing Apps in the Android Market. We showed, that it is possible to acquire a large number of users, e.g. in the case of Tap It we obtained nearly 7,000 participants contributing over 7,000,000 data points in two months only. The Apps enabled us to conduct experiments with real users in real usage situations without the huge effort that conventional methods would require. Two of the studies already yielded results that have been accepted by peer-reviewed HCI conferences (Henze & Boll, 2010; Henze et al., 2010). Furthermore, the five case studies allowed us to get general insight, identify challenges and

provide guidelines for further experiments using mobile application stores.

Our first insight is that the participants we obtained are mostly English-speaking users from the United States. The findings obtained by such a sample cannot be generalized to e.g. the world's population. Thus, when interpreting the data, researchers either need to highlight that the findings are only valid for this certain population, or they have to draw a more representative sub-sample. Furthermore, testing for differences between the different user populations allows checking whether cultural differences exist.

Secondly, when users have to opt-out of the study instead of allowing them to opt-in, a much larger fraction of the users actually takes part in the study. In e.g. the case of the PocketNavigator the opt-in method lead to a hardly useful small number of participants. However, the challenge remains to act legally and ethically, while still attracting as much participants as possible. The most successful approaches are to inform the user about the study on the first start of the application and offer the possibility to opt-out.

Another finding we made is that many users may use the applications only for a short period of time. This might be difficult if the study intends to study experienced users and not beginners. It therefore is important to collect data as early as possible and to motivate long-term use e.g. by providing badges, high-scores, or similar. Furthermore, the App has to offer a user experience that comes close to commercial products to promote extended use.

The qualitative feedback we received from the users was largely useless. The comments from the market were mostly complaints about errors and the application's usability. Mails mostly requested new features. Still, qualitative feedback is the key to understand the collected data. McMillan et al. (McMillan et al., 2010) tested two promising approaches by providing incentives, such as badges or achievements, for giving qualitative feedback and tried to contact users directly.

During the analysis of our data we found plenty of examples where Apps are used in unforeseen ways, such as the user leaving the device lying on the table with the App still running. Thus, it has to be expected that results always contain artefacts or noise, which makes it difficult to obtain valid results from the data. Researchers therefore have to filter the data before analysing it. Furthermore, it should be considered to design the App in a way that makes unforeseen usage less likely, e.g. by providing incentives for the desired usage.

As a bottom line, we found that experiments in the market allow to gain insight that otherwise would only be possible to obtain with an enormous amount of effort. However, our case studies also illustrated that such studies are not automatically successful. Future work should try to uncover pitfalls and establish further guidelines for conducting experiments in mobile application stores successfully. Furthermore, it has to be investigated, how the results obtained by experiments in application stores related to those results obtained by lab studies.

ACKNOWLEDGMENT

This paper is partially supported by the European Commission within the projects InterMedia (FP6-038419) and HaptiMap (FP7-224675). We thank our colleagues for their support and we thank the anonymous users for using our Apps.

REFERENCES

AdMob. (2010a). *AdMob mobile metrics: App usage survey.* Retrieved from http://metrics.admob.com/wp-content/uploads/2010/02/AdMob-Mobile-Metrics-Jan-10.pdf

AdMob. (2010b). *AdMob mobile metrics: Metrics highlights.* Retrieved from http://metrics.admob.com/wp-content/uploads/2010/06/May-2010-AdMob-Mobile-Metrics-Highlights.pdf

Baudisch, P., & Rosenholtz, R. (2003). Halo: A technique for visualizing off-screen objects. In *Proceedings of the SIGCHI Conference on Human Factors in Computing Systems* (pp. 481-488).

Burigat, S., Chittaro, L., & Gabrielli, S. (2006). Visualizing locations of off-screen objects on mobile devices: A comparative evaluation of three approaches. In *Proceedings of the 8th Conference on Human-Computer Interaction with Mobile Devices and Services* (pp. 239-246).

Cooper, A., Reimann, R., & Cronin, D. (2007). *About face 3: The essentials of interaction design.* New Delhi, India: Wiley Publications.

Danziger, K., & Ballantyne, P. (1997). Psychological experiments . In Bringmann, W. G., Luck, H. E., Miller, R., & Early, C. E. (Eds.), *A pictorial history of psychology* (pp. 233–239). Hanover Park, IL: Quintessence Publishing.

Evers, V., & Day, D. (1997). The role of culture in interface acceptance. In *Proceedings of the IFIP TC13 Interantional Conference on Human-Computer Interaction* (pp. 260-267).

Gilbertson, P., Coulton, P., Chehimi, F., & Vajk, T. (2008). Using "tilt" as an interface to control "no-button" 3-D mobile games. *Computers in Entertainment*, 6(3), 1–13. doi:10.1145/1394021.1394031

Girardello, A. (2010). AppAware: Serendipity in mobile applications. In *Proceedings of the Conference on Human-Computer Interaction with Mobile Devices and Services* (pp. 479-480).

Henze, N., & Boll, S. (2010). Push the study to the app store: Evaluating off-screen visualizations for maps in the android market. In *Proceedings of the Conference on Human-Computer Interaction with Mobile Devices and Services* (pp. 373-374).

Henze, N., Poppinga, B., & Boll, S. (2010). Experiments in the wild: Public evaluation of off-screen visualizations in the android market. In *Proceedings of the 6th Nordic Conference on Human-Computer Interaction: Extending Boundaries* (pp. 675-678).

Kellogg, D. (2010). *iPhone vs. Android.* Retrieved from http://blog.nielsen.com/nielsenwire/online_mobile/iphone-vs-android

Kjeldskov, J., & Graham, C. (2003). A review of mobile HCI research methods. In *Proceedings of the Conference on Human-Computer Interaction with Mobile Devices and Services* (pp. 317-335).

Korn, M. (2010). Understanding use situated in real-world mobile contexts. In *Proceedings of the Workshop on Research in the Large.*

McMillan, D., Morrison, A., Brown, O., Hall, M., & Chalmers, M. (2010). Further into the wild: Running worldwide trials of mobile systems. In P. Floréen, A. Krüger, & M. Spasojevic (Eds.), *Proceedings of the 8th International Conference on Pervasive Computing* (LNCS 6030, pp. 210-227).

Michahelles, F. (2010). Getting closer to reality by evaluating released apps? In *Proceedings of the Workshop on Research in the Large.*

Morrison, A., & Chalmers, M. (2010). SGVis: Analysis of mass participation trial data. In *Proceedings of the Workshop on Research in the Large.*

Morrison, A., Reeves, S., McMillan, D., & Chalmers, M. (2010). Experiences of mass participation in Ubicomp research. In *Proceedings of the Workshop on Research in the Large.*

Organisation for Economic Co-operation and Development. (2002). *OECD guidelines on the protection of privacy and transborder flows of personal data.* Paris, France: OECD.

Park, Y., Han, S., Park, J., & Cho, Y. (2008). Touch key design for target selection on a mobile phone. In *Proceedings of the Conference on Human-Computer Interaction with Mobile Devices and Services* (pp. 423-426).

Pielot, M., & Boll, S. (2010). Tactile wayfinder: Comparison of tactile waypoint navigation with commercial pedestrian navigation systems. In P. Floréen, A. Krüger, & M. Spasojevic (Eds.), *Proceedings of the 8th International Conference on Pervasive Computing* (LNCS 6030, pp. 76-93).

Pielot, M., Poppinga, B., & Boll, S. (2010). PocketNavigator: Vibro-tactile waypoint navigation for everyday mobile devices. In *Proceedings of the Conference on Human-Computer Interaction with Mobile Devices and Services* (pp. 423-426).

Schinke, T., Henze, N., & Boll, S. (2010). Visualization of off-screen objects in mobile augmented reality. In *Proceedings of the Conference on Human-Computer Interaction with Mobile Devices and Services* (pp. 313-316).

Simon, S. (2000). The impact of culture and gender on web sites: An empirical study. *ACM SIGMIS Database, 32*(1), 18–37. doi:10.1145/506740.506744

van Erp, J. B. F., van Veen, H. A. H. C., Jansen, C., & Dobbins, T. (2005). Waypoint navigation with a vibrotactile waist belt. *ACM Transactions on Applied Perception, 2*(2), 106–117. doi:10.1145/1060581.1060585

Young, P. (2008). Integrating culture in the design of ICTs. *British Journal of Educational Technology, 39*(1), 6–17.

Zellweger, P. T., Mackinlay, J. D., Good, L., Stefik, M., & Baudisch, P. (2003). City lights: Contextual views in minimal space. In *Proceedings of the SIGCHI Conference on Human Factors in Computing Systems* (pp. 838-839).

Zhai, S., Kristensson, P., Gong, P., Greiner, M., Peng, S., Liu, L., & Dunnigan, A. (2009). Shapewriter on the iPhone: From the laboratory to the real world. In *Proceedings of the SIGCHI Conference on Human Factors in Computing Systems* (pp. 2667-2670).

This work was previously published in the International Journal of Mobile Human Computer Interaction, Volume 3, Issue 4, edited by Joanna Lumsden, pp.71-91, copyright 2011 by IGI Publishing (an imprint of IGI Global).

Compilation of References

Abdul-Rahman, A., & Hailes, S. (1997). A distributed trust model. In *Proceedings of the Workshop on New Security Paradigms* (pp. 48-60).

Abowd, G., & Mynatt, E. (2000). Charting past, present, and future research in ubiquitous computing. *Transactions on Computer-Human Interaction, 7*(1).

Abu-Nasr, D. (2005, August). *Bluetooth takes a bite out of Saudi anti-flirting rules.* Retrieved from http://findarticles.com/p/articles/mi_qn4188/is_20050812/ai_n14886549

acbPocketSoft (n. d.). *acbTaskMan for PocketPC.* Retrieved from http://www.acbpocketsoft.com

Acquaviva, A., Lattanzi, E., & Bogliolo, A. (2004). Power-aware network swapping for wireless palmtop PCs. *IEEE Transactions on Mobile Computing, 5*(5), 571–582. doi:10.1109/TMC.2006.71

AdMob Mobile Metrics. (2010). *Metric highlights.* Retrieved from http://metrics.admob.com/wp-content/uploads/2010/06/May-2010-AdMob-Mobile-Metrics-Highlights.pdf

AdMob. (2010a). *AdMob mobile metrics: App usage survey.* Retrieved from http://metrics.admob.com/wp-content/uploads/2010/02/AdMob-Mobile-Metrics-Jan-10.pdf

AdMob. (2010b). *AdMob mobile metrics: Metrics highlights.* Retrieved from http://metrics.admob.com/wp-content/uploads/2010/06/May-2010-AdMob-Mobile-Metrics-Highlights.pdf

Adomavicius, G., Sankaranarayanan, R., Sen, S., & Tuzhilin, A. (2005). Incorporating contextual information in recommender systems using a multidimensional approach. *ACM Transactions on Information Systems, 23*, 103–145. doi:10.1145/1055709.1055714

Adomavicius, G., & Tuzhilin, A. (2005). Toward the next generation of recommender systems: A survey of the state-of-the-art and possible extensions. *IEEE Transactions on Knowledge and Data Engineering, 17*(6), 734–749. doi:10.1109/TKDE.2005.99

Agarwal, S. K., Chakraborty, D., Kumar, A., Nanavati, A. A., & Rajput, N. (2007). HSTP: Hyperspeech transfer protocol. In *Proceedings of the Eighteenth Conference on Hypertext and Hypermedia* (pp. 67-76). New York, NY: ACM Press.

Agarwal, S. K., Kumar, A., Nanavati, A., & Rajput, N. (2008). The world wide telecom web browser. In *Proceeding of the 17th International Conference on World Wide Web* (pp. 1121-1122). New York, NY: ACM Press.

Agarwal, S., Kumar, A., Nanavati, A. A., & Rajput, N. (2008, April 21-25). The world wide telecom web browser. In *Proceedings of the 17th international Conference on World Wide Web (WWW '08)*, Beijing, China (pp. 1121-1122). New York: ACM. Retrieved from http://doi.acm.org/10.1145/1367497.1367686

Agrocom. (2008). *Agrocom software technologies private limited - achievements.* Retrieved from http://agrocom.co.in/achievements.php

Albrecht, I., Haber, J., & Seidel, H. (2002). Speech synchronization for physicsbased facial animation. In *Proceedings of the 10th International Conference on Computer Graphics, Visualization, and Computer Vision* (pp. 9-16).

Alexa, M., Berner, U., Hellenschmidt, M., & Rieger, T. (2001). An animation system for user interface agents. In *Proceedings of the 6th International European Conference on Computer Graphics, Visualization, and Computer Vision.*

Alexa. (n.d.). *The Web Information Company*. Retrieved August 8, 2008, from http://www.alexa.com/site/ds/top_sites?ts_mode=lang&lang=en

Allen, G. L. (1999). Cognitive abilities in the service of wayfinding: A functional approach. *The Professional Geographer, 51*(4), 554–561. doi:10.1111/0033-0124.00192

Amin, A., Townsend, S., et al. (2009). Fancy a drink in Canary Wharf?: A user study on location-based mobile search. In *Proceedings of the Human-Computer Interaction (INTERACT 2009)*, Uppsala, Sweden (pp. 736-749).

Angell, L., Auflick, J., Austria, P. A., Kochhar, D., Tijerina, L., & Biever, W. (2006). *Driver workload metrics task 2 final report DOT HS 810 635*. Washington, DC: National Highway Traffic Safety Administration.

Anonymous. (2009, February 18). '*AppStore secrets*' - *Pinch media blog*. Retrieved from http://www.pinchmedia.com/blog/appstore-secrets

Anurag, P., Vivek, B., & Sasank, T. (2008). *CERES information services*. Retrieved from http://ceres.co.in

Aoki, P. M., Honicky, R. J., Mainwaring, A., Myers, C., Paulos, E., Subramanian, S., & Woodruff, A. (2009). A vehicle for research: Using street sweepers to explore the landscape of environmental community action. In *Proceedings of the 27th International Conference on Human Factors in Computing Systems* (pp. 375-384).

App Store Metrics. (2011, January 15). *iPhone development news and information for the community, by the community*. Retrieved from http://148apps.biz/app-store-metrics/

Appelt, D., & Martin, D. (1999). Named entity extraction from speech: Approach and results using the textpro system. In *Proceedings of the DARPA Broadcast News Workshop* (pp. 51-54).

Apple Inc. (2011, January 23). *Apple – iTunes – 10 billion app countdown*. Retrieved from http://www.apple.com/itunes/10-billion-app-countdown/

Apple. (2010). *App store review guidelines*. Retrieved from https://developer.apple.com/appstore/resources/approval/guidelines.html

Aptana. (n.d.). *Apple iPhone Emulator*. Retrieved July 10, 2008, from http://www.aptana.com/iphone/

Ark, W. S., & Selker, T. (1999). A look at human interaction with pervasive computers. *IBM Systems Journal, 38*(4), 504–507. doi:10.1147/sj.384.0504

Arter, D., Buchanan, G., Jones, M., & Harper, R. (2007, September 9-12). Incidental information and mobile search. In *Proceedings of the 9th international Conference on Human Computer interaction with Mobile Devices and Services (MobileHCI '07)*, Singapore (Vol. 309, pp. 413-420). New York: ACM. Retrieved from http://doi.acm.org/10.1145/1377999.1378047

Ault, A., Krogmeier, J. V., Dunlop, S. R., & Coyle, E. J. (2008). eStadium: The mobile wireless football experience. In *Proceedings of the Third International Conference on Internet and Web Applications and Services* (pp. 644-649).

Awasthi, U. (2008). *IFFCO Kisan Sanchar Ltd*. Retrieved from http://www.iffco.nic.in/applications/iffcowebr5.nsf/?Open

B&NES Council. (2008). *Bath in focus*. Retrieved from http://www.business-matters.biz/site.aspx?i=pg64

Baeza-Yates, R., Dupret, G., & Velasco, J. (2007). A study of mobile search queries in Japan. In *Proceedings of the WWW 2007 Workshop on Query Log Analysis: Social and Technological Analysis*. Retrieved July 8, 2010, from http://querylogs2007.webir.org/program.htm

Bahuman, A., & Kirthi, R. (2007). *aAqua mini*. Retrieved from www.agrocom.co.in

Bahuman, A., & Ramamritham, K. (2010). *aAQUA mobile - almost all questions answered*. Retrieved from http://www.slideshare.net/bahuman/aaqua-mobile-pilot-to-advise-50000-farmers-over-the-telephone

Baillie, L., Frohlich, P., & Schatz, R. (2007). Exploring social TV. In *Proceedings of the International Conference on Information Technology Interfaces* (pp. 215-220).

Balabanovič, M., & Shoham, Y. (1997). Content-based, collaborative recommendation. *Communications of the ACM, 40*(3), 66–72. doi:10.1145/245108.245124

Balci, K. (2005). Xface: Open source toolkit for creating 3d faces of an embodied conversational agent. In *Proceedings of the International Conference on Smart Graphics* (pp. 263-266).

Balci, K., Not, E., Zancanaro, M., & Pianesi, F. (2007). Xface: Open source project and smil-agent scripting language for creating and animating embodied conversational agents. In *Proceedings of the 15th International Conference on Multimedia* (pp. 1013-1016). New York, NY: ACM Press.

Bamford, W. (2007). *Bombus: One button mobile game.* Retrieved from http://www.youtube.com/watch?v=aVQIUtLzZAg

Bamford, W. (2008). *Boom! A mobile phone game for WidSets.* Retrieved from http://www.youtube.com/watch?v=E5V3L_sVtF8

Banks, K., & Hersman, E. (2009). FrontlineSMS and ushahidi - a demo. In *Proceedings of the 3rd International Conference on Information and Communication Technologies and Development*, Doha, Qatar (pp. 484-484).

Bardzell, J., Bardzell, S., Pace, T., & Karnell, J. (2008). Making user engagement visible: A multimodal strategy for interactive media experience research. In *Proceedings of Extended Abstracts on Human Factors in Computing Systems* (pp. 3663–3668). New York, NY: ACM Press.

Barker, E., Polifroni, J., Walker, M., & Gaizauskas, R. (2009). *Angle-seeking as a scenario for task-based evaluation of information access technology.* Paper presented at the International Workshop on Intelligent User Interfaces, Sanibel Island, FL.

Barker, R., & Molle, F. (2004). *Evolution of irrigation in South and Southeast Asia.* Colombo, Sri Lanka: Comprehensive Assessment Secretariat.

Barnard, E., Davel, M., & van Heerden, C. (2009). ASR corpus design for resource-scarce languages. In *Proceedings of the 10th Annual Conference of the International Speech Communication Association* (pp. 2847-2850).

Barnard, E., Davel, M., & van Huyssteen, G. (2010). Speech technology for information access: A South African case study. In *Proceedings of the AAAI Symposium on Artificial Intelligence*, Palo Alto, CA (pp. 8-13).

Barnard, L., & Yi, J. S. (2005). An empirical comparison of use-in-motion evaluation scenarios for mobile computing devices. *International Journal of Human-Computer Studies, 62*(4), 487–520. doi:10.1016/j.ijhcs.2004.12.002

Basu, P., & Srivastava, P. (2005). *Scaling-up microfinance for India's rural poor.* Washington, DC: World Bank. doi:10.1596/1813-9450-3646

Baudisch, P., & Rosenholtz, R. (2003). Halo: A technique for visualizing off-screen objects. In *Proceedings of the SIGCHI Conference on Human Factors in Computing Systems* (pp. 481-488).

Beatty, P., Reay, I., Dick, S., & Miller, J. (2007). P3P Adoption on E-Commerce Websites: A Survey & Analysis. *IEEE Internet Computing, 11*(2), 65–71. doi:10.1109/MIC.2007.45

Béchet, F., Gorin, A. L., Wright, J. H., & Hakkani-Tür, D. (2004). Detecting and extracting named entities from spontaneous speech in a mixed initiative spoken dialogue context: How may I help you? *Speech Communication, 42*(2), 207–225. doi:10.1016/j.specom.2003.07.003

Becker, R. A., & Cleveland, W. S. (1987). Brushing scatterplots. *Technometrics, 29*, 127–142. doi:10.2307/1269768

Belkin, N., Cool, C., Stein, A., & Thiel, U. (1995). Cases, scripts, and information seeking strategies: On the design of interactive information retrieval systems. *Expert Systems with Applications, 9*(3), 379–395. doi:10.1016/0957-4174(95)00011-W

Bell, M., Chalmers, M., Barkhuus, L., Hall, M., Sherwood, S., Tennent, P., et al. (2006). Interweaving mobile games with everyday life. In *Proceedings of the SIGCHI Conference on Human Factors in Computing Systems* (pp. 417-426).

Bellotti, V., & Edwards, K. (2001). Intelligibility and accountability: Human considerations in context-aware systems. *Human-Computer Interaction, 16*(2), 193–212. doi:10.1207/S15327051HCI16234_05

Benford, S., Rowland, D., Flintham, M., Drozd, A., Hull, R., Reid, J., et al. (2005). Life on the edge: Supporting collaboration in location-based experiences. In *Proceedings of the SIGCHI Conference on Human Factors in Computing Systems* (pp. 721-730). New York, NY: ACM Press.

Bentley, F., & Groble, M. (2009). TuVista: Meeting the multimedia needs of mobile sports fans. In *Proceedings of the 17th ACM International Conference on Multimedia* (pp. 471-480).

Bentley, F., Kaushik, P., Narasimhan, N., & Dhiraj, A. (2006). Ambient mobile communications. In *Proceedings of the ACM SIGCHI Conference on Human Factors in Computing Systems* (pp. 1-3).

Bentley, J. W. (1994). Facts, fantasies, and failures of farmer participatory research. *Agriculture and Human Values, 11*(2), 140–150. doi:10.1007/BF01530454

Bernhaupt, R., IJsselsteijn, W., Mueller, F., Tscheligi, M., & Wixon, D. R. (2008). Evaluating user experiences in games. In *Proceedings of Extended Abstracts on Human Factors in Computing Systems* (pp. 3905–3908). New York, NY: ACM Press.

Bhise, V. D., Forbes, L. M., & Farber, E. I. (1986). *Driver Behavioral Data and Considerations in Evaluating In-vehicle Controls and Displays*. Paper presented at the Transportation Research Board, National Academy of Sciences, 65th Annual Meeting, Washington, DC.

Bickerton, D. (1981). *Roots of language*. Ann Arbor, MI: Karoma Publishers.

Biggs, S. D. (1989). Resource-poor farmer participation in research: A synthesis of experiences from nine national agricultural research systems. *OFCOR Comparative Study Paper, 3*, 1–4.

Black, A., & Lenzo, K. (2001). Flite: A small fast run-time synthesis engine. In *Proceedings of the 4th ISCA Tutorial and Research Workshop on Speech Synthesis* (pp. 20-24).

Blattman, C., Jensen, R., & Roman, R. (2003). Assessing the need and potential of community networking for development in rural India. *The Information Society, 19*(5), 349–364. doi:10.1080/714044683

Bontcheva, K., Tablan, V., Maynard, D., & Cunningham, H. (2004). Evolving GATE to meet new challenges in language engineering. *Natural Language Engineering, 10*(3-4), 349–373. doi:10.1017/S1351324904003468

Bormuth, J. R. (1966). Readability: A new approach. *Reading Research Quarterly, 1*, 79–132. doi:10.2307/747021

Bowman, D., Coquillart, S., Froehlich, B., Hirose, M., Kitamura, Y., & Kiyokawa, K. (2008). 3D user interfaces: New directions and perspectives. *IEEE Computer Graphics and Applications, 28*(6), 20–36. doi:10.1109/MCG.2008.109

Branavan, S. R. K., Chen, H., Eisenstein, J., & Barzilay, R. (2009). Learning document-level semantic properties from free-text annotations. *Journal of Artificial Intelligence Research, 34*(1), 569–603.

Brewster, S., & Brown, L. M. (2004) Tactons: Structured tactile messages for non-visual information display. In *Proceedings of the Fifth Conference on Australasian User Interface,* Darlinghurst, Australia (pp. 15-23).

Brewster, S. (2002). Overcoming the lack of screen space on mobile computers. *Personal and Ubiquitous Computing, 6*(3), 188–205. doi:10.1007/s007790200019

Brignull, H., & Rogers, Y. (2003). Enticing people to interact with large public displays in public places. In *Proceedings of INTERACT* (pp. 17-24).

Brown, L. M., Brewster, S. A., & Purchase, H. C. (2006). Multidimensional tactons for non-visual information display in mobile devices. In *Proceedings of the 8th International Symposium on Human Computer Interaction with Mobile Devices and Services* (pp. 231-238). New York, NY: ACM Press.

Bulling, A., Roggen, D., & Tröster, G. (2009). Wearable eog goggles: Eye-based interaction in everyday environments. In *Proceedings of the 27th International Conference on Human Factors in Computing Systems* (pp. 3259-3264). New York, NY: ACM Press.

Burigat, S., Chittaro, L., & Gabrielli, S. (2006). Visualizing locations of off-screen objects on mobile devices: A comparative evaluation of three approaches. In *Proceedings of the 8th Conference on Human-Computer Interaction with Mobile Devices and Services* (pp. 239-246).

Burnett, G. E. (1998). *'Turn right at the King's Head' driver's requirements for route guidance information*. Unpublished doctoral dissertation, Loughborough University, Leicestershire, UK.

Burnett, G. E., Smith, D., & May, A. (2001). Supporting the navigation task: Characteristics of 'good' landmarks. In Hanson, M. A. (Ed.), *The annual conference of the ergonomics society* (pp. 441–446). London, UK: Taylor & Francis.

Burrell, J., & Gay, G. K. (2002). E-graffiti: Evaluating real-world use of a context-aware system. *Interacting with Computers, 14*(4), 301–312. doi:10.1016/S0953-5438(02)00010-3

Campbell, A. T., Eisenman, S. B., Fodor, K., Lane, N. D., Lu, H., Miluzzo, E., et al. (2008). Transforming the social networking experience with sensing presence from mobile phones. In *Proceedings of the 6th ACM Conference on Embedded Network Sensor Systems*, Raleigh, NC (pp. 367-368).

Card, S. K., Mackinlay, J. D., & Shneiderman, B. (1999). *Readings in information visualization: Using vision to think.* San Diego, CA: Academic Press.

Carenini, G., & Moore, J. D. (2001). A strategy for evaluating generative arguments. In *Proceedings of the First International Conference on Natural Language Generation* (pp. 1307–1314).

Carenini, G., & Rizoli, L. (2009). A multimedia interface for facilitating comparisons of opinions. In *Proceedings of the International Conference on Intelligent User Interfaces* (pp. 325-334).

Carenini, G., Ng, R. T., & Zwart, E. (2005). Extracting knowledge from evaluative text. In *Proceedings of the 3rd International Conference on Knowledge Capture* (pp. 11-18).

Carenini, G., & Moore, J. D. (2006). Generating and evaluating evaluative arguments. *Artificial Intelligence, 170*(11), 925–952. doi:10.1016/j.artint.2006.05.003

Carpineto, C., Mizzaro, S., Romano, G., & Snidero, M. (2009). Mobile information retrieval with search results clustering: Prototypes and evaluations. *Journal of the American Society for Information Science and Technology, 60*(5), 877–895. Retrieved from http://dx.doi.org/10.1002/asi.v60:5. doi:10.1002/asi.21036

Carter, C., & Mankoff, J. (2005). When participants do the capturing: The role of media in diary studies. In *Proceedings of the ACM SIGCHI Conference on Human Factors in Computing Systems* (pp. 1-10).

Carter, S., Mankoff, J., & Heer, J. (2007). Momento: Support for situated Ubicomp experimentation. In *Proceedings of the SIGCHI Conference on Human Factors in Computing Systems* (pp. 125-134). New York, NY: ACM Press.

Carter, J., & Fourney, D. (2005). Research based tactile and haptic interaction guidelines. In Carter, J., & Fourney, D. (Eds.), *Guidelines on tactile and haptics interaction* (pp. 84–92). Saskatoon, SK, Canada: University of Saskatchewan.

Cassell, J., Vilhjalmsson, H. H., & Bickmore, T. (2001). Beat: The behavior expression animation toolkit. In *Proceedings of the 28th Annual Conference on Computer Graphics and Interactive Techniques* (pp. 477-486). New York, NY: ACM Press.

Casson, A. J., Smith, S., Duncan, J. S., & Rodriguez-Villegas, E. (2008). Wearable eeg: What is it, why is it needed and what does it entail? In *Proceedings of the IEEE 30th Annual International Conference on Engineering in Medicine and Biology Society* (pp. 5867-5870). Washington, DC: IEEE Computer Society.

Chae, M., & Kim, J. (2004). Do size and structure matter to mobile users? An empirical study of the effects of screen size, information structure, and task complexity on user activities with standard web phones. *Behaviour & Information Technology, 23*(3), 165–181. doi:10.1080/01449290410001669923

Chall, J. S. (1988). The beginning years. In Zakaluk, B. L., & Samuels, S. J. (Eds.), *Readability: Its past, present, and future.* Newark, DE: International Reading Association.

Chalmers, M. (1996). A linear iteration time layout algorithm for visualising high-dimensional data. In *Proceedings of the IEEE 7th Conference on Vizualisation* (pp. 127-132). Washington, DC: IEEE Computer Society.

Chalmers, M. (2010). A population approach to ubicomp system design. In *Proceedings of the ACM-BCS Visions of Computer Science Conference* (p. 1).

Chan, A., MacLean, K. E., & McGrenere, J. (2005). Learning and identifying haptic icons under workload. In *Proceedings of the 1st Joint Eurohaptics Conference and Symposium on Haptic Interfaces for Virtual Environment and Teleoperator Systems* (pp. 432-439). Washington DC: IEEE Computer Society.

Chang, A., O'Modhrain, S., Jacob, R., Gunther, E., & Ishii, H. (2002). Comtouch: Design of a vibrotactile communication device. In *Proceedings of the 4th Conference on Designing Interactive Systems* (pp. 312-320). New York, NY: ACM Press.

Chang, P. (2010). Drivers and moderators of consumer behaviour in the multiple use of mobile phones. *International Journal of Mobile Communications*, 8(1), 88–105. doi:10.1504/IJMC.2010.030522

Chehimi, F., Coulton, P., & Edwards, R. (2008, July 10-13). 3D motion control of connected augmented virtuality on mobile phones. In *Proceedings of the International Symposium on Ubiquitous Virtual Reality* (pp. 67-70).

Cherubini, M., Gutierrez, A., de Oliveira, R., & Oliver, N. (2010). Social tagging revamped: Supporting the users' need of self-promotion through persuasive techniques. In *Proceedings of the 28th SIGCHI International Conference on Human Factors in Computing Systems* (pp. 985-994). New York, NY: ACM Press.

Cheverst, K., Coulton, P., Bamford, W., & Taylor, N. (2008). Supporting (mobile) user experience at a rural village 'scarecrow festival': A formative study of a geo-located photo mashup utilising a situated display. In *Proceedings of the Mobile HCI Workshop on Mobile Interaction in the Real World* (pp. 27-31).

Cheverst, K., Davies, N., Mitchell, K., & Friday, A. (2000). Experiences of developing and deploying a context-aware tourist guide: The guide project. In *Proceedings of the 6th Annual International Conference on Mobile Computing and Networking* (pp. 20-31). New York, NY: ACM Press.

Chin, A., & Salomaa, J. (2009). A user study of mobile Web services and applications from the 2008 beijing olympics. In *Proceedings of Hypertext 2009*, Turin, Italy.

Chin, A. (2009). Finding cohesive subgroups and relevant members in the Nokia friend view mobile social network. *Computing in Science & Engineering*, 278–283.

Chittaro, L. (2009). Distinctive aspects of mobile interaction and their implications for the design of multimodal interfaces. *Journal on Multimodal User Interfaces*, 3(3), 157–165. doi:10.1007/s12193-010-0036-2

Choi, S.-M., Kim, Y.-G., Lee, D.-S., Lee, S.-O., & Park, G.-T. (2004). Nonphotorealistic 3-d facial animation on the PDA based on facial expression recognition. In A. Butz, A. Kruger, & P. Olivier (Eds.), *Proceedings of the 4th International Symposium on Smart Graphics* (LNCS 3031, pp. 11-20).

Choudhure, T., Consolvo, S., Harrison, B., Hightower, J., LaMarca, A., & LeGrand, L. (2008). The mobile sensing platform: An embedded activity recognition system. *IEEE Pervasive Computing*, 7(2), 32–40. doi:10.1109/MPRV.2008.39

Church, K., & Cherubini, M. (2010, September 26). *Evaluating mobile user experience in-the-wild: Prototypes, playgrounds and contextual experience sampling*. Paper presented at the Workshop Research in the Large: Using App Stores, Markets, and other Wide Distribution Channels in UbiComp Research, Copenhagen, Denmark.

Church, K., & Smyth, B. (2008b). Who, what, where & when: a new approach to mobile search. In *Proceedings of the 13th international Conference on intelligent User interfaces*, Gran Canaria, Spain (pp. 309-312). New York: ACM. Retrieved from http://doi.acm.org/10.1145/1378773.1378817

Church, K., & Smyth, B. (2009). Understanding the intent behind mobile information needs. In *Proceedings of the 13th international Conference on intelligent User interfaces*, Sanibel Island, FL (pp. 247-256). New York: ACM. Retrieved from http://doi.acm.org/10.1145/1502650.1502686

Church, K., Smyth, B., Bradley, K., & Cotter, P. (2008a, September 2-5). A large scale study of European mobile search behaviour. In *Proceedings of the 10th international Conference on Human Computer interaction with Mobile Devices and Services (MobileHCI '08)*, Amsterdam, The Netherlands (pp. 13-22). New York: ACM. Retrieved from http://doi.acm.org/10.1145/1409240.1409243

Churchill, E., Girgensohn, A., Nelson, L., & Lee, A. (2004). Blending digital and physical spaces for ubiquitous community participation. *Communications of the ACM*, 47(2), 38–44. doi:10.1145/966389.966413

CIA. (2009). *The world factbook*. Retrieved from https://www.cia.gov/library/publications/the-world-factbook/geos/in.html

Cleaver, F. (1999). Paradoxes of participation: Questioning participatory approaches to development. *Journal of International Development*, 11(4), 597–612. doi:10.1002/(SICI)1099-1328(199906)11:4<597::AID-JID610>3.0.CO;2-Q

Coleman, E. B., & Blumenfeld, P. J. (1963). Cloze scores of nominalization and their grammatical transformations using active verbs. *Psychological Reports, 13*, 651–654.

Columbia Broadcasting System (CBS). (n.d.). Retrieved March 3, 2008, from http://www.cbsnews.com

Comte, A. (1880). *A general view of positivism* (2nd ed.). London, UK: Reeves & Turner.

Consolvo, S., Arnstein, L., & Franza, B. R. (2002). User study techniques in the design and evaluation of a ubicomp environment. In *Proceedings of the 4th International Conference on Ubiquitous Computing* (pp. 73-90).

Consolvo, S., Klasnja, P., McDonald, D. W., Avrahami, D., Froehlich, J., LeGrand, L., et al. (2008). Flowers or a robot army? Encouraging awareness & activity with personal, mobile displays. In *Proceedings of the 10ᵗʰ International Conference on Ubiquitous Computing* (pp. 54-63).

Consolvo, S., Klasnja, P., McDonald, D. W., Avrahami, D., Froehlich, J., LeGrand, L., et al. (2009). Flowers or a robot army? Encouraging awareness & activity with personal, mobile displays. In *Proceedings of the Conference on Ubiquitous Computing* (pp. 54-63).

Consolvo, S., McDonald, D. W., & Landay, J. A. (2009). Theory-driven design strategies for technologies that support behavior change in everyday life. In *Proceedings of the 27ᵗʰ International Conference on Human Factors in Computing Systems* (pp. 405-414).

Consolvo, S., McDonald, D. W., Toscos, T., Chen, M. Y., Froehlich, J., Harrison, B., et al. (2008). Activity sensing in the wild: a field trial of ubifit garden. In *Proceedings of the 26ᵗʰ Annual SIGCHI Conference on Human Factors in Computing System* (pp. 1797-1806).

Consolvo, S., Harrison, B. L., Smith, I. E., Chen, M. Y., Everitt, K., & Froehlich, J. (2007). Conducting in situ evaluations for and with ubiquitous computing technologies. *International Journal of Human-Computer Interaction, 22*(1-2), 103–118. doi:10.1207/s15327590ijhc2201-02_6

Consolvo, S., & Walker, M. (2003). Using the experience sampling method to evaluate ubicomp applications. *IEEE Pervasive Computing, 2*(2), 24–31. doi:10.1109/MPRV.2003.1203750

Cooke, B., & Kothari, U. (2001). *Participation: The new tyranny?* London, UK: Zed Books.

Cooper, A., Reimann, R., & Cronin, D. (2007). *About face 3: The essentials of interaction design*. New Delhi, India: Wiley Publications.

Coppens, T., Trappeniers, L., & Godon, M. (2004). AmigoTV: Towards a social TV experience. In *Proceedings of European Interactive TV Conference* (pp.1-3).

Corporate Solutions Consulting (UK) Limited. (2007). *Research Report Fair Processing Notifications: Current Effectiveness and Opportunities for Improvement*. Retrieved from http://www.ico.gov.uk/upload/documents/library/corporate/research_and_reports/ic_final_report_version_1.1_final.pdf

Costanza, E., Panchard, J., Zufferey, G., Nembrini, J., Freudiger, J., Huang, J., et al. (2010). SensorTune: A mobile auditory interface for DIY wireless sensor networks. In *Proceedings of the 28th International Conference on Human Factors in Computing Systems*, Atlanta, GA. (pp. 2317-2326).

Coulton, P., Bamford, W., & Edwards, R. (2008). Mud, mobiles and a large interactive display. In *Proceedings of the OzCHI Conference on Public and Situated Displays to Support Communities*.

Coulton, P., Copic Pucihar, K., & Bamford, W. (2008, November). Mobile social gaming. In *Proceedings of the International Workshop on Social Interaction and Mundane Technologies* Cambridge, UK (pp. 20-21).

Coulton, P., Rashid, O., Edwards, R., & Thompson, R. (2005). Creating entertainment applications for cellular phones. *Computers in Entertainment, 3*(3), 3. doi:10.1145/1077246.1077254

Cramer, H., Mentis, H., & Fernaeus, Y. (2010). Serious work on playful experiences: A preliminary set of challenges. In *Proceedings of the 'Fun, Seriously?' Workshop at CSCW*. Savannah, GA.

Cramer, H., Rost, M., Belloni, N., Chincholle, D., & Bentley, F. (2010). Research in the large: Using app stores, markets and other wide distribution channels in UbiComp research. In *Proceedings of the 12ᵗʰ ACM International Conference Adjunct Papers on Ubiquitous Computing* (p. 511-514). New York, NY: ACM Press.

Cranor, L., Langheinrich, M., Marchiori, M., Presler-Marshall, M., & Reagle, J. (2002). *The Platform for Privacy Preferences 1.0 (P3P1.0) Specification*. Retrieved from http://www.w3.org/TR/P3P/#intro_example

Cranor, L., Egelman, S., Sheng, S., McDonald, A., & Chowdhury, A. (2008). P3P Deployment on Websites. *Electronic Commerce Research and Applications*, *7*(3), 274–293. doi:10.1016/j.elerap.2008.04.003

Crossan, A., Murray-Smith, R., Brewster, S., & Musizza, B. (2009). Instrumented usability analysis for mobile devices. *International Journal of Mobile Human Computer Interaction*, *1*(1), 1–19. doi:10.4018/jmhci.2009010101

Crum, C. (2010). *Consumer Demographics and Their Wireless Devices. B2B Publications*. Retrieved July 21, 2010 from http://www.webpronews.com/topnews/2010/02/25/consumer-demographics-and-their-wireless-devices

Csikszentmihalyi, M. (1990). *Flow: The psychology of optimal experience*. New York, NY: Harper Perennial.

Cui, Y., & Roto, V. (2008). How people use the Web on mobile divices. In *Proceedings of the 17th international conference on World Wide Web,* Beijing, China (pp. 905-914). New York: ACM.

Danihelka, J., Kencl, L., & Zara, J. (2010). Reduction of animated models for embedded devices. In *Proceedings of the 18th International European Conference on Computer Graphics, Visualization, and Computer Vision* (pp. 89-95).

Danziger, K., & Ballantyne, P. (1997). Psychological experiments . In Bringmann, W. G., Luck, H. E., Miller, R., & Early, C. E. (Eds.), *A pictorial history of psychology* (pp. 233–239). Hanover Park, IL: Quintessence Publishing.

Dave, K., Lawrence, S., & Pennock, D. M. (2003). Mining the peanut gallery: Opinion extraction and semantic classification of product reviews. In *Proceedings of the 12th International Conference on World Wide Web* (pp. 519-528).

Davison, A. (1984). *Readability formulas and comprehension. Comprehension instruction: Perspectives and suggestions*. New York: Longman.

de Leon, M. P., Balasubramaniam, S., & Donnelly, W. (2006). Creating a distributed mobile networking testbed environment-through the living labs approach. In *Proceeding on Testbeds and Research Infrastructures for the Development of Networks and Communities* (pp. 134-139).

Demberg, V., & Moore, J. (2006). Information presentation in spoken dialogue systems. In *Proceedings of the 11th International Conference of the European Chapter of the Association for Computational Linguistics*.

Deng, L., Wang, Y., Wang, K., Acero, A., Hon, H., & Droppo, J. (2004). Speech and language processing for multimodal human-computer interaction. *The Journal of VLSI Signal Processing*, *36*(2), 161–187. doi:10.1023/B:VLSI.0000015095.19623.73

Department for Transport. (2007). *Road Casualties Great Britain*. Retrieved from http://www.dft.gov.uk/pgr/statistics/datatablespublications/accidents/casualtiesmr/rcgbmainresults2007

Derks, D., Bos, A. E., & von Grumbkow, J. (2008). Emoticons in computer-mediated communication: Social motives and social context. *Cyberpsychology & Behavior*, *11*(1), 99–101. doi:10.1089/cpb.2007.9926

Devevey, P., Lorenzon, N., & Tambary, C. (2005). *Measuring wireless energy consumption on PDAs and on laptops* (Tech. Rep. No. 2005). Genoa, Italy: University of Genoa.

Dey, A. K. (2001). Understanding and using context. *Personal and Ubiquitous Computing*, *5*(1), 4–7. doi:10.1007/s007790170019

Dey, A. K., Abowd, G. D., & Salber, D. (2001). A conceptual framework and a toolkit for supporting the rapid prototyping of context-aware applications. *Human-Computer Interaction*, *16*(2-4), 97–166. doi:10.1207/S15327051HCI16234_02

Di Fabbrizio, G., Gupta, N., Besana, S., & Mani, P. (2010). Have2eat: A restaurant finder with review summarization for mobile phones. In *Proceedings of the 23rd International Conference on Computational Linguistics: Demonstrations* (pp. 17-20).

Diakopoulos, N. A., & Shamma, D. A. (2010). Characterizing debate performance via aggregated Twitter sentiment. In *Proceedings of the ACM SIGCHI Conference on Human Factors in Computing Systems* (pp. 1195-1198).

Dix, A., Finlay, J., Abowd, G. D., & Beale, R. (2004). *Human-Computer Interaction* (3rd ed.). Upper Saddle River, NJ: Pearson Education Limited, Prentice Hall.

Dourish, P. (2001). *Where the action is: The foundations of embodied interaction*. Cambridge, MA: MIT Press.

Dourish, P. (2004). What we talk about when we talk about context. *Personal and Ubiquitous Computing, 8*(1), 19–30. doi:10.1007/s00779-003-0253-8

Dowman, M., Tablan, V., Cunningham, H., Ursu, C., & Popov, B. (2005). *Semantically enhanced television news through web and video integration.* Paper presented at the Second European Semantic Web Conference Workshop, Crete, Greece.

Dryer, D. C. (1999). Getting personal with computers: How to design personalities for agents. *Applied Artificial Intelligence, 13*(3), 273–295. doi:10.1080/088395199117423

Duistermaat, M. (2005). *Tactile land in night operations* (Tech. Rep. No. TNO-DV3 2005 M065). Soesterberg, Netherlands: Netherlands Organisation for Applied Scientific Research

Duyree, T. (2010, May 10). *Beyond AdMob: There's plenty more mobile ad networks to go around.* Retrieved from http://moconews.net/article/419-beyond-admob-theres-plenty-more-mobile-ad-networks-to-go-around/

Eades, P. (1984). A heuristic for graph drawing. *Congressus Numerantium, 42*, 149–160.

Eagle, N., & Pentland, A. S. (2006). Reality mining: Sensing complex social systems. *Personal and Ubiquitous Computing, 10*(4), 255–268. doi:10.1007/s00779-005-0046-3

EDS IDG Shopping Report. (2007). *Shopping choices: Attraction or distraction?* Retrieved from http://www.eds.com/industries/cir/downloads/EDSIDGReport_aw_final.pdf

Edwards, G. T., Liu, L. S., Moulic, R., & Shea, D. G. (2008). Proxima: a mobile augmented-image search system. In *Proceeding of the 16th ACM international Conference on Multimedia*, Vancouver, BC (pp. 921-924). New York: ACM. Retrieved from http://doi.acm.org/10.1145/1459359.1459522

Eichhorn, E., Wettach, R., & Hornecker, E. (2008). A stroking device for spatially separated couples. In *Proceedings of the 10th International Conference on Human Computer Interaction with Mobile Devices and Services* (pp. 303-306). New York, NY: ACM Press.

Ekman, P. (1999). Basic emotions . In Dalgleish, T., & Power, M. (Eds.), *Handbook of cognition and emotion* (pp. 45–60). Chichester, UK: John Wiley & Sons.

El Ali, A., & Nack, F. (2009). Touring in a living lab: Some methodological considerations. In *Proceedings of the Mobile Living Labs Workshop*, Enschede, The Netherlands (pp. 23-26).

El Ali, A., Nack, F., & Hardman, L. (2010). Understanding contextual factors in location-aware multimedia messaging. In *Proceedings of the 12th International Conference on Multimodal Interfaces* (p. 22). New York, NY: ACM Press.

Enriquez, M. J., & MacLean, K. E. (2003). The hapticon editor: A tool in support of haptic communication research. In *Proceedings of the 11th International Symposium on Haptic Interfaces for Virtual Environment and Teleoperator Systems* (pp. 356-362). Washington, DC: IEEE Computer Society.

Eriksson, M., Niitamo, V., & Kulkki, S. (2005). *State-of-the-art in utilizing living labs approach to user-centric ICT innovation – a European approach.* Stromsund, Sweden: Luleå University of Technology.

Esbjörnsson, M., Brown, B., Juhlin, O., Normark, D., Östergren, M., & Laurier, E. (2006). Watching the cars go round and round: Designing for active spectating. In *Proceedings of the ACM SIGCHI Conference on Human Factors in Computing Systems* (pp. 1221-1224).

Evaluation of Online Privacy Notices. (2004). In *Proceedings of ACM Conference on Human Factors in Computing Systems (CHI 2004)*, Vienna, Austria (pp. 471-478).

Evers, V., & Day, D. (1997). The role of culture in interface acceptance. In *Proceedings of the IFIP TC13 Interantional Conference on Human-Computer Interaction* (pp. 260-267).

Falaki, H., Mahajan, R., Kandula, S., Lymberopoulos, D., Govindan, R., & Estrin, D. (2010). Diversity in smartphone usage. In *Proceedings of the 8th International Conference on Mobile Systems, Applications and Services* (pp. 179-194).

Fanguy, B., Kleen, B., & Soule, L. (2004). Privacy policies: Cloze test reveals readability concerns. *Issues in Information Systems, 5*(1), 117–123.

FAO. (2005). *Database*. Retrieved from http://faostat.fao.org/

Fasolo, B., McClelland, G. H., & Todd, P. M. (2007). Escaping the tyranny of choice: When fewer attributes make choice easier. *Marketing Theory*, 7(1), 13–26. doi:10.1177/1470593107073842

Favre, B., Béchet, F., & Nocéra, P. (2005). Robust named entity extraction from large spoken archives. In *Proceedings of the Conference on Human Language Technology and Empirical Methods in Natural Language Processing* (pp. 491-498).

Fawcett, J., Beyer, B., Hum, D., Ault, A., & Krogmeier, J. (2009). Rich immersive sports experience: A hybrid multimedia system for content consumption. In *Proceedings of the 6th IEEE Conference on Consumer Communications and Networking* (pp. 1-5).

Federal Trade Commission. (2007). *Fair Information Practice Principles*. Retrieved from http://www.ftc.gov/reports/privacy3/fairinfo.shtm

Feng, J., Bangalore, S., & Gilbert, M. (2009). Role of natural language understanding in voice local search. In *Proceedings of the 10th Annual Conference of the International Speech Communication Association* (pp. 1859-1862).

Feunekes, G., Gortemaker, I., Willems, A., Lion, R., & van den Kommer, M. (2008). Front-of-pack nutrition labelling: Testing effectiveness of different nutrition labelling formats front-of-pack in 4 European countries. *Appetite*, 50, 57–70. doi:10.1016/j.appet.2007.05.009

Figo, D., Diniz, P., Ferreira, D., & Cardoso, J. (2010). Preprocessing techniques for context recognition from accelerometer data. *Personal and Ubiquitous Computing*, 14(7), 645–662. doi:10.1007/s00779-010-0293-9

Fisher, C., & Sanderson, P. (1996). Exploratory sequential data analysis: Exploring continuous observational data. *Interaction*, 3(2), 25–34. doi:10.1145/227181.227185

Flurry. (2010). *Flurry homepage*. Retrieved from http://www.flurry.com

Flurry. (2010, October 8). *Mobile application analytics, iPhone analytics, android analytics*. Retrieved from http://www.flurry.com/

Fotos, S. S. (2006). The Cloze Test as an Integrative Measure of EFL Proficiency: A Substitute for Essays on College Entrance Examinations? *Language Learning*, 41(3), 313–336. doi:10.1111/j.1467-1770.1991.tb00609.x

Froehlich, J., Chen, M. Y., Consolvo, S., Harrison, B., & Landay, J. A. (2007). Myexperience: A system for in situ tracing and capturing of user feedback on mobile phones. In *Proceedings of the 5th International Conference on Mobile Systems, Applications and Services* (pp. 57-70). New York, NY: ACM Press.

Froehlich, J., Chen, M. Y., Smith, I. E., & Potter, F. (2006). Voting with your feet: An investigative study of the relationship between place visit behavior and preference. In *Proceedings of the 8th International Conference on Ubiquitous Computing* (pp. 333-350).

Frohlich, D. M., Rachovides, D., Riga, K., Bhat, R., Frank, M., Edirisinghe, E., et al. (2009). StoryBank: mobile digital storytelling in a development context. In *Proceedings of the 27th international Conference on Human Factors in Computing Systems*, Boston (pp. 1761-1770). New York: ACM. Retrieved from http://doi.acm.org/10.1145/1518701.1518972

Galantucci, B. (2005). An experimental study of the emergence of human communication systems. *Cognitive Science: A Multidisciplinary Journal*, 29(5), 737-767.

Gallace, A., Tan, H. Z., & Spence, C. (2006). Numerosity judgments for tactile stimuli distributed over the body surface. *Perception*, 35(2), 247–266. doi:10.1068/p5380

Garner, P., Coulton, P., & Edwards, R. (2006, June 29-July 1). XEPS – Enabling card-based payment for mobile terminals. In *Proceedings of the Tenth IEEE International Symposium on Consumer Electronics*, St. Petersburg, Russia (pp. 375-380).

Garzonis, S., Jones, S., Jay, T., & O'Neill, E. (2009). Auditory icon and Earcon mobile service notifications: Intuitiveness, learnability, memorability and preference. In *Proceedings of the SIGCHI Conference on Human Factors in Computing Systems* (pp. 1513-1522). New York, NY: ACM Press.

Geerts, D. (2006). Comparing voice chat and text chat in a communication tool for interactive television. In *Proceedings of the Nordic Conference on Human-Computer Interaction: Changing Roles* (pp. 461-464).

Geerts, D., & De Grooff, D. (2009). Supporting the social uses of television: Sociability heuristics for social TV. In *Proceedings of the ACM SIGCHI Conference on Human Factors in Computing Systems* (pp. 595-604).

Gelau, C., & Krems, J. F. (2004). The occlusion technique: A procedure to assess the HMI of in-vehicle information and communication systems. *Applied Ergonomics, 35*(3), 185–187. doi:10.1016/j.apergo.2003.11.009

Gemoets, D., Rosemblat, G., Tse, T., & Logan, R. (2004). Assessing readability of consumer health information: An exploratory study. In *Proceedings of the 11th World Congress on Medical Informatics*, San Francisco, CA (pp. 869-874). Amsterdam, The Netherlands: IOS Press.

Geoghegan, E. (2009). *Experian Simmons Fall 2009 Consumer Study/National Hispanic Study.*

Gigerenzer, G., & Selten, R. (Eds.). (2001). *Bounded rationality: The adaptive toolbox.* Cambridge, MA: MIT Press.

Gigerenzer, G., & Todd, P. M.ABC Research Group. (1999). *Simple heuristics that make us smart.* New York, NY: Oxford University Press.

Gilbertson, P., Coulton, P., Chehimi, F., & Vajk, T. (2008). Using "tilt" as an interface to control "no-button" 3-D mobile games. *Computers in Entertainment, 6*(3), 1–13. doi:10.1145/1394021.1394031

Girardello, A. (2010). AppAware: Serendipity in mobile applications. In *Proceedings of the Conference on Human-Computer Interaction with Mobile Devices and Services* (pp. 479-480).

Girardello, A., & Michahelles, F. (2010). AppAware: Which mobile applications are hot? In *Proceedings of the 12th International Conference on Human Computer Interaction with Mobile Devices & Services*, Lisbon, Portugal (pp. 431-434).

Gitau, S., Marsden, G., & Donner, J. (2010). After access: challenges facing mobile-only internet users in the developing world. In *Proceedings of the 28th international Conference on Human Factors in Computing Systems*, Atlanta, GA (pp. 2603-2606). New York: ACM. Retrieved from http://doi.acm.org/10.1145/1753326.1753720

Glass, J., Hazen, T., Cyphers, S., Malioutov, I., Huynh, D., & Barzilay, R. (2007). Recent progress in the MIT spoken lecture processing project. In *Proceeding of the 8th Annual Conference of the International Communication Association* (pp. 2553-2556).

Goldberg, A. B., & Zhu, X. (2006). Seeing stars when there aren't many stars: Graph-based semi-supervised learning for sentiment categorization. In *Proceedings of TextGraphs: The First Workshop on Graph Based Methods for Natural Language Processing* (pp. 45-52)

Goldin-Meadow, S., & Mylander, C. (1998). Spontaneous sign systems created by deaf children in two cultures. *Nature, 391*, 279–280. doi:10.1038/34646

Goldstein, D. G., & Gigerenzer, G. (2002). Models of ecological rationality: The recognition heuristic. *Psychological Review, 109*, 75–90. doi:10.1037/0033-295X.109.1.75

Golledge, R. G. (1999). Human wayfinding and cognitive maps . In Golledge, R. G. (Ed.), *Wayfinding behavior: Cognitive mapping and other spatial process* (pp. 5–45). Baltimore, MD: Johns Hopkins University Press.

Good, N., Schafer, J. B., Konstan, J. A., Borchers, A., Sarwar, B., Herlocker, J., et al. (1999). Combining collaborative filtering with personal agents for better recommendations. In *Proceedings of the Sixteenth National Conference on Artificial Intelligence* (pp. 439-446).

Google, Inc. (2010) *Android market.* Retrieved from http://www.android.com/market

Gorin, A. L., Parker, B. A., Sachs, R. M., & Wilpon, J. G. (1997). How may I help you? *Speech Communication, 23*, 113–127. doi:10.1016/S0167-6393(97)00040-X

Gorlenko, L., & Merrick, R. (2003). No wires attached: Usability challenges in the connected mobile world. *IBM Systems Journal, 42*(4), 639–651. doi:10.1147/sj.424.0639

Grabler, F., Agrawala, M., Sumner, R. W., & Pauly, M. (2008). Automatic generation of tourist maps. In *Proceedings of the SIGGRAPH 35th International Conference and Exhibition on Computer Graphics and Interaction Techniques* (pp. 1-11). New York, NY: ACM Press.

GREATDANE. (n. d.). Generic reality trace data analysis engine. Retrieved from http://www.loevborg.com/tools/greatdane

Greaves, A., & Akerman, M. (2009). Exloring user reaction to personal projection when used in shared public places: a formative study. In *Proceedings of the Context-Aware Mobile Media and Mobile Social Networks Workshop in MobileHCI 2009 Conference*, Germany.

Green, P. (1994). *Measures and Methods Used to Assess the Safety and Usability of Driver Information Systems* (Tech. Rep. No. UMTRI-93-12). Ann Arbor, MI: The University of Michigan.

Green, P. (1999b). The 15-Second Rule for Driver Information Systems. In *Proceedings of the Intelligent Transportation Society of America Conference*.

Greenberg, S., & Buxton, B. (2008). Usability evaluation considered harmful (some of the time). In *Proceeding of the Twenty-Sixth Annual SIGCHI Conference on Human Factors in Computing Systems* (pp. 111-120). New York, NY: ACM Press.

Greenberg, S., & Rounding, M. (2001). The notification collage: Posting information to public and personal displays. In *Proceedings of the SIGCHI Conference on Human Factors in Computing Systems* (pp. 514-521).

Greenfield, A. (2006). *Everyware: The dawning age of ubiquitous computing*. Berkeley, CA: New Riders Publishing.

Greenhalgh, C., French, A., Tennent, P., & Humble, J. (2007, October). *From ReplayTool to digital replay system*. Paper presented at the 3rd International Conference on e-Social Science, Ann Arbor, MI.

Green, P. (1999a). *Visual and Task Demands of Driver Information Systems (Tech. Rep. No. UMTRI- 98-16)*. Ann Arbor, MI: The University of Michigan.

Gupta, N., Di Fabbrizio, G., & Haffner, P. (2010). Capturing the stars: Predicting rankings for service and product reviews. In *Proceedings of the NAACL HLT Workshop on Semantic Search* (pp. 36-43).

Hall, M., Bell, M., Morrison, A., Reeves, S., Sherwood, S., & Chalmers, M. (2009). Adapting Ubicomp software and its evaluation. In *Proceedings of the 1st ACM SIGCHI Symposium on Engineering Interactive Computing Systems* (pp.143-148). New York, NY: ACM Press.

Hamann, B. (1994). A data reduction scheme for triangulated surfaces. *Computer Aided Geometric Design, 11*(2), 197–214. doi:10.1016/0167-8396(94)90032-9

Hansen, F. A., & Grønbæk, K. (2008). Social web applications in the city: A lightweight infrastructure for urban computing. In *Proceedings of the 19th ACM Conference on Hypertext and Hypermedia* (pp. 175-180).

Harboe, G., Metcalf, C. J., Bentley, F., Tullio, J., Massey, N., & Romano, G. (2008). Ambient social TV: Drawing people into a shared experience. In *Proceedings of the ACM SIGCHI Conference on Human Factors in Computing Systems* (pp. 1-10).

Harboe, G., Massey, N., Metcalf, C., Wheatley, D., & Romano, G. (2008). The uses of social television. *Computers in Entertainment, 6*(1), 1–15. doi:10.1145/1350843.1350851

Harris, Z. S. (1962). *String Analysis of Sentence Structure*. The Hague, The Netherlands: Mouton.

Harr, J., & Kosack, S. (1990). Employee benefit packages: How understandable are they? *Journal of Business Communication, 27*(2), 185–200. doi:10.1177/002194369002700205

Harsh, S. (2010, June 30). *May 2010 mobile metrics report*. Retrieved from http://metrics.admob.com/2010/06/may-2010-mobile-metrics-report/

Hassenzahl, M., Diefenbach, S., & Göritz, A. (2010). Needs, affect, and interactive products – Facets of user experience. *Interacting with Computers, 22*(5), 353–362. doi:10.1016/j.intcom.2010.04.002

Hassenzahl, M., & Sandweg, N. (2004). From mental effort to perceived usability: transforming experiences into summary assessments . In *Proceedings of Extended Abstracts on Human Factors in Computing Systems* (pp. 1283–1286). New York: ACM Press.

Hassenzahl, M., & Tractinsky, N. (2006). User experience - a research agenda. *Behaviour & Information Technology, 25*(2), 91–97. doi:10.1080/01449290500330331

Healey, P., Swoboda, N., Umata, I., & King, J. (2007). Graphical language games: Interactional constraints on representational form. *Cognitive Science: A Multidisciplinary Journal, 31*(2), 285-309.

Hearst, M. (1999). User Interfaces and visualisation . In Baeza-Yates, R., & Ribeiro-Neto, B. (Eds.), *Modern Information Retrieval*. New York: ACM Press.

Heikkinen, J., Olsson, T., & Vaananen-Vainio-Mattila, K. (2009). Expectations for user experience in haptic communication with mobile devices. In *Proceedings of the 11th International Conference on Human-Computer Interaction with Mobile Devices and Services* (pp. 1-10). New York, NY: ACM Press.

Heimonen, T. (2009). Information needs and practices of active mobile internet users. In *Proceedings of the 6th International Conference on Mobile Technology, Application & Systems*, Nice, France (pp. 1-8).

Heimonen, T., & Käki, M. (2007). Mobile findex: supporting mobile web search with automatic result categories. In *Proceedings of the 9th international Conference on Human Computer interaction with Mobile Devices and Services*, Singapore (Vol. 309, pp. 397-404). New York: ACM. Retrieved from http://doi.acm.org/10.1145/1377999.1378045

Henze, N., & Boll, S. (2010). Push the study to the app store: Evaluating off-screen visualizations for maps in the android market. In *Proceedings of the Conference on Human-Computer Interaction with Mobile Devices and Services* (pp. 373-374).

Henze, N., Poppinga, B., & Boll, S. (2010). Experiments in the wild: Public evaluation of off-screen visualizations in the android market. In *Proceedings of the 6th Nordic Conference on Human-Computer Interaction: Extending Boundaries* (pp. 675-678).

HERMES. (n. d.). *HERMES tool*. Retrieved from http://www.loevborg.com/tools/hermes

Hertwig, R., & Todd, P. M. (2003). More is not always better: The benefits of cognitive limits . In Hardman, D., & Macchi, L. (Eds.), *Thinking: Psychological perspectives on reasoning, judgment and decision making* (pp. 213–231). Chichester, UK: John Wiley & Sons.

Hilbert, D. M., & Redmiles, D. F. (2000). Extracting usability information from user interface events. *ACM Computing Surveys, 32*(4), 384–421..doi:10.1145/371578.371593

Hillard, D., Huang, Z., Ji, H., Grishman, R., Hakkani-Tür, D., Harper, M., et al. (2006). Impact of automatic comma prediction on POS/name tagging of speech. In *Proceedings of the IEEE/ACL Workshop on Spoken Language Technology* (pp. 58-61).

Hoggan, E., & Brewster, S. A. (2010). Crosstrainer: Testing the use of multimodal interfaces in situ. In Proceedings of the 28th International Conference on Human Factors in Computing Systems (pp. 333-342).

Horlock, J., & King, S. (2003). Discriminative methods for improving named entity extraction on speech data. In *Proceedings of the 8th European Conference on Speech Communication and Technology* (pp. 2765-2768).

Horvath, R. (2008). Innovation - mobile services. In *Proceedings of the CII/GIS Conference.*

Hu, M., & Liu, B. (2004). Mining opinion features in customer reviews. In *Proceedings of the 19th National Conference on Artificial Intelligence* (pp. 755-760).

Huang, E. M., Harboe, G., Tullio, J., Novak, A., Massey, N., Metcalf, C. J., & Romano, G. (2009). Of social television comes home: A field study of communication choices and practices in TV-based text and voice chat. In *Proceedings of the ACM SIGCHI Conference on Human Factors in Computing Systems* (pp. 585-594).

Huang, J., Zweig, G., & Padmanabhan, M. (2001). Information extraction from voicemail. In *Proceedings of the Conference of the Association for Computational Linguistics* (pp. 290-297).

Huggins-Daines, D., Kumar, M., Chan, A., Black, A., Ravishankar, M., & Rudnicky, A. (2006). Pocketsphinx: A free, real-time continuous speech recognition system for hand-held devices. In *Proceedings of the IEEE International Conference on Acoustics, Speech and Signal Processing* (p. 1). Washington, DC: IEEE Computer Society.

Hurst, M., & Nigam, K. (2004). Retrieving topical sentiments from online document collections. *Document Recognition and Retrieval 11, 5296*, 27-34.

Husserl, E. (1893-1917). *On the phenomenology of the consciousness of internal time*. New York, NY: Springer.

Inselberg, A., & Dimsdale, B. (1990). Parallel coordinates: A tool for visualizing multi-dimensional geometry. In *Proceedings of the 1st Conference on Visualization* (pp. 361-378). Washington, DC: IEEE Computer Society.

International Organization for Standardization. (2010). *ISO DIS 9241-210: Ergonomics of human system interaction - part 210: Human-centred design for interactive systems.* Retrieved from http://www.iso.org/iso/catalogue_detail.htm?csnumber=52075

International Telecommunications Union. (2009). *ICT statistics.* Retrieved from http://www.itu.int/ITU-D/ict/statistics/

iOS Dev Center. (2010). *Apple developer, developing for iOS.* Retrieved from http://developer.apple.com/devcenter/ios/index.action

Irune, A., & Burnett, G. E. (2007). Locating in-car controls: Predicting the effects of varying design layout. In *Proceedings of Road Safety and Simulation conference (RSS2007)*, Rome, Italy.

Irune, A. (2009). Evaluating the visual demand of in-vehicle information systems: The development of a novel method.

Ishii, H., Wisneski, C., Brave, S., Dahley, A., Gorbet, M., Ullmer, B., & Yarin, P. (1998). ambientROOM: Integrating ambient media with architectural space. In *Proceedings of the International Conference on Human Factors in Computing Systems* (pp. 173-174).

ISO. (2007). *Road vehicles – Ergonomic aspects of transport information and control systems – Occlusion method to assess visual demand due to the use of in-vehicle systems.* Geneva, Switzerland: ISO International Standard.

Ivory, M. Y., & Hearst, M. A. (2001). The state of the art in automating usability evaluation of user interfaces. *ACM Computing Surveys, 33*(4), 470–516. .doi:10.1145/503112.503114

Jacucci, G., Oulasvirta, A., Ilmonen, T., Evans, J., & Salovaara, A. (2007). CoMedia: Mobile group media for active spectatorship. In *Proceedings of the ACM SIGCHI Conference on Human Factors in Computing Systems* (pp. 1273-1282).

Jafarinaimi, N., Forlizzi, J., Hurst, A., & Zimmerman, J. (2005). Breakaway: An ambient display designed to change human behavior. In *Proceedings of the 7th International Conference on Human Factors in Computing Systems* (pp. 1945-1948).

Jansche, M., & Abney, S. P. (2002). Information extraction from voicemail transcripts. In *Proceedings of the ACL Conference on Empirical Methods in Natural Language Processing* (Vol. 10).

Jensen, C., & Potts, C. (2004). *Privacy Policies as Decision-Making Tools: an evaluation of online privacy notices.*

Jensen, K. L., & Larsen, L. B. (2007). Evaluating the usefulness of mobile services based on captured usage data from longitudinal field trials. In Proceedings of the 4th International Conference on Mobile Technology, Application, and Systems and the 1st International Symposium on Computer Human Interaction in Mobile Technology (pp. 675-682).

Jensen, K. L., Krishnasamy, R., & Selvadurai, V. (2010). Studying PH. A. N. T. O. M. in the wild: a pervasive persuasive game for daily physical activity. In *Proceedings of the 22nd Conference of the Computer-Human Interaction Special Interest Group of Australia on Computer-Human Interaction* (pp 17-20).

Jiang, X., Chen, N. Y., Hong, J. I., Wang, K., Takayama, L., & Landay, J. A. (2004). Siren: Context-aware computing for firefighting. In A. Ferscha & F. Mattern (Eds.), *Proceedings of the International Conference on Pervasive and Ubiquitous Computing* (LNCS 3001, pp. 87-105).

John, B. E. (2003). Information processing and skilled behaviour . In Carroll, J. M. (Ed.), *HCI Models, Theories and Frameworks: Toward a Multidisciplinary Science.* London: Morgan-Kaufmann. doi:10.1016/B978-155860808-5/50004-6

Johnson, K. (2004). *Readability.* Retrieved July 17, 2010, from http://www.timetabler.com/reading.html

Jones, M., Buchanan, G., Harper, R., & Xech, P. (2007). Questions not answers: a novel mobile search technique. In *Proceedings of the SIGCHI Conference on Human Factors in Computing Systems*, San Jose, CA (pp. 155-158). New York: ACM. Retrieved from http://doi.acm.org/10.1145/1240624.1240648

Jones, M., Marsden, G., Mohd-Nasir, N., Boone, K., & Buchanan, G. (1999). Improving Web interaction on small displays. In P. H. Enslow (Ed.), *Proceedings of the Eighth international Conference on World Wide Web*, Toronto, Canada (pp. 1129-1137). New York: Elsevier.

Jones, M., Buchanan, G., Cheng, T., & Jain, P. (2006c). Changing the pace of search: Supporting "background" information seeking. *Journal of the American Society for Information Science and Technology*, *57*(6), 838–842. Retrieved from http://dx.doi.org/10.1002/asi.v57:6. doi:10.1002/asi.20304

Jones, M., Buchanan, G., & Thimbleby, H. (2003). Improving web search on small screen devices. *Interacting with Computers*, *15*(4), 479–495. doi:10.1016/S0953-5438(03)00036-5

Jones, M., Jain, P., Buchanan, G., & Marsden, G. (2003). Using a mobile device to vary the pace of search. In . *Proceedings of the Mobile HCI, 2003*, 90–94.

Jones, M., & Jones, S. (2006a). The music is the message. *Interaction*, *13*(4), 24–27. Retrieved from http://doi.acm.org/10.1145/1142169.1142190. doi:10.1145/1142169.1142190

Jones, M., Jones, S., Bradley, G., Warren, N., Bainbridge, D., & Holmes, G. (2008). ONTRACK: Dynamically adapting music playback to support navigation. *Personal and Ubiquitous Computing*, *12*(7), 513–525. Retrieved from http://dx.doi.org/10.1007/s00779-007-0155-2. doi:10.1007/s00779-007-0155-2

Jones, M., & Marsden, G. (2006b). *Mobile Interaction Design*. New York: John Wiley & Sons.

Jones, S., Jones, M., & Deo, S. (2004). Using keyphrases as search result surrogates on small screen devices. *Personal and Ubiquitous Computing*, *8*(1), 55–68. Retrieved from http://dx.doi.org/10.1007/s00779-004-0258-y. doi:10.1007/s00779-004-0258-y

Juul, J. (2009). *A casual revolution: Reinventing video games and their players*. Cambridge, MA: MIT Press.

Kadous, M., & Sammut, C. (2002). *Mobile conversational characters. Virtual conversational characters: Applications, methods, and research challenge*. Paper presented at the Joint HF/OZCHI Workshop on Human Factors and Human-Computer Interaction, Melbourne, Australia.

Kahneman, D., Slovic, P., & Tversky, A. (Eds.). (1982). *Judgment under uncertainty: Heuristics and biases*. Cambridge, UK: Cambridge University Press.

Kaikkonen, A. (2008). Full or tailored mobile Web - where and how do people browse on their mobile phones. In *Proceedings of the International Conference on Mobile Technology, Applications, and Systems*, Yilan, Taiwan (pp. 1-8).

Kaikkonen, A., Kekäläinen, A., Cankar, M., Kallio, T., & Kankainen, A. (2005). Usability testing of mobile applications: A comparison between laboratory and field testing. *Journal of Usability Studies*, *1*(1), 4–17.

Kamvar, M., & Baluja, S. (2006). A large scale study of wireless search behavior: Google mobile search. In R. Grinter, T. Rodden, P. Aoki, E. Cutrell, R. Jeffries, & G. Olson (Eds.), *Proceedings of the SIGCHI Conference on Human Factors in Computing Systems*, Montréal, Québec, Canada (pp. 701-709). New York: ACM. Retrieved from http://doi.acm.org/10.1145/1124772.1124877

Kamvar, M., & Baluja, S. (2007b). The role of context in query input: using contextual signals to complete queries on mobile devices. In *Proceedings of the 9th international Conference on Human Computer interaction with Mobile Devices and Services*, Singapore (Vol. 309, pp. 405-412). New York: ACM. Retrieved from http://doi.acm.org/10.1145/1377999.1378046

Kamvar, M., & Baluja, S. (2008). Query suggestions for mobile search: understanding usage patterns. In *Proceeding of the Twenty-Sixth Annual SIGCHI Conference on Human Factors in Computing Systems*, Florence, Italy (pp. 1013-1016). New York: ACM. Retrieved from http://doi.acm.org/10.1145/1357054.1357210

Kamvar, M., Kellar, M., Patel, R., & Xu, Y. (2009). Computers and iphones and mobile phones, oh my!: a logs-based comparison of search users on different devices. In *Proceedings of the 18th international Conference on World Wide Web*, Madrid, Spain (pp. 801-810). New York: ACM. Retrieved from http://doi.acm.org/10.1145/1526709.1526817

Kamvar, M., & Baluja, S. (2007a). Deciphering Trends in Mobile Search. *Computer*, *40*(8), 58–62. doi:10.1109/MC.2007.270

Kane, S., Wobbrock, J., et al. (2008). TrueKeys: Identifying and Correcting Typing Errors for People with Motor Impairments. In *Proceedings of the 13th International Conference on Intelligent User Interfaces* (pp. 349-352). New York: ACM.

Kanis, H., & Wendel, I. E. M. (1990). Redesigned use, a designer's dilemma. *Ergonomics, 33*(4), 459–464. doi:10.1080/00140139008927151

Karlson, A. K., Robertson, G. G., Robbins, D. C., Czerwinski, M. P., & Smith, G. R. (2006). FaThumb: a facet-based interface for mobile search. In R. Grinter, T. Rodden, P. Aoki, E. Cutrell, R. Jeffries, & G. Olson (Eds.), *Proceedings of the SIGCHI Conference on Human Factors in Computing Systems*, Montréal, Québec, Canada (pp. 711-720). New York: ACM. Retrieved from http://doi.acm.org/10.1145/1124772.1124878

Karpischek, S., & Michahelles, F. (2010, November-December). *my2cents - Digitizing consumer opinions and comments about retail products*. Paper presented at the Internet of Things Conference, Tokyo, Japan.

Katsikopoulos, K. V., & Fasolo, B. (2006). New tools for decision analysts. *IEEE Transactions on Systems, Man, and Cybernetics . Part A, 36*, 960–967.

Kawahara, T., Lee, C.-H., & Juang, B.-H. (1997). Combining key-phrase detection and subword-based verification for flexible speech understanding. In . *Proceedings of the International Conference on Acoustics, Speech, and Signal Processing, 2*, 1159–1162.

Kegl, J. (1994). The Nicaraguan sign language project: An overview. *Signpost, 7*(1), 24–31.

Kellar, M., Reilly, D., Hawkey, K., Rodgers, M., MacKay, B., & Dearman, D. (2005). It's a jungle out there: Practical considerations for evaluation in the city . In *Proceedings of Extended Abstracts on Human Factors in Computing Systems* (pp. 1533–1536). New York, NY: ACM Press.

Kelley, P. G., Bresee, J., Cranor, L. F., & Reeder, R. W. (2009). A "Nutrition Label" for Privacy. In *Proceedings of the Symposium on Usable Privacy and Security.*

Kellogg, D. (2010). *iPhone vs. Android.* Retrieved from http://blog.nielsen.com/nielsenwire/online_mobile/iphone-vs-android

Keskin, C., Balci, K., Aran, O., Sankur, B., & Akarun, L. (2007). A multimodal 3D healthcare communication system. In [Washington, DC: IEEE Computer Society.]. *Proceedings of the Conference on, 3DTV*, 1–4.

Kim, H., Kim, J., et al. (2002). An empirical study of the use contexts and usability problems in mobile Internet. In *Proceedings of the 35th Annual Hawaii international conference on system Sciences (HICSS'02)* (pp. 132). Washingotn, DC: IEEE Computer Society.

Kindberg, T., Chalmers, M., & Paulos, E. (2007). Guest editors' introduction: Urban computing. *IEEE Pervasive Computing / IEEE Computer Society [and] IEEE Communications Society, 6*(3), 46–51. doi:10.1109/MPRV.2007.57

Kishonti Informatics. (2003). *GL benchmark.* Retrieved from http://glbenchmark.com

Kjeldskov, J., & Graham, C. (2003). A review of mobile HCI research methods. *Human-Computer Interaction with Mobile Devices and Services,* 317-335.

Kjeldskov, J., & Graham, C. (2003). A review of mobile HCI research methods. In *Proceedings of the Conference on Human-Computer Interaction with Mobile Devices and Services* (pp. 317-335).

Kjeldskov, J., & Graham, C. (2003). A review of mobilehci research methods. In L. Chittaro (Ed.), *Proceedings of the 5th International Symposium on Human Computer Interaction with Mobile Devices and Services* (LNCS 2795, pp. 8-11).

Kjeldskov, J., Skov, M. B., Als, B. S., & Høegh, R. T. (2004). Is it worth the hassle? Exploring the added value of evaluating the usability of context-aware mobile systems in the field. In S. Brewster & M. Dunlop (Eds.), *Proceedings of the 6th International Symposium on Mobile Human-Computer Interaction* (LNCS 3160, pp. 529-535).

Kjeldskov, J., Skov, M., Als, B., & Høegh, R. (2004). Is it worth the hassle? Exploring the added value of evaluating the usability of context-aware mobile systems in the field. In S. Brewster & M. Dunlop (Eds.), *Proceedings of the 6th International Symposium on Mobile Human-Computer Interaction* (LNCS 3160, pp. 529-535).

Kjeldskov, J., & Skov, M. B. (2007). Studying usability in sitro: Simulating real world phenomena in controlled environments. *International Journal of Human-Computer Studies, 22*, 7–37.

Kjeldskov, J., & Stage, J. (2004). New techniques for usability evaluation of mobile systems. *International Journal of Human-Computer Studies, 60*(5-6), 599–620. doi:10.1016/j.ijhcs.2003.11.001

Klare, G. R. (1963). *The Measurement of Readability.* Ames, Iowa: Iowa State University Press.

Klare, G. R. (1975). Assessing readability. *Reading Research Quarterly, 10*, 62–102. doi:10.2307/747086

Kleimann Communication Group, Inc. (2006). *Evolution of a Prototpye Financial Privacy Notice.* Retrieved from http://www.ftc.gov/privacy/privacyinitiatives/ftcfinalreport060228.pdf

Klippel, A., & Winter, S. (2005). Structural salience of landmarks for route directions. In A. G. Cohn & D. M. Mark (Eds.), *Proceedings of the International Conference of Spatial Information Theory* (LNCS 3693, pp. 347-362).

Knoche, H., Prabhakar, T., Jamadagni, H., Pittet, A., Sheshagiri Rao, P., et al. (2010). Common sense net 2.0 - minimizing uncertainty of rain-fed farmers in semi-arid India with sensor networks. In *UNESCO Technologies for Development.* Lausanne, Switzerland.

Kolář, Ŝ., & Psutka, J. (2004), Automatic punctuation annotation in Czech broadcast news speech. In *Proceedings of the International Speech Communication Association* (pp. 319-325).

Koommey, J. G., Berard, S., Sanchez, M., & Wong, H. (2009). Assessing trends in the electrical effciency of computation over time. *IEEE Annals of the History of Computing.*

Korn, M. (2010). Understanding use situated in real-world mobile contexts. In *Proceedings of the Workshop on Research in the Large.*

Kostakos, V., Nicolai, T., Yoneki, E., O'Neill, E., Kenn, H., & Crowcroft, J. (2009). Understanding and measuring the urban pervasive infrastructure. *Personal and Ubiquitous Computing, 13*(5), 355–364. doi:10.1007/s00779-008-0196-1

Kostakos, V., O'Neill, E., Penn, A., Roussos, G., & Papadongonas, D. (2010). Brief encounters: Sensing, modeling and visualizing urban mobility and copresence networks. *ACM Transactions on Computer-Human Interaction, 17*(1), 1–38. doi:10.1145/1721831.1721833

Kotkar, P., Thies, W., & Amarasinghe, S. (2008). An audio wiki for publishing user-generated content in the developing world. In *Proceedings of the HCI Workshop for Community and International Development*, Florence, Italy.

Kristoffersen, S., & Ljungberg, F. (1999). "Making place" to make it work: emprical explorations of hci for mobile cscw. In *Proceedings of the international ACM SIGGROUP conference on supporting group work* (pp. 276-285). New York: ACM.

Kronous Groups. (n. d.). *OpenGL ES - the standard for embedded accelerated 3D graphics.* Retrieved from http://www.khronos.org/opengles/

Kshirsagar, S., Magnenat-Thalmann, N., Guye-Vuilleme, A., Thalmann, D., Kamyab, K., & Mamdani, E. (2002). Avatar markup language. In *Proceedings of the Workshop on Virtual Environments*, Aire-la-Ville, Switzerland (pp. 169-177).

Kuang, C. (2009). *Better choices through technology.* Retrieved from http://www.good.is/post/better-choices-through-technology/

Kukkonen, J., Lagerspetz, E., Nurmi, P., & Andersson, M. (2009). BeTelGeuse: A platform for gathering and processing situational data. *IEEE Pervasive Computing / IEEE Computer Society [and] IEEE Communications Society, 8*(2), 49–56. doi:10.1109/MPRV.2009.23

Kumar, A., Agarwal, S. K., & Manwani, P. (2010). The spoken web application framework: User generated content and service creation through low-end mobiles. In *Proceedings of the International Cross Disciplinary Conference on Web Accessibility*, Raleigh, NC (pp. 1-10).

Kumar, A., Rajput, N., Chakraborty, D., Agarwal, S. K., & Nanavati, A. A. (2007). WWTW: The world wide telecom web. In *Proceedings of the Workshop on Networked Systems for Developing Regions* (pp. 1-6). New York, NY: ACM Press.

Kumlin, K. (2007). *Mobilize your campaign.* Nokia Emerging Business Unit, WidSets.

Kunc, L., & Kleindienst, J. (2007). ECAF: Authoring language for embodied conversational agents. In V. Matousek & P. Mautner (Eds.), *Proceedings of the 10th International Conference on Text, Speech, and Dialogue* (LNCS 4629, pp. 206-213).

Kunc, L., Slavik, P., & Kleindienst, J. (2008). Talking head as life blog. In P. Sojka, A. Horak, I. Kopecek, & K. Pala (Eds.), *Proceedings of the 11th International Conference on Text, Speech, and Dialogue* (LNCS 5246, pp. 365-372).

Kuniavsky, M. (2003). *Observing the user experience: A practitioner's guide to user research (Morgan Kauffman series in interactive technologies).* San Francisco, CA: Morgan Kauffman.

Kurti, A., Milrad, M., & Spikol, D. (2007). Designing innovative learning activities using ubiquitous computing. In *Proceedings of the Seventh IEEE International Conference on Advanced Learning Technologies* (pp. 386-390).

Lalji, Z., & Good, J. (2008). Designing new technologies for illiterate populations: A study in mobile phone interface design. *Interacting with Computers, 20*(6), 574–586. doi:10.1016/j.intcom.2008.09.002

Larsen, L. B., Jensen, K. L., Larsen, S., & Rasmussen, M. H. (2007). A paradigm for mobile speech-centric services. *Proceedings of the Abstract Interspeech, 1*(4), 2344–2347.

Law, E. L. C., Roto, V., Hassenzahl, M., Vermeeren, A. P., & Kort, J. (2009). Understanding, scoping and defining user experience: A survey approach. In *Proceedings of the 27th International Conference on Human Factors in Computing Systems* (pp. 719-728). New York, NY: ACM Press.

Lee, H., & Kim, J. (2006). Privacy threats and issues in mobile RFID. In *Proceedings of the First International Conference on Availability, Reliability and Security.*

Lee, I., & Kim, J. (2005). Use contexts for the mobile internet: a longitudinal study monitoring actual use of mobile Internet Services. *International Journal of Human-Computer Interaction, 18*(3), 269–292. doi:10.1207/s15327590ijhc1803_2

Lermos, R. (2005). *MSN sites get easy-to-read privacy label, CNET News.* Retrieved from http://news.cnet.com/2100-1038_3-5611894.html

Licoppe, C., & Inada, Y. (2006). Emergent uses of a multiplayer location-aware mobile game: The interactional consequences of mediated encounters. *Mobilities, 1*(1), 39–61. doi:10.1080/17450100500489221

Lim, M. Y., & Aylett, R. (2009). An emergent emotion model for an affective mobile guide with attitude. *Applied Artificial Intelligence, 23*(9), 835–854. doi:10.1080/08839510903246518

Lin, J. J., Mamykina, L., Lindtner, S., Delajoux, G., & Strub, H. (2006). Fish 'n' Steps: Encouraging physical activity with an interactive computer game. In *Proceedings of the International Conference on Ubiquitous Computing* (pp. 261-278).

Lin, M., & Goldman, R. (2007). How do people tap when walking? An empirical investigation of nomadic data entry. *International Journal of Human-Computer Studies, 65*(9), 759–769. doi:10.1016/j.ijhcs.2007.04.001

Liu, J., & Seneff, S. (2009). Review sentiment scoring via a parse-and-paraphrase paradigm. In *Proceedings of the Conference on Empirical Methods in Natural Language Processing* (pp. 161-169).

Liu, Y., Shriberg, E., Stolcke, A., Hillard, D., Ostendorf, M., & Harper, M. (2006). Enriching speech recognition with automatic detection of sentence boundaries and disfluencies. *IEEE Transactions on Audio, Speech, and Language Processing, 14*(5), 1526–1540. doi:10.1109/TASL.2006.878255

Llamas, R. (2010). *Worldwide converged mobile device market grows 39.0% year over year in fourth quarter.* Retrieved from http://www.idc.com/getdoc.jsp?containerId=prUS22196610

Loomis, J. M., & Lederman, S. J. (1986). Tactual perception. In K. R. Boff, L. Kaufman, & J. P. Thomas (Eds.), *Handbook of perception and human performance volume II: Cognitive process and performance* (pp. 1-41). Hoboken, NJ: John Wiley & Sons.

Lumsden, J., & Brewster, S. (2003). A paradigm shift: alternative interaction techniques for use with mobile wearable devices. In *Proceedings of the 13th annual IBM centres for advanced studies conference* (pp. 197-210). IBM Press.

Luyten, K., Thys, K., Huypens, S., & Coninx, K. (2006). Social stitching with interactive television . In *Proceedings of Extended Abstracts of Human Factors in Computing Systems* (pp. 1049–1054). Telebuddies. doi:10.1145/1125451.1125651

MacKenzie, I. S., Kober, H., Smith, D., Jones, T., & Skepner, E. (2001). LetterWise: prefix-based disambiguation for mobile text input. In *Proceedings of the 14th Annual ACM Symposium on User interface Software and Technology*, Orlando, FL (pp. 111-120). New York: ACM. Retrieved from http://doi.acm.org/10.1145/502348.502365

MacLean, K., & Enriquez, M. (2003). Perceptual design of haptic icons. In *Proceedings of the EuroHaptics Meeting*, Dublin, Ireland (pp. 351-363).

Mäkelä, K., Belt, S., Greenblatt, D., & Häkkilä, J. (2007). Mobile interaction with visual and RFID tags: A field study on user perceptions. In *Proceedings of the SIGCHI Conference on Human Factors in Computing Systems* (pp. 991-994).

Malioutov, I., & Barzilay, R. (2006). Minimum cut model for spoken lecture segmentation. In *Proceedings of the 21st International Conference on Computational Linguistics and the 44th Annual Meeting of the Association for Computational Linguistics* (pp. 25-32).

Mandryk, R., Inkpen, K., & Calvert, T. (2006). Using psychophysiological techniques to measure user experience with entertainment technologies. *Behaviour & Information Technology*, 25(2), 141–158. doi:10.1080/01449290500331156

MarkerClusterer. (2011). *Gmaps utility library.* Retrieved from http://gmaps-utility-library-dev.googlecode.com/svn/tags/markerclusterer/

Marketing Vox. (2008). *The Latest Data about Demographics of iPhone Users.* Retrieved from http://www.futurelab.net/blogs/marketingstrategyinnovation/2010/01/latest_data_about_demographics.html

Marsden, G. (2003). Using HCI to leverage communication technology. *Interaction*, 10(2), 48–55. doi:10.1145/637848.637862

Marsden, G., Gillary, P., Thimbleby, H., & Jones, M. (2002). The Use of Algorithms in Interface Design. *International Journal of Personal and Ubiquitous Technologies*, 6(2), 132–140. doi:10.1007/s007790200012

Massa, P., & Avesani, P. (2004). Trust-aware collaborative filtering for recommender systems. In *Proceedings of the Federated International Conference on the Move to Meaningful Internet: CoopIS, DOA, ODBASE* (pp. 492-508).

Mathew, D. (2005) *vSmileys: Imaging emotions through vibration patterns.* Paper presented at the Forum on Alternative Access: Feelings & Games.

Matthews, R. B., & Stephens, W. (2002). *Crop-soil simulation models: Applications in developing countries.* New York, NY: CABI. doi:10.1079/9780851995632.0000

May, A. J., Ross, T., Bayer, S., & Tarkiainen, M. (2003). Pedestrian navigation aids: Information requirements and design implications. *Personal and Ubiquitous Computing*, 7(6), 331–338. doi:10.1007/s00779-003-0248-5

McConnell, C. R. (1982). Readability formulas as applied to college economics textbooks. *Journal of Reading*, 14–17.

McDonald, A. M., & Cranor, L. F. (2008). The Cost of Reading Privacy Policies. *Journal of Law and Policy.*

McDonald, A. M., Reeder, R. W., Kelley, P. G., & Cranor, L. F. (2009). A comparative study of online privacy policies and formats. In *Proceedings of the Privacy Enhancing Technologies Symposium.*

McGwin, G., & Brown, D. B. (1999). Characteristics of traffic crashes among young, middle-aged, and older drivers. *Accident; Analysis and Prevention*, 31, 181–198. doi:10.1016/S0001-4575(98)00061-X

McKee, H., & Porter, J. (2009). *The ethics of Internet research a rhetorical, case-based process.* New York, NY: Peter Lang.

McMillan, D. (2010, September). *iPhone software distribution for mass participation.* Paper presented at the Research in the Large Workshop, Ubicomp, Copenhagen, Denmark.

McMillan, D., Morrison, A., Brown, O., Hall, M., & Chalmers, M. (2010). Further into the wild: Running worldwide trials of mobile systems. In P. Floréen, A. Krüger, & M. Spasojevic (Eds.), *Proceedings of the 8th International Conference on Pervasive Computing* (LNCS 6030, pp. 210-227).

Mehra, A. (2007). *Reuters market light now available in local post offices across Maharastra.* Retrieved from http://news.thomasnet.com/companystory/Reuters-Market-Light-Now-Available-in-Local-Post-Offices-across-Maharashtra-808401

Melax, S. (1998). A simple, fast, and effective polygon reduction algorithm. *Game Developer, 11,* 44–49.

Metcalf, C., Harboe, G., Tullio, J., Massey, N., Romano, G., Huang, E. M., & Bentley, F. (2008). Examining presence and lightweight messaging in a social television experience. *ACM Transactions on Multimedia Computing . Communications and Applications, 4*(4), 1–16.

Michahelles, F. (2010). Getting closer to reality by evaluating released apps? In *Proceedings of the Workshop on Research in the Large.*

Michon, P., & Denis, M. (2001). When and why are visual landmarks used in giving directions? In D. R. Montello (Ed.), *Proceedings of the 5th International Conference on Spatial Information Theory: Foundations of Geographic Information Science* (LNCS 2205, pp. 202-305).

Millard, P., & Saint-Andre, P. (2010). *XEP-060: Publish-subscribe.* Retrieved from http://xmpp.org/extensions/xep-0060.html

Millar, S., & Al-Attar, Z. (2004). External and body-centered frames of reference in spatial memory: Evidence from touch. *Perception & Psychophysics, 66*(1), 51–59. doi:10.3758/BF03194860

Miller, D., Schwartz, R., Weischedel, R., & Stone, R. (1999). Named entity extraction from broadcast news. In *Proceedings of the DARPA Broadcast News Workshop* (pp. 37-40).

Millonig, A., & Schechtner, K. (2005). Developing landmark-based pedestrian navigation systems. In *Proceedings of the 8th International IEEE Conference on Intelligent Transportation Systems* (pp. 197-202). Washington, DC: IEEE Computer Society.

Miluzzo, E., Lane, N. D., Fodor, K., Peterson, R., Lu, H., Musolesi, M., et al. (2008). Sensing meets mobile social networks: The design, implementation and evaluation of the CenceMe application. In *Proceedings of the 6th ACM Conference on Embedded Network Sensor Systems* (pp. 337-350).

Miluzzo, E., Lane, N. D., Lu, H., & Campbell, A. T. (2010). Research in the app store era: Experiences from the CenceMe app deployment on the iPhone. In *Proceedings of the Workshop on Research in the Large.*

Miluzzo, E., Lane, N., Lu, H., & Campbell, A. (2010, September). *Research in the app store era: Experiences from the CenceMe app deployment on the iPhone.* Paper presented at the Research in the Large Workshop, Ubicomp, Copenhagen, Denmark.

Miyamori, H., Nakamura, S., & Tanaka, K. (2005). Generation of views of TV content using TV viewers' perspectives expressed in live chats on the web. In *Proceedings of the International Multimedia Conference* (pp. 853-861).

Mizobuchi, S., Chignell, M., et al. (2005). Mobile text entry: relationship between walking speed and text input task difficulty. In *Proceedings of the 7th international conference on Human computer interaction with mobile devices & services* (pp. 122-128). New York: ACM.

Mobile Radicals. (2008). *3D Bombus: WidSets game.* Retrieved from http://www.youtube.com/watch?v=RSi-LndShE8

Mochocki, B., Lahiri, K., & Cadambi, S. (2006). Power analysis of mobile 3d graphics. In *Proceedings of the Conference on Design, Automation and Test in Europe,* Leuven, Belgium (pp. 502-507).

Morrison, A., & Chalmers, M. (2010). SGVis: Analysis of mass participation trial data. In *Proceedings of the Workshop on Research in the Large.*

Morrison, A., Bell, M., & Chalmers, M. (2009). Visualisation of spectator activity at stadium events. In *Proceedings of the 13th International Conference on Information Visualisation* (pp. 219-226). Washington, DC: IEEE Computer Society.

Morrison, A., Reeves, S., McMillan, D., & Chalmers, M. (2010). Experiences of mass participation in Ubicomp research. In *Proceedings of the Workshop on Research in the Large*.

Morrison, A., Tennent, P., & Chalmers, M. (2006). Co-ordinated visualisation of video and system log data. In *Proceedings of the 4th International Conference on Co-ordinated & Multiple Views in Exploratory Visualization* (pp. 91-102). Washington, DC: IEEE Computer Society.

Morrison, A., Tennent, P., Chalmers, M., & Williamson, J. (2007). Using location, bearing and motion data to filter video and system logs. In *Proceedings of the 5th International Conference on Pervasive Computing* (pp. 109-126).

Morrison, A., Ross, G., & Chalmers, M. (2003). Fast multidimensional scaling through sampling, springs and interpolation. *Information Visualization*, 2(1), 68–77. doi:10.1057/palgrave.ivs.9500040

Moss-Pultz, S., & Chen, T. (2010). *Openmoko*. Retrieved from http://www.openmoko.com/

Moss-Pultz, S., Welte, H., Lauer, M., Almesberg, W., Hung, T., Lai, W., et al. (2010). *Freerunner*. Retrieved from http://www.openmoko.com/freerunner.html

Muhanna, A. (2007). *Exploration of human-computer interaction challenges in designing software for mobile devices*. Unpublished Master's thesis, University of Nevada, Reno, USA.

Müller, C., Grossmann-Hutter, B., Jameson, A., Rummer, R., & Wittig, F. (2001) Recognizing time pressure and cognitive load on the basis of speech: An experimental study. In *Proceedings of the 8th International Conference on User Modeling* (pp. 24-33).

Mun, I. K., Kantrowitz, A. B., Carmel, P. W., Mason, K. P., & Engels, D. W. (2007). Active RFID system augmented with 2D barcode for asset management in a hospital setting. In *Proceedings of the IEEE International Conference on RFID* (pp. 205-211).

Munteanu, C., Penn, G., & Zhu, X. (2009). Improving automatic speech recognition for lectures through transformation-based rules learned from minimal data. In *Proceedings of the Joint Conference of the 47th Annual Meeting of the ACL and the 4th International Joint Conference on Natural Language Processing of the AFNLP* (pp. 764-772).

Murray-Smith, R., Ramsay, A., Garrod, S., Jackson, M., & Musizza, B. (2007). Gait alignment in mobile phone conversations. In *Proceedings of the 9th International Conference on Human Computer Interaction with Mobile Devices and Services* (pp. 214-221). New York, NY: ACM Press.

Mustonen, T., Olkkonen, M., et al. (2004). Examning mobile phone text legibility while walking. In *Proceedings of Conference on Human Factors in Computing Systems* (pp. 1243-1246). New York: ACM.

Nack, F. (2003). Capturing experience - a matter of contextualising events. In *Proceedings of the ACM SIGMM Workshop on Experiential Telepresence* (pp. 53-64). New York, NY: ACM Press.

Nacke, L. E., Drachen, A., Kuikkaniemi, K., Niesenhaus, J., Korhonen, H. J., & van den Hoogen, W. M. (2009). Playability and player experience research . In *Proceedings of Breaking New Ground*. Innovation in Games, Play, Practice and Theory.

Nagamine, K. (2010, February 4). *Worldwide converged mobile device market grows 39.0% year over year in fourth quarter, says IDC*. Retrieved from http://www.idc.com/getdoc.jsp?containerId=prUS22196610

Nakamura, S., Shimizu, M., & Tanaka, K. (2008). Can social annotation support users in evaluating the trustworthiness of video clips? In *Proceedings of the 2nd Workshop on Information Credibility on the Web* (pp. 59-62).

Nass, C., Moon, Y., Fogg, B. J., Reeves, B., & Dryer, C. (1995). Can computer personalities be human personalities? In *Proceedings of the Conference Companion on Human Factors in Computing Systems* (pp. 228-229). New York, NY: ACM Press.

Neal, V. L., Dingus, T. A., Klauer, S. G., Sudweeks, J., & Goodman, M. J. (2005). *An overview of the 100-car naturalistic study and findings* (Paper No. 05-0400). Washington, DC: National Highway Traffic Safety Administration.

Nielsen, C. M., Overgaard, M., Pedersen, M. B., Stage, J., & Stenild, S. (2006). Its worth the hassle! The added value of evaluating the usability of mobile systems in the field. In *Proceedings of the 4th Nordic Conference on Human-Computer Interaction: Changing Roles* (pp. 272-280).

Nielson, J. (2009). *Mobile usability: Mobile usability test.* Retrieved from http://www.useit.com/alertbox/mobile-usability.html

Nigam, K., & Hurst, M. (2004). *Towards a robust metric of opinion.* Paper presented at the AAAI Spring Symposium on Exploring Attitude and Affect in Text, Stanford, CA.

Nilsson, A. (2004). Using IT to make place in space: evaluating mobile technology support for sport spectators. In *Proceedings of the 12th European Conference on Information Systems* (p. 127).

Nokia Siemens Networks, Nokia and Commonwealth Telecommunications Organization (CTO). (2008). *Towards effective e-governance: The delivery of public services through local e-content.* Retrieved from http://www.e-agriculture.org/en/news/towards-effective-e-governance-delivery-public-services-through-local-e-content

Nokia. (2006). *Turn limitation into strength: Design one-button games version 1.0.* Retrieved from http://www.forum.nokia.com/info/sw.nokia.com/id/8dff4326-3979-4149-96c0-5fa95a14a3cb/Turn_Limitation_into_Strength_Design_One-Button_Games_v1_0_en.pdf.html

Norman, D. (2009). *The invisible computer: Why good products can fail, the personal computer is so complex, and information appliances are the solution.* Cambridge, MA: MIT Press.

Novotney, S., & Callison-Burch, C. (2010). Cheap, fast and good enough: Automatic speech recognition with non-expert transcription. In *Proceedings of the Annual NAACL Conference on Human Language Technologies* (pp. 207-215).

O'Brien, S., & Mueller, F. F. (2006). Holding hands over a distance: Technology probes in an intimate, mobile context. In *Proceedings of the 18th Australia Conference on Computer-Human Interaction* (pp. 293-296). New York, NY: ACM Press.

O'Donovan, J., & Smyth, B. (2005). Trust in recommender systems. In *Proceedings of the 10th International Conference on Intelligent User Interfaces* (pp. 167-174).

O'Hara, K., & Kindberg, T. (2007). Understanding user engagement with barcoded signs in the 'coast' location-based experience. *Journal of Location Based Services, 1*(4), 256–273. doi:10.1080/17489720802183423

O'Neill, E., Thompson, P., Garzonis, G., & Warr, A. (2007). Reach out and touch: Using NFC and 2D barcodes for service discovery and interaction with mobile devices. In A. LaMarca, M. Langheinrich, & K. N. Truong (Eds.), *Proceedings of the 5th International Conference on Pervasive Computing* (LNCS 4480, pp. 19-36).

Ochsner, K. N., & Schacter, D. L. (2000). A social cognitive neuroscience approach to emotion and memory . In Borod, J. C. (Ed.), *The neuropsychology of emotion* (pp. 163–193). Oxford, UK: Oxford University Press.

Oehlberg, L., Ducheneaut, N., Thornton, J., Moore, R. J., & Nickell, E. (2006). Social TV: Designing for distributed, sociable television viewing. In *Proceedings of the European Conference on Interactive TV* (pp. 25-26).

Okabe, A., Boots, B., Sugihara, K., & Chiu, S. N. (2000). *Spatial tessellations: Concepts and applications of Voronoi diagrams* (2nd ed.). Chichester, UK: John Wiley & Sons.

Oliver, E. (2010). The challenges in large-scale smartphone user studies. In *Proceedings of the 2nd ACM International Workshop on Hot Topics in Planet-scale Measurement* (p. 5).

Oliver, E. (2010). The challenges in large-scale smartphone user studies. In *Proceedings of the International Conference on Mobile Systems, Applications and Services.*

Oliver, E. (2009). A survey of platforms for mobile networks research. *ACM SIGMOBILE Mobile Computing and Communications Review, 12*(4), 53–69. doi:10.1145/1508285.1508292

O'Neill, E., Kostakos, V., Kindberg, T., Schiek, A. F. G., Penn, A., Fraser, D. S., & Jones, T. (2006). Instrumenting the city: Developing methods for observing and understanding the digital cityscape. In *Proceedings of the 8th International Conference on Ubiquitous Computing* (pp. 315-332).

Öquist, G., Hein, A. L., Ygge, J., & Goldstein, M. (2004). Eye Movement Study of Reading on a Mobile Device Using the Page and RSVP Text Presentation Formats. In *Proceedings of the Mobile Human-Computer Interaction (MobileHCI 2004)* (pp. 108- 119).

Organisation for Economic Co-operation and Development. (2002). *OECD guidelines on the protection of privacy and transborder flows of personal data.* Paris, France: OECD.

Ortiz, A., del Puy Carretero, M., Oyarzun, D., Yanguas, J., Buiza, C., Gonzalez, M., et al. (2007). Elderly users in ambient intelligence: Does an avatar improve the interaction? In C. Stephanidis & M. Pieper (Eds.), *Proceedings of the 9th Conference on User Interfaces for All* (LNCS 4397, pp. 99-114).

Oulasvirta, A., Tamminen, S., et al. (2005). Interaction in 4-second bursts: the fragmented nature of attentional resources in mobile HCI. In *Proceedings of the SIGCHI conference on Human factors in computing systems* (pp. 919-928). New York: ACM.

Oulasvirta, A., Petit, R., Raento, M., & Tiita, S. (2007). Interpreting and acting on mobile awareness cues. *Human-Computer Interaction, 21*, 97–135.

Oviatt, S. (2006). Human-centered design meets cognitive load theory: Designing interfaces that help people think. In *Proceedings of the 14th Annual ACM International Conference on Multimedia* (pp. 871-880).

Ozowa, V. N. (1997). Information needs of small scale farmers in Africa: The Nigerian example. *Consultative Group on International Agricultural Research News, 4*(3).

Paek, T., Thiesson, B., Ju, Y., & Lee, B. (2008). Search Vox: leveraging multimodal refinement and partial knowledge for mobile voice search. In *Proceedings of the 21st Annual ACM Symposium on User interface Software and Technology*, Monterey, CA (pp. 141-150). New York: ACM. Retrieved from http://doi.acm.org/10.1145/1449715.1449738

Paksima, T., Georgila, K., & Moore, J. D. (2009). Evaluating the effectiveness of information presentation in a full end-to-end dialogue system. In *Proceedings of the SIGDIAL Conference on the 10th Annual Meeting of the Special Interest Group on Discourse and Dialogue* (pp. 1-10).

Pandzic, I. S. (2002). Facial animation framework for the web and mobile platforms. In *Proceedings of the Seventh International Conference on 3D Web Technology* (pp. 27-34). New York, NY: ACM Press.

Pandzic, I. S., Ahlberg, J., Wzorek, M., Rudol, P., & Mosmondor, M. (2003). *Faces everywhere: Towards ubiquitous production and delivery of face animation.* Paper presented at the 2nd International Conference on Mobile and Ubiquitous Multimedia, Norrkoping, Sweden.

Pandzic, I. S., & Forchheimer, R. (2002). *Mpeg-4 facial animation: The standard, implementation and applications* (pp. 15–61). Chichester, UK: John Wiley & Sons. doi:10.1002/0470854626.part2

Pang, B., & Lee, L. (2005). Seeing stars: Exploiting class relationships for sentiment categorization with respect to rating scales. In *Proceedings of the 43rd Annual Meeting on Association for Computational Linguistics* (pp. 115-124).

Pang, B., Lee, L., & Vaithyanathan, S. (2002). Thumbs up? Sentiment classification using machine learning techniques. In *Proceedings of the ACL Conference on Empirical Methods in Natural Language Processing* (pp. 79-86).

Pang, B., & Lee, L. (2008). Opinion mining and sentiment analysis. *Foundations and Trends in Information Retrieval, 2*(1-2), 1–135. doi:10.1561/1500000011

Park, Y., Han, S., Park, J., & Cho, Y. (2008). Touch key design for target selection on a mobile phone. In *Proceedings of the Conference on Human-Computer Interaction with Mobile Devices and Services* (pp. 423-426).

Pascoe, J., & Ryan, N. (2000). Using while moving: HCI issues in fieldwork environments. *ACM Transactions on Computer-Human Interaction, 7*(3), 417–437. doi:10.1145/355324.355329

Patel, N., Agarwal, S., Rajput, N., Nanavati, A., Dave, P., & Parikh, T. S. (2009). A comparative study of speech and dialed input voice interfaces in rural India. In *Proceedings of the 27th international Conference on Human Factors in Computing Systems*, Boston (pp. 51-54). New York: ACM. Retrieved from http://doi.acm.org/10.1145/1518701.1518709

Patel, N., Chittamuru, D., Jain, A., Dave, P., & Parikh, T. S. (2010). Avaaj Otalo—a field study of an interactive voice forum for small farmers in rural India. In *Proceedings of the 28th International Conference on Human Factors in computing systems*, Atlanta, GA (pp. 733-742).

Payne, J. W., Bettman, J. R., & Johnson, E. J. (1993). *The adaptive decision maker*. Cambridge, UK: Cambridge University Press.

Peltonen, P., Salovaara, A., Jacucci, G., Ilmonen, T., Ardito, C., Saarikko, P., et al. (2007). Extending large-scale event participation with user-created mobile media on a public display. In *Proceedings of the 6th International Conference on Mobile and Ubiquitous Multimedia* (pp. 131-138).

Pentikousis, K. (2010). In search of energy-effcient mobile networking. *IEEE Communications Magazine*, *48*(1), 95–103. doi:10.1109/MCOM.2010.5394036

Persson, P., & Fagerberg, P. (2002). *Geonotes: A real-use study of a public location-aware community system* (Tech. Rep. No. T2002:27). Gothenburg, Sweden: University of Göteburg.

Petrie, H., Johnson, V., et al. (1998). Design lifecycles and wearable computers for users with disabilities. In *Proceedings of the First Workshop on Human-Computer Interaction with Mobile Devices*.

Pettitt, M. A. (2007). *Visual demand evaluation methods for in-vehicle interfaces*. Unpublished doctoral dissertation, University of Nottingham, UK.

Pielot, M., & Boll, S. (2010). Tactile wayfinder: Comparison of tactile waypoint navigation with commercial pedestrian navigation systems. In P. Floréen, A. Krüger, & M. Spasojevic (Eds.), *Proceedings of the 8th International Conference on Pervasive Computing* (LNCS 6030, pp. 76-93).

Pielot, M., Poppinga, B., & Boll, S. (2010). PocketNavigator: Vibro-tactile waypoint navigation for everyday mobile devices. In *Proceedings of the Conference on Human-Computer Interaction with Mobile Devices and Services* (pp. 423-426).

Pirhonen, A., Brewster, V., et al. (2002). Gestural and audio metaphors as a means of control in mobile devices. In *Proceedings of the SIGCHI conference on Human factors in computing systems: Changing our world, changing ourselves* (pp. 291-298). New York: ACM.

Plauche, M., Nallasamy, U., Pal, J., Wooters, C., & Ramachandran, D. (2006). Speech recognition for illiterate access to information and technology. In *Proceedings of the International Conference on Information and Communications Technologies and Development* (pp. 83-92).

Poels, K., de Kort, Y., & Ijsselsteijn, W. (2007). It is always a lot of fun!: Exploring dimensions of digital game experience using focus group methodology. In *Proceedings of the Conference on Future Play* (pp. 83-89). New York, NY: ACM Press.

Polifroni, J., & Seneff, S. (2010). Combining word based features, statistical language models, and parsing for named entity recognition. In *Proceedings of the 11th Annual Conference of the International Speech Communication Association* (pp. 1289-1292).

Polifroni, J., & Walker, M. (2008). Intentional summaries as cooperative responses in dialogue: Automation and evaluation. In *Proceedings of the ACL Conference on Human Language Technologies* (pp. 479-487).

Polifroni, J., Seneff, S., Branavan, S. R. K., Wang, C., & Barzilay, R. (2010). *Good grief, I can speak it! Preliminary experiments in audio restaurant reviews*. Paper presented at the IEEE Workshop on Spoken Language Technology, Berkley, CA.

Poller, P., & Muller, J. (2002). Distributed audio-visual speech synchronization. In *Proceedings of the Seventh International Conference on Spoken Language Processing* (pp. 205-208).

Prante, T., Rocker, C., Streitz, N. A., Stenzel, R., Magerkurth, C., van Alphen, D., & Plewe, D. A. (2003). Hello. Wall – beyond ambient displays. In *Proceedings of the International Conference on Ubiquitous Computing* (pp. 277-278).

Preece, J., & Shneiderman, B. (2009). The reader-to-leader framework: Motivating technology-mediated social participation. *AIS Transactions on Human-Computer Interaction*, *1*(1), 13–32.

Qiu, L., & Benbasat, I. (2005). An investigation into the effects of text-to-speech voice and 3D avatars on the perception of presence and flow of live help in electronic commerce. *ACM Transactions on Computer-Human Interaction*, *12*(4), 329–355. doi:10.1145/1121112.1121113

Qt Software. (2008). *Qt cross-platform application framework.* Retrieved from http://qt.nokia.com/products

Raento, M., Oulasvirta, A., Petit, R., & Toivonen, H. (2005). ContextPhone - a prototyping platform for context-aware mobile applications. *IEEE Pervasive Computing*, *4*(2), 51–59. doi:10.1109/MPRV.2005.29

Rainie, L. (2010). *Internet, broadband, and cell phone statistics.* PewResearchCenter.

Ramakrishnan, I. V., Stent, A., & Yang, G. (2004). Hearsay: Enabling audio browsing on hypertext content. In *Proceedings of the 13th International Conference on World Wide Web* (pp. 80-89). New York, NY: ACM Press.

Rao, K. V., & Sonar, R. M. (2009). M4D applications in agriculture: Some developments and perspectives in India. *Defining the 'D' in ICT4D*, 104-111.

Rao, S., Gadgil, M., Krishnapura, R., Krishna, A., Gangadhar, M., & Gadgil, S. (2004). *Information needs for farming and livestock management in semi-arid tracts of Southern India (Tech. Rep. No. AS 2).* Bangalore, India: CAOS.

Raubal, M., & Winter, S. (2002). Enriching wayfinding instructions with local landmarks. In M. J. Egenhofer & D. M. Mark (Eds.), *Proceedings of the 2nd International Conference on Geographic Information Science* (LNCS 2478, pp. 243-259).

Reades, J., Calabrese, F., Sevtsuk, A., & Ratti, C. (2007). Cellular census: Explorations in urban data collection. *Pervasive Computing*, *6*(3), 30–38. doi:10.1109/MPRV.2007.53

Reay, I., Beatty, P., Dick, S., & Miller, J. (2007). A Survey and Analysis of the P3P Protocol's Agents, Adoption, Maintenance, and Future. *IEEE Transactions on Dependable and Secure Computing*, *5*(2), 151–164. doi:10.1109/TDSC.2007.1004

Reay, I., Dick, S., & Miller, J. (2009). An Analysis of Privacy Signals on the World Wide Web: Past, Present and Future. *Information Sciences*, *179*(8), 1102–1115. doi:10.1016/j.ins.2008.12.012

RECON. (n. d.). *Remote controlled reconnaissance in real contexts.* Retrieved from http://www.loevborg.com/tools/recon

Reddy, M. (2003). SCROOGE: Perceptually-driven polygon reduction. *Computer Graphics Forum*, *15*(4), 191–203. doi:10.1111/1467-8659.1540191

Reuters. (2011). *Apple's app store hits 10 billion downloads.* Retrieved from http://www.reuters.com/article/idUSTRE70L0Y220110122

Ricker, T. (2007, July 10). *iPhone hackers: We have owned the filesystem.* Retrieved from http://www.engadget.com/2007/07/10/iphone-hackers-we-have-owned-the-filesystem/

Roberts, J. (2005, October 2). Poll: Privacy Rights under Attack. *CBS News.* Retrieved from http://www.cbsnews.com/

Robinson, S., Eslambolchilar, P., & Jones, M. (2009). Sweep-Shake: finding digital resources in physical environments. In *Proceedings of the 11th international Conference on Human-Computer interaction with Mobile Devices and Services*, Bonn, Germany (pp. 1-10). New York: ACM. Retrieved from http://doi.acm.org/10.1145/1613858.1613874

Robinson, S., Eslambolchilar, P., & Jones, M. (2010). Exploring Casual Point-and-Tilt Interactions for Mobile Geo-Blogging. *Personal and Ubiquitous Computing*, *4*(14), 363–379. doi:10.1007/s00779-009-0236-5

Rockwell, T. H. (1988). Spare visual capacity in driving revisited: New empirical results for an old idea . In Gale, A. G., Freeman, M. H., Haslegrave, C. M., Smith, P., & Taylor, S. P. (Eds.), *Vision in Vehicles II*. London: Elsevier Science.

Rogers, Y., Connelly, K., Tedesco, L., Hazlewood, W., Kurtz, A., Hall, B., et al. (2007). Why it's worth the hassle: The value of in-situ studies when designing UbiComp. In J. Krumm, G. D. Abowd, A. Seneviratne, & T. Strang (Eds.), *Proceedings of the 9th International Conference on Ubiquitous Computing* (LNCS 4717, pp. 336-353).

Rogers, Y., Hazlewood, W., Marshall, P., Dalton, N. S., & Hertrich, S. (2010). Ambient influence: Can twinkly lights lure and abstract representations trigger behavioral change? In *Proceedings of the 12th International Conference on Ubiquitous Computing*, Copenhagen, Denmark (pp. 291-300).

Rogers, E. M. (1995). *Diffusion of innovations* (4th ed.). New York, NY: Free Press.

Rogers, Y., Lim, Y., Hazlewood, W., & Marshall, P. (2009). Equal opportunities: Do shareable interfaces promote more group participation than single users displays? *Human-Computer Interaction, 24*(2), 79–116. doi:10.1080/07370020902739379

Rohs, M., Kratz, S., Schleicher, R., Sahami, A., & Schmidt, A. (2010, September). *WorldCupinion: Experiences with an Android app for real-time opinion sharing during World Cup soccer games*. Paper presented at the Research in the Large Workshop, Ubicomp, Copenhagen, Denmark.

Ross, D. A., & Blasch, B. B. (2000). Wearable interfaces for orientation and wayfinding. In *Proceedings of the 4th International ACM Conference on Assistive Technologies* (pp. 193-200). New York, NY: ACM Press.

Ross, G., Morrison, A., & Chalmers, M. (2004). Coordinating components for visualisation and algorithmic profiling. In *Proceedings of the 2nd International Conference on Coordinated and Multiple Views in Exploratory Visualization* (pp. 3-14). Washington, DC: IEEE Computer Society.

Ross, G., & Chalmers, M. (2003). A visual workspace for constructing hybrid MDS algorithms and coordinating multiple views. *Information Visualization, 2*(4), 247–257. doi:10.1057/palgrave.ivs.9500056

Rovers, A., & van Essen, H. (2004). HIM: A framework for haptic instant messaging. In *Proceedings of the International Conference on Extended Abstracts on Human factors in Computing Systems* (pp. 1313-1316). New York, NY: ACM Press.

Saint-Andre, P. (2004). *Extensible messaging and presence protocol (XMPP): Core*. Retrieved from http://datatracker.ietf.org/doc/rfc3920/

Salen, K., & Zimmerman, E. (2003). *Rules of play: Game design fundamentals*. Cambridge, MA: MIT Press.

Schapire, R. E., & Singer, Y. (2000). Boostexter: A boosting-based system for text categorization. *Machine Learning, 39*(2-3), 135–168. doi:10.1023/A:1007649029923

Schiffman, H. R. (1976). *Sensation and perception: An integrated approach*. New York, NY: John Wiley & Sons.

Schinke, T., Henze, N., & Boll, S. (2010). Visualization of off-screen objects in mobile augmented reality. In *Proceedings of the Conference on Human-Computer Interaction with Mobile Devices and Services* (pp. 313-316).

Schleicher, R., & Trösterer, S. (2009). The 'joy-of-use'-button: Recording pleasant moments while using a PC. In T. Gross, J. Gulliksen, P. Kotzé, L. Oestreicher, P. Palanque, R. O. Prates, & M. Winckler (Eds.), *Proceedings of the 12th IFIP TC 13 International Conference on Human-Computer Interaction* (LNCS 5727, pp. 630-633).

Schmidmayr, P., Ebner, M., & Kappe, F. (2008). What's the power behind 2D barcodes? Are they the foundation of the revival of print media? In *Proceeding of the International Conference on New Media Technology* (pp. 234-242).

Schmidt, A., Beigl, M., & Gellersen, H. W. (1999). There is more to context than location. *Computers & Graphics Journal, 23*(6), 893–902. doi:10.1016/S0097-8493(99)00120-X

Schmiedl, G., Seidl, M., et al. (2009). Mobile Phone Web Browsing - a study on usage and usability of the mobile web. In *Proceedings of the 11th International Conference on Human-Computer Interaction with Mobile Devices and Services*, Bonn, Germany (pp. 1-2).

Schusteritsch, R., Rao, S., & Rodden, K. (2005). Mobile search with text messages: designing the user experience for google SMS. In *Proceedings of CHI '05 Extended Abstracts on Human Factors in Computing Systems*, Portland, OR (pp. 1777-1780). New York: ACM. Retrieved from http://doi.acm.org/10.1145/1056808.1057020

Seneff, S. (1992). TINA: A natural language system for spoken language applications. *Computational Linguistics, 18*(1), 61–86.

Senghas, A., Kita, S., & Ozyurek, A. (2004). Children creating core properties of language: Evidence from an emerging sign language in Nicaragua. *Science, 305*(5691), 1779–1782. doi:10.1126/science.1100199

Sherwani, J., Ali, N., Rosé, C. P., & Rosenfeld, R. (2009). Orality-grounded HCI: Understanding the oral user. *Information Technologies & International Development, 5*(4).

Sherwani, J., Palijo, S., Mirza, S., Ahmed, T., Ali, N., & Rosenfeld, R. (2009). Speech vs. touch-tone: Telephony interfaces for information access by low literate users. In *Proceedings of Information and Communications Technologies and Development* (pp. 447-457).

Shrestha, S. (2007). Mobile web browsing: Usability study. In *Proceedings of the 4th International Conference on Mobile Technology, Applications, and Systems and the 1st International Symposium on Computer Human Interaction in Mobile Technology* (pp. 187-194). New York, NY: ACM Press.

Shultz, W., Nolan, J., Cialdini, R., Goldstein, N., & Griskevicius, V. (2007). The constructive, destructive and reconstructive power of social norms. *Psychological Science, 18,* 429–434. doi:10.1111/j.1467-9280.2007.01917.x

Silven, O., & Jyrkka, K. (2007). Observations on power-effciency trends in mobile communication devices. *EURASIP Journal on Embedded Systems,* (1): 17.

Simon, H. A. (1990). Invariants of human behavior. *Annual Review of Psychology, 41,* 1–19. doi:10.1146/annurev.ps.41.020190.000245

Simon, S. (2000). The impact of culture and gender on web sites: An empirical study. *ACM SIGMIS Database, 32*(1), 18–37. doi:10.1145/506740.506744

Singh, V. K., Jain, R., & Kankanhalli, M. S. (2009). Motivating contributors in social media networks. In *Proceedings of the First SIGMM Workshop on Social Media* (pp. 11-18). New York: ACM Press.

Singh, R. I., & Miller, J. (2010). Empirical Knowledge Discovery by Triangulation in Computer Science. *Advances in Computers, 80,* 163–190. doi:10.1016/S0065-2458(10)80004-X

Singular Inversion. (2010). *FaceGen.* Retrieved from http://www.facegen.com

Sivak, M. (1996). The information that drivers use; is it indeed 90% visual? *Perception, 25*(9), 1081–1089. doi:10.1068/p251081

Smith, I. (2005). Social-mobile applications. *IEEE Computer, 38*(4), 84–85.

Snyder, B., & Barzilay, R. (2007). Multiple aspect ranking using the good grief algorithm. In *Proceedings of the Human Language Technology Conference of the North American Chapter of the Association of Computational Linguistics* (pp. 300-307).

Society of Automotive Engineers. (2000). *SAE J2364 Recommended practice: Navigation and route guidance function accessibility while driving.*

Sohn, T., Li, K. A., Griswold, W. G., & Hollan, J. D. (2008). A diary study of mobile information needs. In *Proceeding of the Twenty-Sixth Annual SIGCHI Conference on Human Factors in Computing Systems,* Florence, Italy (pp. 433-442). New York: ACM. Retrieved from http://doi.acm.org/10.1145/1357054.1357125

Sorrows, M. E., & Hirtle, S. C. (1999). The nature of landmarks for real and electronic spaces. In C. Freksa & D. Mark (Eds.), *Proceedings of the International Conference on Spatial Information Theory* (LNCS 1661, pp. 37-50).

Srikulwong, M., & O'Neill, E. (2010a). *A direct experimental comparison of back array and waist-belt tactile interfaces for indicating direction.* Paper presented at Workshop on Multimodal Location Based Techniques for Extreme Navigation in conjunction with the 8th International Conference on Pervasive Computing, Helsinki, Finland.

Srikulwong, M., & O'Neill, E. (2010b). A comparison of two wearable tactile interfaces with a complementary display in two orientations. In *Proceedings of the 5th International Workshop on Haptic and Audio Interaction Design* (pp. 139-148).

Statistics Canada. (2010). *Internet use by individuals, by Internet privacy concern and age.* Retrieved from http://www40.statcan.gc.ca/l01/cst01/comm31a-eng.htm

Stephen, P. (1996). *Bodyspace: Anthropometry, Ergonomics and the Design of Work.* London: Taylor & Francis.

Stephens, W., & Middleton, T. (2002). Why has the uptake of decision support systems been so poor. *Crop-Soil Simulation Models,* 129-147.

Stevens, K. T., & Stevens, K. C. (1992). Measuring the Readability of Business Writing: The Cloze Procedure vs. Readability Formulas. *Journal of Business Communication, 29*, 367–382. doi:10.1177/002194369202900404

Stone, P., & Hochman, Z. (2004). If interactive decision support systems are the answer, have we been asking the right questions? In *Proceedings of the 4th International Crop Science Congress on New Directions for a Diverse Planet*, Brisbane, Australia.

Story, L. (2007, November 1). FTC to review online ads and privacy. *The New York Times*. Retrieved from http://www.nytimes.com/

Strachan, S., Eslambolchilar, P., Murray-Smith, R., Hughes, S., & O'Modhrain, S. (2005). GpsTunes: controlling navigation via audio feedback. In *Proceedings of the 7th international Conference on Human Computer interaction with Mobile Devices &Amp; Services*, Salzburg, Austria (Vol. 111, pp. 275-278). New York: ACM. Retrieved from http://doi.acm.org/10.1145/1085777.1085831

Stroud, D. (2010). The Latest Data about Demographics of iPhone Users. *FUTURELAB*. Retrieved July 21, 2010 from http://www.futurelab.net/blogs/marketing-strategy-innovation/2010/01/latest_data_about_demographics.html

Stutts, J. C., Reinfurt, D. W., Staplin, L., & Rodgman, E. A. (2001). *The role of driver distraction in traffic crashes*. Washington, DC: AAA Foundation for Traffic Safety.

Sumeeth, M., & Miller, J. (2009). Are on-line privacy policies readable? *International Journal of Information Security and Privacy*.

Sun, Z., Stent, A., & Ramakrishnan, I. V. (2006). Dialog generation for voice browsing. In *Proceedings of the International Crossdisciplinary Workshop on Web Accessibility* (pp. 49-56). New York, NY: ACM Press.

Sun, X., & May, A. (2009). The role of spatial contextual factors in mobile personliszation at large sports events. *Personal and Ubiquitous Computing, 13*(4), 293–302. doi:10.1007/s00779-008-0203-6

Suzuki, Y., Fukumoto, F., & Sekiguchi, Y. (1998). Keyword extraction using term-domain interdependence for dictation of radio news. In *Proceedings of the 17th International Conference on Computational Linguistics* (pp. 1272-1276).

Swedberg, C. (2009). *Ohio music festival sings RFID's praises*. Retrieved from http://www.rfidjournal.com/article/articleview/4985

Sweeney, S., & Crestani, F. (2006). Effective search results summary size and device screen size: is there a relationship? *Information Processing & Management, 42*(4), 1056–1074. doi:10.1016/j.ipm.2005.06.007

Swiss Agency for Development and Cooperation SDC. (2008). *Annual report 2008 Switzerland's international cooperation*. Retrieved from http://www.cosude.ch/de/Home/Dokumentation/ressources/resource_en_181617.pdf

Synovate. (2005a). *Quantitative Evaluation of Alternative Food Signposting Concepts, Report of the U.K. Food Standards Agency, No. 265087*. Retrieved from http://www.food.gov.uk/multimedia/pdfs/signpostquanresearch.pdf

Synovate. (2005b). *Qualitative Signpost Labelling Refinement Research, Report of the U.K. Food Standards Agency, No. 951968*. Retrieved from http://www.food.gov.uk/multimedia/pdfs/signpostqualresearch.pdf

Tan, H. Z., Gray, R., Young, J. J., & Traylor, R. (2003). A haptic back display for attentional and directional Cueing. *Haptics-e: Electronic Journal of Haptics Research, 3*(1).

Taylor, W. L. (1953). Cloze procedure: A new tool for measuring readability. *The Journalism Quarterly, 30*, 415–433.

TED. (2007). *Evelyn Glennie shows how to listen*. Retrieved from http://www.ted.com/

ter Hofte, G. H. (2007a). *What's that hot thing in my pocket? SocioXensor, a smartphone data collector*. Paper presented at the Third International Conference on e-Social Science.

ter Hofte, G. H. (2007b). Xensible interruptions from your mobile phone. In *Proceedings of the 9th International Conference on Human Computer Interaction with Mobile Devices and Services* (pp. 178-181).

Ternes, D., & MacLean, K. E. (2008). Designing large sets of haptic icons with rhythm. In M. Ferre (Ed.), *Proceedings of the 6th International Conference on Haptics: Perception, Devices, and Scenarios* (LNCS 5024, pp. 199-208).

Tero-Markus, S., & Lumivuori, M. (2007). *WidSets platform for mobile Web 2.0 users and developers.* Retrieved from http://fruct.org/sem2/WidSets%20Platform%20 for%20Mobile%20Web%202.0%20Users%20and%20 Developers.pdf

Thaler, R. H., & Sunstein, C. R. (2008). *Nudge: Improving decisions about health, wealth, and happiness.* New York, NY: Penguin.

The Center for Information Policy Leadership. (2009). *Multi-layered Notices Explained.* Retrieved from http://www.hunton.com/files/tbl_s47Details/FileUpload265/1303/CIPL-APEC_Notices_White_Paper.pdf

The New York Times. (n.d.). Retrieved March 3, 2008, from http://www.nytimes.com/

Thom-Santelli, J. (2007). Mobile social software: Facilitating serendipity or encouraging homogeneity? *IEEE Pervasive Computing / IEEE Computer Society [and] IEEE Communications Society, 6*(3), 46–51. doi:10.1109/MPRV.2007.60

Thong, J.-M. D., Goddeau, D., Litvinova, A., Logan, B., Moreno, P., & Swain, M. (2000). Speechbot: A speech recognition based audio indexing system for the web. In *Proceedings of the 6th International Conference on Computer-Assisted Information Retrieval* (pp. 106-115).

Tijerina, L., Palmer, E., & Goodman, M. J. (1998). *Driver workload assessment of route guidance system destination entry while driving* (Tech. Rep. No. UMTRI-96-30). Ann Arbor, MI: The University of Michigan Transportation Research Institute.

Titov, I., & McDonald, R. (2008). Modeling online reviews with multi-grain topic models. In *Proceeding of the 17th International Conference on World Wide Web* (pp. 111-120).

Todd, P. M., Gigerenzer, G., & the ABC Research Group (in press). *Ecological rationality: Intelligence in the world.* New York, NY: Oxford University Press.

Todd, P. M. (2007). How much information do we need? *European Journal of Operational Research, 177,* 1317–1332. doi:10.1016/j.ejor.2005.04.005

Todd, P. M., & Gigerenzer, G. (2007). Environments that make us smart: Ecological rationality. *Current Directions in Psychological Science, 16*(3), 167–171. doi:10.1111/j.1467-8721.2007.00497.x

Todd, P. M., & Miller, G. F. (1999). From pride and prejudice to persuasion: Satisficing in mate search . In Gigerenzer, G., & Todd, P. M. ABC Research Group (Eds.), *Simple heuristics that make us smart* (pp. 287–308). New York, NY: Oxford University Press.

Toscos, T., Faber, A. M., An, S., & Gandhi, M. (2006). Chick clique: Persuasive technology to motivate teenage girls to exercise. In *Proceedings of Extended Abstracts on Human Factors in Computing Systems* (pp. 1873-1878).

Trafton, J. G., Altmann, E. M., Brock, D. P., & Mintz, F. (2003). Preparing to resume an interrupted task: Effects of prospective goal encoding and retrospective rehearsal. *International Journal of Human-Computer Studies, 58,* 583–603. doi:10.1016/S1071-5819(03)00023-5

Trelease, R. B. (2008). Diffusion of innovations: smartphones and wireless anatomy learning resources. *Anatomical Sciences Education, 1,* 233–239. doi:10.1002/ase.58

Trisorio-Liuzzi, G., & Hamdy, A. (2008). Rain-fed agriculture improvement: Water management is the key challenge. Paper presented at the 13th IWRA World Water Congress, Montpellier, France.

Tsukada, K., & Yasumura, M. (2004). ActiveBelt: Belt-type wearable tactile display for directional navigation. In *Proceedings of the 6th International Conference on Ubiquitous Computing* (LNCS 3205, pp. 384-399).

Tulving, E. (1993). What is episodic memory? *Current Directions in Psychological Science,* 67–70. doi:10.1111/1467-8721.ep10770899

Tulving, E. (2002). Episodic memory: From mind to brain. *Annual Review of Psychology, 53,* 1–25. doi:10.1146/annurev.psych.53.100901.135114

Tuulos, V. H., Scheible, J., & Nyholm, H. (2007). Combining web, mobile phones and public displays in large-scale: Manhattan story mashup. In A. LaMarca, M. Langheinrich, & K. N. Truong (Eds.), *Proceedings of the 5ᵗʰ International Conference on Pervasive Computing* (LNCS 4480, pp. 37-54).

Tweedie, L., Spence, B., Williams, D., & Bhogal, R. (1994). The attribute explorer. In *Proceedings of the Conference Companion on Human Factors in Computing Systems* (pp. 435-436).

U.S. Copyright Office, Librarian of Congress. (2010). *Statement of the librarian of congress relating to section 1201 rulemaking.* Retrieved from http://www.copyright.gov/1201/2010/Librarian-of-Congress-1201-Statement.html

Van Erp, J. B. F., Van Veen, H. A. H. C., Jansen, C., & Dobbins, T. (2005). Waypoint navigation with a vibrotactile waist belt. *ACM Transactions on Applied Perception, 2*(2), 106–117. doi:10.1145/1060581.1060585

van Heerden, C., Barnard, E., & Davel, M. (2009). Basic speech recognition for spoken dialogues. In *Proceedings of the 10ᵗʰ Annual Conference of the International Speech Communication Association* (pp. 3003-3006).

van Mensvoort, K. (2005). *Datafountain.* Retrieved from http://infosthetics.com/archives/2005/08/datafountain.html

Vetere, F., Gibbs, M. R., Kjeldskov, J., Howard, S., Mueller, F. F., Pedell, S., et al. (2005). Mediating intimacy: Designing technologies to support strong-tie relationships. In *Proceedings of the SIGCHI Conference on Human factors in Computing Systems* (pp. 471-480). New York, NY: ACM Press.

Vila, T. R., & Greenstadt, D. Molnar. (2003). *Why we can't be bothered to read privacy policies models of privacy economics as a lemon market.* New York: ACM International

von Neumann, J., & Morgenstern, O. (1944). *Theory of games and economic behavior.* Princeton, NJ: Princeton University Press.

Wagner, D., Billinghurst, M., & Schmalstieg, D. (2006). How real should virtual characters be? In *Proceedings of the ACM SIGCHI International Conference on Advances in Computer Entertainment Technology* (p. 57). New York, NY: ACM Press.

Walker, M., Whittaker, S., Stent, A., Maloor, P., Moore, J., Johnston, M., & Vasireddy, G. (2004). Generation and evaluation of user tailored responses in multimodal dialogue. *Cognitive Science, 28*(5), 811–840. doi:10.1207/s15516709cog2805_8

Walton, M. Vukovic', V., & Marsden, G. (2002). 'Visual literacy' as challenge to the internationalisation of interfaces: a study of South African student web users. In *CHI '02 Extended Abstracts on Human Factors in Computing Systems*, Minneapolis, MN (pp. 530-531). New York: ACM. Retrieved from http://doi.acm.org/10.1145/506443.506465

Wang, A., Emmi, M., & Faloutsos, P. (2007). Assembling an expressive facial animation system. In *Proceedings of the ACM SIGGRAPH Symposium on Video Games* (pp. 21-26). New York, NY: ACM Press.

Ward, W. (1989). Understanding spontaneous speech. In *Proceedings of the Workshop on Speech and Natural Language* (pp. 137-141).

Warren, N., Jones, M., Jones, S., & Bainbridge, D. (2005). Navigation via continuously adapted music. In *CHI '05 Extended Abstracts on Human Factors in Computing Systems*, Portland, OR (pp. 1849-1852). New York: ACM. Retrieved from http://doi.acm.org/10.1145/1056808.1057038

Waters, K., & Levergood, T. (1993). *DECface: An automatic lipsynchronization algorithm for synthetic faces* (Tech. Rep. No. 93/4). Cambridge, MA: Cambridge Research Laboratory.

Webb, E. J. (2000). *Unobtrusive measures* (Rev. ed.). Thousand Oaks, CA: Sage Publications.

Weilenmann, A. (2001). Negotiating use: making sense of mobile technology. *Personal and Ubiquitous Computing, 5*(2), 137–145. doi:10.1007/PL00000015

Weiser, M. (1991). The computer for the 21st century. *Scientific American, 265*(3), 66–75. doi:10.1038/scientificamerican0991-94

Weisz, J. D., Kiesler, S., Zhang, H., Ren, Y., Kraut, R. E., & Konstan, J. A. (2007). Watching together: Integrating text chat with video. In *Proceedings of the SIGCHI Conference on Human Factors in Computing Systems* (pp. 877-886).

Whitaker, S., Hirschberg, J., Amento, B., Stark, L., Bacchiani, M., Isenhour, P., et al. (2002). SCANMail: A voicemail interface that makes speech browsable, readable, and searchable. In *Proceedings of the Conference on Human Factors in Computing Systems* (pp. 275-282).

WidSets. (2008). *WidSets business model.* Retrieved from http://research.nokia.com/publication/11214

Wigdor, D., & Balakrishnan, R. (2003, November 2-5). TiltText: Using tilt for text input to mobile phones. In *Proceedings of the 16th Annual ACM Symposium on User Interface Software and Technology*, Vancouver, BC, Canada (pp. 81-90).

Wigelius, H., & Väätäjä, H. (2009). Dimensions of context affecting user experience in mobile work. In T. Gross, J. Gulliksen, P. Kotzé, L. Oestreicher, P. Palanque, R. O. Prates et al. (Eds.), *Proceedings of the 12th IFIP TC 13 International Conference on Human-Computer Interaction* (LNCS 5727, pp. 604-617).

Williams, A., & Dourish, P. (2006). Imagining the city: The cultural dimensions of urban computing. *IEEE Pervasive Computing / IEEE Computer Society [and] IEEE Communications Society*, 38–43.

Wisneski, C., Ishii, H., Dahley, A., Gorbet, M., Brave, S., Ullmer, B., & Yarin, P. (1998). Ambient displays: Turning architectural space into an interface between people and digital information. In *Proceedings of the First International Workshop on Cooperative Building, Integrating Information, Organization, and Architecture* (pp. 22-32).

Wittenburg, K., Lanning, T., Heinrichs, M., & Stanton, M. (2001). Parallel bargrams for consumer-based information exploration and choice. In *Proceedings of the 14th Annual ACM Symposium on User Interface Software and Technology* (pp. 51-60).

Wobbrock, J. O. (2006). The future of mobile device research in HCI. In *Proceedings of the Workshop on What is the Next Generation of Human-Computer Interaction?* Montréal, Canada (pp. 131-134).

Wright, M., & Freed, A. (1997). Open sound control: A new protocol for communicating with sound synthesizers. In *Proceedings of the International Computer Music Conference* (pp. 101-104).

Yardley, G. (2009). *App store secrets: What we've learned from 300,000,000 downloads.* Retrieved from http://www.slideshare.net/360conferences/pinch-media-app-store-secrets

Yin, M., & Zhai, S. (2006). The benefits of augmenting telephone voice menu navigation with visual browsing and search. In *Proceedings of the SIGCHI Conference on Human Factors in Computing Systems* (pp. 319-328). New York, NY: ACM Press.

Young, P. (2008). Integrating culture in the design of ICTs. *British Journal of Educational Technology*, 39(1), 6–17.

Zakaluk, B. L., & Samuels, S. J. (1988). *Readability: Its past, present and future.* Newark, DE: International Reading Association.

Zellweger, P. T., Mackinlay, J. D., Good, L., Stefik, M., & Baudisch, P. (2003). City lights: Contextual views in minimal space. In *Proceedings of the SIGCHI Conference on Human Factors in Computing Systems* (pp. 838-839).

Zeni, N., Kiyavitskaya, N., Barbera, S., Oztaysi, B., & Mich, L. (2009). RFID-based action tracking for measuring the impact of cultural events on tourism . In Höpken, W., Gretzel, U., & Law, R. (Eds.), *Information and communication technologies in tourism* (pp. 223–235). New York, NY: Springer.

Zhai, L., Fung, P., Schwartz, R., Carpuat, M., & Wu, D. (2004). Using *N*-best lists for named entity recognition from Chinese speech. In *Proceedings of HLT-NAACL: Short Papers* (pp. 37-40).

Zhai, S., Kristensson, P., Gong, P., Greiner, M., Peng, S., Liu, L., & Dunnigan, A. (2009). Shapewriter on the iPhone: From the laboratory to the real world. In *Proceedings of the SIGCHI Conference on Human Factors in Computing Systems* (pp. 2667-2670).

Zwahlen, H. T. (1988). Safety aspects of cellular telephones in automobiles. In *Proceedings of the International Symposium on Automotive Technology and Automation (ISATA)*, Florence, Italy.

About the Contributors

Joanna Lumsden, PhD., is a senior lecturer/researcher in the School of Engineering & Applied Sciences at Aston University (Birmingham, UK) and is the Lab Manager of the Aston Interactive Media (AIM) Lab. Prior to moving to Aston in 2009, Joanna was a researcher with the National Research Council of Canada (NRC) and the designer and lab manager for a state-of-the-art mobile human computer interaction (HCI) evaluation lab within the NRC facility. Joanna is also an adjunct professor with the Faculty of Interdisciplinary Studies at the University of New Brunswick (Canada). She obtained her BSc in software engineering (Hons) from the University of Glasgow (Scotland, 1996), where she also later achieved her PhD in HCI in 2001. Her research interests and expertise are mainly in mobile HCI and associated evaluation techniques. She has served on program committees for several international HCI/general computer science conferences and was also editor of the Handbook of Research on User Interface Design and Evaluation for Mobile Technology.

* * *

Ainojie Alexander Irune is a Human Factors Researcher Fellow at the Human Factors Research Group (HFRG) and Horizon Digital Economy Research at the University of Nottingham. His work involves the investigation of Human Factors and Ergonomic issues within transportation research. Currently, he is involved in providing human factors guidance within a Horizon research project exploring ways of harnessing social networks to support sustainable transport. Prior to this, his research focused on human-centred design and evaluation issues for in-vehicle information systems - the impacts of in-vehicle information systems on the driver; the potential for novel user-interfaces in a driving context; and the methods appropriate for use in the design and evaluation of such technology. He holds a BSc. and a PhD. in Computer Science and Human Factors from the University of Nottingham.

Abdallah El Ali is a PhD student in Mobile Human-Computer Interaction at the Information & Language Processing Systems (ILPS) group part of the University of Amsterdam (UvA). His current research interests include attention and interruption in mobile and ubiquitous environments, adaptive and predictive user modeling, and multimodal interaction. He received his M.S. in Cognitive Science from the University of Amsterdam, where he also worked as a research assistant in Artificial Intelligence & Law at the Leibniz Center for Law. His master's thesis was on the psychophysics of reproduced duration estimates in humans. He also worked as a freelance English copywriter at several design and advertising agencies. Website: http://staff.science.uva.nl/~elali/

Will Bamford is a Ph.D. research student from Lancaster University in the UK. He has been studying various forms of contextual user-generated content on mobile phones. The main focus of his research is in developing real-world/deployed systems that use the web together with emerging mobile sensor and other input technologies such as integrated cameras, near field communications, positioning systems, touch, digital compasses and accelerometers, to create new ways of interacting with the mobile phone, other mobile users, and their environment.

Susanne Boll is Professor for Media Informatics and Multimedia Systems at the University of Oldenburg, Germany. Also since 2002 she is a member of executive board of the OFFIS Institute for Information Technology where she is scientific head of many international and national research projects in the field of semantic retrieval and intelligent user interfaces. In 2001, Susanne Boll received her doctorate with distinction from the Technical University of Vienna. Susanne Boll's research interests lie in the field of multimodal user interfaces, mobile and pervasive media and semantic retrieval of digital media. She is a member of the editorial board of the IEEE Multimedia Magazine and Associate Editor of the ACM Transactions on Multimedia Computing, Communications and Applications (ACM TOMCCAP) and Springer Multimedia Tools and Applications. As an active member of the scientific community, Susanne served on program committees for many international conferences and also co-organized several international events.

Matthew Chalmers is a Reader in Computer Science, and a researcher in HCI, ubiquitous computing (ubicomp) and information visualisation. Half of his career has been in industrial research labs, including work at Xerox on the first ubicomp systems, Active Badges, and work at UBS Ubilab on fast non-linear multidimensional scaling algorithms. His current ubicomp research combines theory, application design and infrastructure, and using applications such mobile multiplayer games, health and fitness, and cultural tourism as vehicles for more general computer science research.

Tianyi Chen is a PhD student at School of Computer Science, University of Manchester. He is interested in HCI, the Mobile Web and Web Accessibility. His PhD topic is reciprocal interoperability of Accessible and Mobile Webs with regard to user input where he investigates whether small mobile device users experience same input problems with motor-impaired desktop users, and whether solutions to these problems can be migrated between different user domains. Further information about Tianyi Chen can be found at: http://homepages.cs.manchester.ac.uk/~chent.

Paul Coulton has over 15 years' research experience in mobile and is Senior Lecturer at Lancaster University and leads the Mobile Experiences Group for the Nokia Innovation Network. Paul has published extensively in mobile both in terms of both academic papers was selected as one of 50 most talented mobile developers worldwide from a community of over 2 million to be a founding Nokia Champion in 2006 and has been re-selected every year since. The main focus of his research surrounds innovative mobile experiences design with an emphasis on mobile entertainment in the form of games and play. Paul was the first academic invited to speak at the Mobile section of the Game Developers Conference and seen as a leading innovator in the field both within industry and academia. Paul's long term relationship with Nokia has meant research projects encompass novel uses of the latest technologies n mobile phones and many of his projects have won a number of international awards for innovation.

Jiri Danihelka received his MSc. degree in Computer Science from the Faculty of Faculty of Mathematics and Physics, Charles University in Prague in 2008. Currently he is a PhD student at the Department of Computer Graphics and Interaction at the Faculty of Electrical Engineering, Czech Technical University in Prague. His supervisor is prof. Jiri Zara. Since May 2006 he is a senior developer of the R&D Centre for Mobile Applications (RDC) at Czech Technical University in Prague. He teaches seminars on virtual reality. His research focuses on mobile graphics and new model reduction algorithms for devices with limited resources.

Roman Hak received his MSc in 2009 from the Faculty of Electrical Engineering, Czech Technical University in Prague. Currently he is a PhD student at the Department of Telecommunications Engineering. His research interests are multimodal interaction and algorithms. His work mainly focuses on mobile devices and environments.

Lynda Hardman is the head of the Interactive Information Access research group at the Centrum voor Wiskunde en Informatica (CWI) in Amsterdam, The Netherlands. Current research includes creating linked-data driven, user-centric applications for exploring media content related to events, in the context of the EventMedia project within the PetaMedia network of excellence. Previous projects include (K-Space NoE and MultimediaN/E-Culture) and document models for hypermedia and synchronized multimedia on the Web (SMIL). Lynda is part-time professor of multimedia interaction at the University of Amsterdam in the Informatics Institute. She co-edited a special issue of the Multimedia Systems Journal on the canonical processes of media production. She is a member of the editorial board for the Journal of Web Semantics and the New Review of Hypermedia and Multimedia. She runs the Interaction Design for the Semantic Web module of User System Interaction at the TU/e. Website: http://homepages.cwi.nl/

Simon Harper works in the Web Ergonomics Lab in the School of Computer Science at the University of Manchester. His main research area is Web accessibility, including those situational impairments in which the user is not conventionally disabled, but is handicapped by the technology or the environment in which a technology is used. Further information about Simon can be found at: http://homepages. cs.manchester.ac.uk/~sharper.

Niels Henze works as a researcher and doctoral student in the Media Informatics and Multimedia Systems Group of the University of Oldenburg. He teaches Human-Computer interaction and is involved in the Intelligent User Interfaces group at the research institute OFFIS. Niels worked for some national and European research projects such as the European project InterMedia. His main research interests are advances in accessing digital information using physical objects to interlink the physical word and digital content. Among his other research interests is interaction with multimedia using mobile devices, tactile and auditory interfaces, and accessibility.

Thomas Hermann studied physics at Bielefeld University. From 1998 to 2001 he was a member of the interdisciplinary Graduate Program "Task-oriented Communication". He started the research on sonification and auditory display in the Neuroinformatics Group and received a Ph.D. in Computer Science in 2002 from Bielefeld University (thesis: Sonification for Exploratory Data Analysis). After research stays at the Bell Labs (NJ, USA, 2000), GIST (Glasgow University, UK, 2004) and McGill

University (Montreal, 2008), he is currently assistant professor and head of the Ambient Intelligence Group within CITEC, the Center of Excellence in Cognitive Interaction Technology, Bielefeld University. Thomas Hermann is Member of the ICAD Board of Directors and vice-chair of the EU COST Action IC0601 Sonic Interaction Design. His research focus is sonification, ambient intelligence, data mining, human-computer interaction and cognitive interaction technology.

Jeffrey Huang is the Director of the Media x Design Laboratory and a Full Professor in Architecture and Information Systems at EPFL, with joint appointments at the Faculty of Computer and Communication Sciences, and the Faculty of Architecture, Civil and Environmental Engineering. His research explores the convergence of physical and virtual environments. Current research projects include the integration of computing into architecture and cities, novel processes of design (parametric design, algorithmic design), mobile applications, and more generally, design-centered approaches to human-computer interaction. Prior to EPFL, Huang was a faculty member at Harvard's Graduate School of Design. Huang received his DiplArch from the ETHZ, and his Masters and Doctoral Degrees from Harvard University, where he was awarded the Gerald McCue medal.

Kasper Løvborg Jensen is an assistant professor in the Department of Architecture, Design and Media Technology at Aalborg University. He is also affiliated with the Section for Multimedia Information and Signal Processing at the Department of Electronic systems at the same university where he did his PhD. Besides researching methods and tools for evaluating mobile and ubiquitous applications in the field, his broader research interests lie within the cross field of science and engineering within mobile, ubiquitous and pervasive computing, HCI, context-awareness, multimodal interfaces and artificial intelligence. He received a M.Sc. from Aalborg University in 2005.

Matt Jones is a Professor of Computer Science at the Future Interaction Technology Lab at Swansea University. He has worked on mobile interaction issues for the past fifteen years and has published a large number of articles in this area. He has had many collaborations and interactions with handset and service developers including Orange, IBM Research, Microsoft Research, Reuters, BT Cellnet and Adaptive Info. He has been a Visiting Fellow at Nokia Research and is currently a member of its Scientific Advisory Board (Tampere, Finland). He is an editor of the International Journal of Personal and Ubiquitous Computing and on the steering committee for the Mobile Human Computer Interaction conference series. Matt is the co-author of "Mobile Interaction Design", John Wiley & Sons Nov 2005. See www.undofuture.com for more.

Lukas Kencl received a Ph.D. degree in Communication Networks from Ecole Polytechnique Federale de Lausanne (EPFL), Switzerland, in 2003, and a MSc. degree in Computer Science from Charles University, Prague, in 1995. Since May 2007 he is director of the R&D Centre for Mobile Applications (RDC) at Czech Technical University in Prague. Previously, he was a Senior Researcher at Intel Research, Cambridge, UK (2003-6) and a Pre-Doc at the IBM Zurich Research Laboratory (1999-2003). His research focuses on novel interfaces, services and applications in mobile wireless networks and on architecture and performance optimization of networking systems. He regularly publishes at established networking conferences and journals, is a co-inventor of multiple network technology patents and holder of several industrial grants and awards (Vodafone, IBM, Microsoft).

Imre Kiss is a Principal Researcher with Nokia Research Center in Cambridge, MA, USA. His research interests are in pattern recognition, machine learning and voice and multimodal human computer interaction (HCI) for mobile systems. Prior to joining the Cambridge, MA lab of Nokia Research in 2007, he has held various research and managerial positions at Nokia Research in Tampere, Finland, where he has worked extensively on various aspects of voice user interfaces for embedded mobile platforms. He received an M.Sc. degree in Electrical Engineering from the Technical University of Budapest, Hungary, in 1996, and a Ph.D. degree in Information Technology from Tampere University of Technology, Tampere, Finland, in 2001. Dr. Kiss is a member of the IEEE, the International Speech Communication Association (ISCA) and the Applied Voice Input/Output Society (AVIOS).

Hendrik Knoche is a post-doctoral researcher in the Media x Design Laboratory at EPFL. He has worked worked on interaction design and user experience both on desktop and mobile devices for the last ten years in industry and academia. He holds a Diploma in computer science from the University of Hamburg and pursued his PhD in computer science in Angela Sasse's Human Centered Systems Group at University College London. Prior to his PhD studies he worked with David Kirsh at the Interactive Cognition Lab of UC San Diego on an ONR funded project on distributed collaboration. He has worked on European projects on interactive TV centric services and mobile TV and various user studies relating to mobile social networking, automated content adaptation, biometric systems, video quality and visual experience. His research interests include socio-technical systems, human-centered design, HCI and mediated experiences.

Sven Kratz is a research assistant with Deutsche Telekom Laboratories at TU Berlin, where he is also currently pursuing his Phd studies. His primary research focus is on sensor-based mobile interfaces that allow for novel types of interaction. This includes gesture recognition, interaction based on distance sensing and pressure, tabletop applications for mobile devices and optical tracking technologies. Sven also conducts research into mobile interfaces providing efficient navigation of large information spaces such as maps, 3D environments or databases. He received his Diplom degree in Computer Science from RWTH-Aachen University in 2008. In 2007, while working at the Media Computing Group at RWTH Aachen, Kratz participated in the development, deployment and the user interface design of the mobile pervasive city game REXplorer, the first mobile city game released in public.

Jakob Eg Larsen, Ph.D. is Associate Professor at the Technical University of Denmark, DTU Informatics at the Cognitive Systems Section where he is managing the Mobile Informatics Laboratory (milab). His research interests include mobile systems, human-computer interaction, personalized user interfaces and experiences, context-awareness, and personal information management. He has published papers on these topics at international conferences and workshops. At DTU Informatics he is teaching M.Sc. level courses in the Digital Media Engineering M.Sc. program. He received his Ph.D. (2005) from the Technical University of Denmark and his M.Sc. in Computer Science (1999) University of Copenhagen, Dept. of Computer Science and human computer interaction & cognitive psychology from University of Copenhagen, Dept. of Psychology.

Christian Leichsenring studied science informatics at Bielefeld University, spending one year abroad at the University of Paris XI and the University of Marne-la-Vallée. During his studies he worked at the Neuroinformatics Group of Bielefeld University as a student worker and held several positions as a student representative. He now does his doctorate in the Ambient Intelligence Group and the Applied Informatics Group of the Center of Excellence for Cognitive Interaction Technology (CITEC). Christian Leichsenring's research focus lies with context-aware mobile devices and augmented reality.

Donald McMillan is a Ph.D. student at the University of Glasgow, UK and a member of the Social Ubiquitous Mobile (SUM) research group. His research focus is on the utilisation of 'mass participation' trials for ubicomp research. This has involved running several trials involving of tens of thousands of users in order to study the method and the factors that affect results, use and participant engagement during trials. He is co-organiser of CHI 2011 workshop on ethics in large-scale distribution of research applications and of the SICSA CHI networking event.~lynda/

James Miller received the B.Sc. and Ph.D. degrees in Computer Science from the University of Strathclyde, Scotland. Subsequently, he worked at the United Kingdom's National Electronic Research Initiative on Pattern Recognition as a Principal Scientist, before returning to the University of Strathclyde to accept a lectureship, and subsequently a senior lectureship in Computer Science. Since 1993, his research interests have been in Software and Systems Engineering. In 2000, he joined the University of Alberta. He is the principal investigator in a number of research projects that investigate software verification, validation and evaluation issues across various domains, including embedded, web-based and ubiquitous environments. He has published over one hundred refereed journal and conference papers (see www.steam.ualberta.ca for details on recent directions); and sits on the editorial board of the Journal of Empirical Software Engineering.

Alistair Morrison is a Research Associate at the University of Glasgow. His background is in information visualisation and his PhD focussed on methods for exploring high dimensional data sets. In his PhD thesis he described novel dimensional reduction algorithms that offered a significant reduction in computational complexity of earlier techniques, while offering comparable accuracy of produced layouts. His subsequent research turned to tools and techniques for analysing data collected from trials of ubiquitous computing systems, including the Replayer software for combining recorded trial video data with system logs and automating filtering of video to show salient information. More recently he has studied mass participation ubiquitous computing trials, running several studies involving tens of thousands of users and examining the issues surrounding the release of trial software through public 'App Store'-style software repositories. He is co-organiser of CHI 2011 workshop on ethics in large-scale distribution of research applications.

Frank Nack is tenure assistant professor at the Information and Language Processing Systems group (ILPS) of the Informatics Institute of the University of Amsterdam (UvA). The main thrust of his research is on representation, retrieval and reuse of media in hypermedia systems, context and process aware media knowledge spaces, representation and adaptation of experiences, hypermedia systems that enhance human communication and creativity, interactive storytelling, and computational humour theory. He has published more than 100 papers on these topics. He is on the editorial board of IEEE Multimedia, where he edits the Media Beat column and also serves as associated editor in chief. He also serves on the board of IEEE Transactions on Computational Intelligence and AI in Games. Frank is member of the ACM, ACM SIGMM, ACM SIGCHI and ACM SIGWEB. Website: http://fnack.wordpress.com/

Eamonn O'Neill is a Reader and Royal Society Industry Fellow in the Department of Computer Science at the University of Bath. He spends half his time working with Vodafone Group Research and Development (R&D) on intelligent mobile services research. His research interests include bringing a human-computer interaction perspective to research on mobile and pervasive systems. The research has two main strands. The first is in the development of mobile and pervasive computing systems, i.e. computing and communication devices, applications and resources distributed throughout our environment and carried in our pockets and on our clothing. The second is in participatory design, i.e. enabling and supporting the active collaboration of users and designers in the production of human-computer systems.

Stephen Payne is professor of human-centric systems in the Department of Computer Science at the University of Bath. Before moving to Bath, Payne was a Professor of Psychology in Cardiff University and (briefly) a Professor in Manchester Business School. Previously he worked at IBM T.J. Watson Research Center. Payne has worked on cognitive approaches to Human-Computer Interaction since his PhD, on Task-Action Grammars. Most recently he has worked on theories of discretionary task switching and of skim-reading. He is a member of the Editorial Board of Human-Computer Interaction.

Martin Pielot is an associate researcher in the Intelligent User Interfaces Group at the OFFIS – Institute for Information Technology, Germany. At the same time he is a doctoral student in the Media Informatics and Multimedia Systems Group at the University of Oldenburg, Germany. Coming from a background of building mobile applications, his research focuses on ubiquitous computing and mobile human-computer interaction in general. In particular he is interested in space & location, and making both "graspable" by conveying spatial information via the skin. He sees research in the large as a fantastic opportunity to bring new insights to this topic.

Joe Polifroni has had over 25 years experience in building spoken dialogue systems. He has done work in automating dialogue design, response generation, and user modelling. He has been a Research Scientist in the Spoken Dialogue Systems Group at MIT's Laboratory for Computer Science, a Research Associate at the University of Sheffield, and a software lead at Unveil Technologies. He is currently a Principal Scientist at Nokia Research Center in Cambridge, Massachusetts, where he works on spoken interfaces to mobile devices.

Benjamin Poppinga is a researcher in the Intelligent User Interfaces Group at the OFFIS – Institute for Information Technology, Germany. At the same time he is a doctoral student at the University of Oldenburg, Germany. Benjamin's research focuses on mobile 'in the wild' context sensing and evaluation techniques, especially in the domain of mobile navigation applications. Benjamin is involved in the European project HaptiMap, where he's continuing the development of the PocketNavigator.

Yvonne Rogers is a professor of Human-Computer Interaction in the Computing Department at the Open University, where she directs the Pervasive Interaction Lab. From 2003-2006 she had a joint position in the School of Informatics and Information Science at Indiana University (US). Prior to that she was a professor in the former School of Cognitive and Computing Sciences at Sussex University. She has also been a Visiting Professor at Stanford, Apple, Queensland University and UCSD. Her research focuses on augmenting and extending everyday, learning and work activities with a diversity of interactive and novel technologies. She was one of the principal investigators on the UK Equator project (2000-2007), where she pioneered and experimented with ubiquitous learning. She has published widely, beginning with her PhD work on graphical interfaces in the early 80s to her most recent work on public visualizations and behavioral change. She is one of the authors of the bestselling textbook *"Interaction Design; Beyond Human-Computer Interaction"* and more recently *"Being Human: Human Computer Interaction in the Year 2020"*.

Michael Rohs is an assistant professor at the Institute for Media Informatics, Department of Informatics, Ludwig-Maximilians-Universitat (LMU) Munchen, Germany, heading the Mobile Human-Computer Interaction group. His primary research interests are in mobile human-computer interaction, mobile interactive media, and pervasive computing. My work focuses on novel interaction techniques for mobile devices, applications of computer vision techniques in mobile HCI, the usage of sensors for mobile interactions, and the integration of physical and virtual resources in the user's environment. Michael received a Diplom in Computer Science from the Technische Universitat Darmstadt, Germany, a Master's degree in Computer Science from the University of Colorado at Boulder, USA, and a Ph.D. in Computer Science from ETH Zurich, Switzerland. In the past, he worked as a research assistant in the Distributed Systems Group at ETH Zurich and as a senior research scientist at Deutsche Telekom Laboratories, Technische Universitat Berlin, Germany.

Alireza Sahami is a doctoral candidate and member of the researcher staff at Plauno, The Ruhr Institute for Software Technology, at the University of Duisburg-Essen and Human Computer Interaction at the university of Stuttgart. His research interests include human computer interaction, ubiquitous computing, and mobile computing. He received his M.Sc. degree in media informatics from RWTH Aachen University, Germany.

Torben Schinke is a co-owner of Worldiety Adrian Macha & Torben Schinke GbR (WDY) a company located in Oldenburg, Lower Saxony and was established in 2010. Worldiety is a startup company working together with known Stock Corporations in Germany and is specializing in the research, design and development of mobile applications and embedded systems. Torben graduated with a diploma at the University of Oldenburg where he was involved in the Intelligent User Interface group at the research institute OFFIS. There he made several contributions to the European project InterMedia. Amongst others, his research interests include mobile human computer interaction.

Robert Schleicher is a senior researcher at the Quality & Usability Lab of the Deutsche Telekom Laboratories (TU Berlin) since 2006. He studied psychology in Bonn and New York City. In 2008, he finished his PhD thesis on psychophysiology, eye movements and emotions at the University of Cologne. His research interests in Human Computer Interaction are Affective Computing/psychophysiology, user experiences, and eyetracking.

Albrecht Schmidt is a professor for Human Computer Interaction at the University of Stuttgart. Previously he was a Professor at University of Duisburg-Essen, had a joined position between the University of Bonn and the Fraunhofer Institute for Intelligent Analysis and Information Systems (IAIS). He studied computer science in Ulm and Manchester and received in 2003 a PhD from the Lancaster University in the UK. His research interest is in human computer interaction beyond the desktop, including user interfaces for mobile devices and cars. Albrecht published well over 100 refereed archival publications and his work is widely cited. He is co-founder of the ACM conference on Tangible and Embedded Interaction (TEI) and initiated the conference on Automotive User Interfaces (auto-ui.org). He is an area editor of the IEEE Pervasive Computing Magazine and edits a column on invisible Computing in the IEEE Computer Magazine.

Stephanie Seneff received her SB, SM, and PhD degrees from MIT. Her research in speech and language has spanned nearly thirty years at MIT, covering many aspects of the speech chain, from signals to symbols to meaning. Her early research, at the MIT Lincoln Laboratory in the seventies, included synthesis, coding, feature extraction, and recognition. Her doctoral thesis, completed in 1985, concerned a model for human auditory processing of speech, and its application to computer speech recognition. Since 1989, her research has primarily focused on natural language processing for speech, encompassing phonology, morphology, syntax, semantics, speech generation, discourse and dialogue. She has consistently pursued a research paradigm that involves testing theoretical ideas in the context of real conversational systems in a number of different domains and languages. She is especially interested in the concept of Web-based interactive dialogue as an aid for second language acquisition.

P. R. Sheshagiri Rao is a Farmer & Researcher, residing in a semi arid village in South India. For research he has worked with Indian Institute of Science, Bangalore (through its field station located in his farm) for over twenty years in areas of – Natural resource management, Biodiversity, Agro-ecology, adaptation to Climate variability & change and application of ICT. He has co-authored a book on Biodiversity and conservation policy and published in peer reviewed journals in each of his areas of research interest and served as member of several Regional and National policy making bodies. In recent years, his applied research work has led to consultancy assignments with- Earth Institute, Columbia University; Corporate firms and NGOs. He leads a small NGO, CK Trust, and several community based natural resource management efforts. He has academic training in both Agriculture and Ecology with a Masters degree in Plant breeding and Genetics from University of Agricultural sciences, Bangalore.

Ravi I Singh received his B.Sc in Computer Science from the University of Alberta, Canada. In 2008, began his M.Sc. in Software Engineering at The University of Alberta. He is also a member of IEEE and two non government organizations whom he has instructed in Software Engineering practices. His major areas of research interest include triangulation, persuasion and machines, and electronic government. He also has a background in linguistics and speaks eight languages..

Mayuree Srikulwong received an MSc in Information Systems from Asian Institute of Technology (AIT), Thailand. She is a final year PhD student in the Department of Computer Science at the University of Bath, UK. Her research is funded by the University of Thai Chamber of Commerce, Thailand. Her research interests include tactile interfaces, vibrotactile feedback, multimodal interactions, and wearable computing. Her current project involves evaluating tactile pedestrian navigation system that provides necessary information such as directions and landmarks for navigation in urban environments. The project aims to develop a theoretical understanding of tactile psychophysics in relation to human haptic perception for pedestrian navigation tasks.

Arek Stopczynski is concluding his Master of Science in Telecommunications degree at Technical University of Denmark (DTU). His research interests include applications of mobile technologies such as utilizing mobile phones as cognitive systems, mobile context awareness and building social networks in physical environments. He has worked as Research Assistant on several research projects in this area, designing and implementing novel mobile applications. Furthermore as Laboratory Assistant at the Mobile Informatics Lab at the Cognitive Systems Section at DTU Informatics he has assisted in the development of mobile applications as part of research and teaching activities. He plans to enroll the Ph.D. program within his research interests at the Cognitive Systems Section at DTU informatics.

Manasa Sumeeth received a Masters in Software Engineering from The University of Alberta in 2009. She is now a Systems Analyst with IBM Canada in Edmonton, Alberta, Canada.

Peter M. Todd is professor of cognitive science, informatics, and psychology at Indiana University, Bloomington. In 1995 he moved to Germany to help found the Center for Adaptive Behavior and Cognition (ABC), based at the Max Planck Institute for Human Development in Berlin, to study the simple ways that humans and other animals can make good decisions in appropriate environments. The Center's work was captured in the book Simple Heuristics That Make Us Smart (Gigerenzer, Todd, and the ABC Research Group; Oxford, 1999); the sequel, Ecological Rationality: Intelligence in the World, covering information-environment structures and their impact on decision making, is in press. Todd moved to Indiana University in 2005 and set up the ABC-West lab there. His ongoing research interests span the interactions between and co-evolution of decision making and decision environments, focusing on the ways that people and other animals search for resources—including mates, information, and food—in space and time.

René Tünnermann is a research associate at the Ambient Intelligence Group at the Cognitive Interaction Technology Center of Excellence at Bielefeld University (CITEC). He studied science informatics at Bielefeld University. During his studies he worked as a student worker at the Neuroinformatics Group of Bielefeld University and the Alignment in AR-based cooperation project of the CRC673-Alignment in Communication. His research focus lies with tangible interfaces and interactive surfaces.

Yeliz Yesilada is a lecturer at the Middle East Technical University Northern Cyprus Campus (METU NCC) in the Computer Engineering programme and an Honorary Research Fellow in the School of Computer Science at the University of Manchester. Her primary research interest is centred around the Human Centred Web; in particular Web accessibility, the mobile Web and using Semantic Web technologies to improve user experience. Further information about Dr. Yesilada can be found at: http://www.metu.edu.tr/~yyeliz.

Jiri Zara is a professor and a head of the Department of Computer Graphics and Interaction at the Czech Technical University in Prague. He teaches courses on 2D and 3D graphics algorithms and virtual reality. He is an author/co-author of three books on Computer Graphics and Virtual Reality (in Czech language) that are used as basic literature at most Czech universities. He worked as a Czech expert for the ISO organization in the field of Computer Graphics (JTC1/SC24) and Multimedia (JTC1/SC29) in previous 10 years. He has led a number of research projects on virtual and augmented reality, including also a digitization of the Langweil paper model of Prague.

Index

2D Barcodes 169

A

Aptana 64
Augmented Reality 100
Authoring Language for Embodied Conversational
    Agents (ECAF) 133
Automatic Speech Recognition (ASR) 114-115, 139
Avatar Markup Language (AML) 133

C

Capture Context 80
μCARS 91
    HERMES 92
CenceMe 260
Cloze Procedure 60-61
    Scoring 61
CO2PENHAGEN Festival 170
Common Sense Net 2.0 162
Contextual Software Project 261

D

Decision Making Strategies 101
    Bounded Rationalities 101
    Ecological Rationalities 102
    Unbounded Rationalities 101
Decision Support Systems (DSS) 156, 158
Distractions 2

E

Electroencephalogram (EEG) 213
Electromyography (EMG) 213
Embodied Conversational Agents (ECA) 132

F

Facial Animation Parameters (FAPs) 139

G

Galvanic Skin Response (GSR) 213
Generic REAlity Trace Data ANalysis Engine
    (GREATDANE) 89
    Components 90
    Usability 91
Geotagging 173
Graphic User Interface (GUI) 92

H

Human-Centered Design (HCD) 156
Human Computer Interaction (HCI) 1, 39, 124, 310
Human Factors (HF) 3, 15
Human Navigation
    Landmarks 187
Hybrid Information Visualisation Environment
    (HIVE) 271
Hyperspeech Transfer Protocol (HSTP) 132

I

Informational Nudges 102
Information and Communication Technologies and
    Development (ICTD) 158
Information and Communication Technologies for
    Development (ICT4D) 156
Information Visualization 103
    Cumulative Display 108
    Fusing Display 108
    Glanceability 104
    Points Of Interest (POI) 295
Interaction Designs 26
    Query Entry 26
    Search Results 28
International Mobile Equipment Identity (IMEI) 311
In the Large 277
In-Vehicle Information Systems (IVIS) 1, 4

J

Jailbreaking 226
Java Micro Edition (J2ME) 278

K

Kruskal-Wallis Test 190

L

Lane Change Test (LCT) 2
Location-aware Multimedia Messaging (LMM) 204,
    206
    GeoNotes 212

M

Mass Participation 223
Millenium Development Goal (MDG) 155
Mircoblogging 173
Mobile Devices
    Android 52, 151, 211
    AppAware 225
    AppAware>Unlocking 226
    Apple iPhone 64
    Attention Switching 46
    Flesch-Kincaid Readability Formula 60
    Keyboard Type 48
    Readability 58
    Typing 46
    Web Usage 50
    While Walking 44
Mobile Search 23
    Developing World 33
    Human Logs 25
    Incidental Search 29
    Machine Logs 24
Multimodal Searching 32

N

Natural Language Processing (NLP) 114-115
Navigation System 10
    Point of Interest (POI) 10
NonPhotorealistic Rendering (NPR) 133

O

Organisation of Economic Cooperation and Devel-
    opment (OECD) 311
Over The Air (OTA) 282

P

Peer-to-Peer (P2P) 268
Peripheral Detection Test (PDT) 2
Phenomenological Theories 209
Phrase Spotting 117
    How May I Help You (HMIHY) System 118
Platform for Privacy Preferences Project (P3P) 69
Positivist Theories 209
Primary Loading Task (PLT) 6
Privacy Policy 57
    Organisation for Economic Co-operation and
        Development (OECD) 58
    TRUSTe 62

R

Radio Frequency IDentification (RFID) 169
Reality Traces 80
Recommender Systems 121
Reflective Search 30
REmote CONtrolled RECONnaissance in REal
    CONtexts (RECON) 84
    Components 85
    PHysical Activity through NewTechnology on
        Mobiles (PH.A.N.T.O.M.) 88
    Usability 88
Rich Immersive Sports Experience (RISE) 242

S

Sentiment Detection 119
SGVis 259
Shopping Study
    Collaborative Tool 110
    Comparative Tool 110
    Cumulative Tool 110
Short Message Service (SMS) 278
Simple Heuristics 102
Single Board Computer (SBC) 159
Social Media
    Facebook 254
    Real-Time Emotion Sharing 242
    Social TV (STV) 241
    Sports 242
Spatial Display 187
Speech Recognition 135
Speech Synthesis 134
Spoken Language Systems 124
    SpeechBot 118
Synchronized Multichannel Integration Language
    for Synthetic Agents (SMIL-Agent) 133

T

Tactile Communication
    Distinguishability 195-196
    feelabuzz 145
    Haptic Display 146, 153
    Memorability 196
    Neo FreeRunner 150-151, 154
    PocketNavigator 296
    Symbolic Mapping 196
    Tactile Representation 194
    Vibrotactile Interaction 146
Telecommunication Companies
    Apple 223, 278
    AT&T 118
    Blackberry 225
    Google 62
    HTC 136
    IDC 223, 258
    Nokia 122, 288
Text-to-Speech (TTS) 140

U

Ubicomp 224
Uninterrupted Power Source (UPS) 159

Unique Device ID (UDID) 267
Urban Pervasive Infrastructure (UPI) 214
    Augmented Gate Count 214
    Experience Sampling Methods (ESMs) 214
User Generated Content (UGC) 282
User Interface Design 19

V

Video Streaming 135
Voice Browser 132
Voiceover IP (VoIP) 274

W

What You See Is Almost What You Hear
    (WYSIAWYH) 119
WidSets 280
    Application Programming Interface (API) 286
    Bombus 280
    Four-In-A-Row 287
Wireless Application Portal (WAP) 278
Wireless Sensor Network (WSN) 159
World Cupinion 245
    Vuvuzela App 253
World Wide Telecom Web (WWTW) 132